THE MATRILINES
of the
Egyptian Arabian Horse

Volume I

Dr. William M. Hudson Jr.

The Matrilines of the Egyptian Arabian Horse
Volume I
Dr. William M. Hudson Jr.

Product Liability: The author and publisher can give no guarantee for the validity or accuracy of any or all of the information contained in this book. This statement also refers to information about the use of DNA data herein presented. The author and publisher specifically state that any use of the DNA data herein presented is prohibited. The author and publisher accept no responsibility or liability for injuries or financial losses resulting from the use of the mDNA data herein published in specific or general animal identification.

© 2018, First Edition, published by Zandai Publications
Printed in the U.S.A. ISBN 978-0-9882939-4-6

Acknowledgments

The author wishes to express his gratitude to the many people who helped in the creation of this book. The generosity of Beatrice Funes and her mother Esperanza Raswan made this book possible.

Joan Kellogg was the graphics artist and design editor for the entire project. She is an InDesign specialist, and cast the text from its rough drafts and drawings into a coherent and artfully constructed whole. She created and optimized many of the images and photographs, and was responsible for the arrangement and flow of the book. She edited the book and rendered it into a format suitable for digital printing.

No book about the Egyptian Arabian horse is complete without recognizing the importance of Judith Forbis whose genius and commitment to this breed is unsurpassed. The author wishes to thank her on behalf of Egyptian Arabian horse breeders around the world. The superior quality of the Egyptian Arabian horse of today is due in large part to her energetic and creative efforts.

Hansi Heck was invaluable in providing an insider's view of the subject at hand. Her PINDEX was an essential source of primary data, including pedigrees and original reproductions of international Arabian horse registrations.

Many photographers contributed archival images of Arabian horses, especially Polly Knoll, Johnny Johnston, and Jerry Sparagowski.

The author was aided by the works of W.S. Maugham and Arthur Conan Doyle. They served as excellent sources of literary instruction.

Rosemary Archer kindly gave permission for use of photographs from her publications.

Interviews with Jane LLewellyn Ott encapsulated views and insights into the early days of the Arabian horse in America.

Scott Langdon, Ph. d. performed the mitochondrial DNA analysis for the study in Chapter 11. He is the director of the Duke University DNA facility.

Special thanks to Nextgen Editing, London, England, for final editing of the mitochondrial DNA study found in Chapter 10. Their persistence was essential in the study being published in the scientific journal PLOSONE.

Thanks to Michael Bowling for reviewing the study and providing useful feedback.

Joe Ferris was a helpful resource in proving old, rare and obscure photographs. As an expert in the field of the Egyptian Arabian horse, he provided many vital facts and information.

Cynthia Culbertson is due special thanks for advice and review of the project.

The work of Ferman Ansel was essential by providing videos of some of the outstanding Egyptian Arabian horses of his day.

W. Georg Olms of the Asil Club, Germany graciously gave permission for the use of several images from Asil Araber, Volume 1 - 6

The author wishes to thank those horse owners who generously donated blood samples from their horses for the mDNA study which forms the core of this book. Without their trust and encouragement this book would not have been written.

Merrill DePetrillo Hudson, beloved wife of the author, contributed substantially to the style and content of this book during the ten years of its creation.

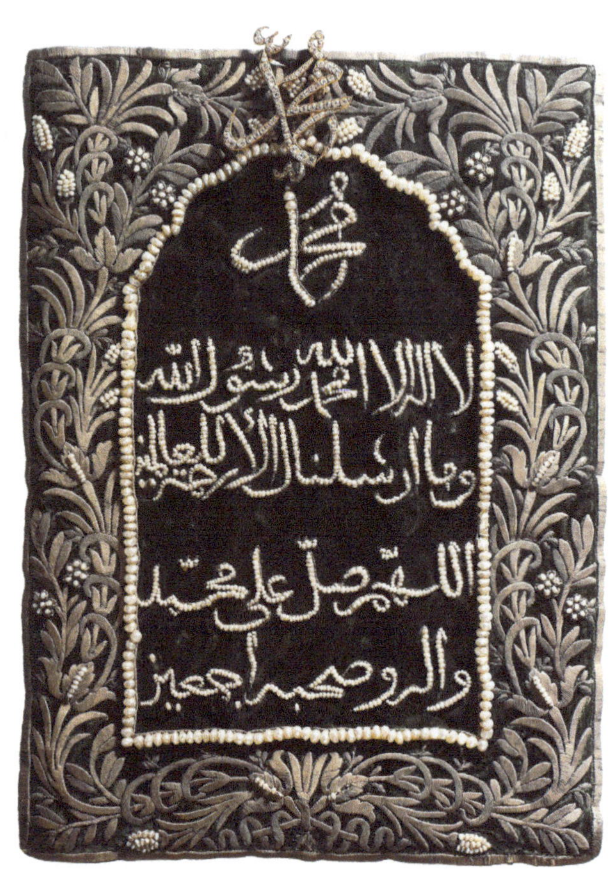

Published by Zandai Publications
Published in 2018 in the U.S.A.

The Matrilines of the Egyptian Arabian Horse
Volume I

Table of Contents

DEDICATION

THIS BOOK IS DEDICATED TO ESPERANZA RASWAN

Esperanza Raswan, wife of Carl Raswan, worked tirelessly to make The Raswan Index a reality. She had a deep commitment to him and an unwavering belief in his works. She met her future husband in 1947 while she was working for the Mexican Consular Service in New Mexico. Prior to this, she had worked for the Consular Service in Washington D.C. and in Dallas Texas. They were married a few years later, in 1951. These became crucial years during which Carl and Esperanza wrote the six volumes of The Index, the last volume being completed just before his death in 1966.

Esperanza felt very privileged to have been a part of this collaborative effort to publish The Raswan Index and The Arab and His Horse. After Carl Raswan died, Esperanza was left with the responsibility of publishing the last volume of The Raswan Index and raising two young daughters by herself.

She always found strength and consolation knowing that, in some way, he is still with her. His written word lives on, as well as his accompanying presence in her life. She fondly recalls that her father Carl once described Esperanza as "a substance made from the angels".

These remembrances were contributed for this dedication by Beatrice Funes, daughter of Carl and Esperanza Raswan.

Figure 1 - *Carl Raswan.*

Figure 2 - *Esperanza Raswan.*

October 15, 2010

Dear Dr. Hudson,

I wish you great success with your book. May it serve to educate and inspire people to appreciate the beauty of the Arabian horse.

Esperanza M. Raswan

يرحب المؤلف باحترام بمحدثي اللغة العربية المهتمين بالخيول العربية الاصيلة و يأمل ان يكون محتوي هذا الكتاب مفيدا لهم. حيث ان الحصان العربي هو من ترات الشعوب العربية و ان الخيول العربية الاصيلة هي التي تولد و تنشأ في الصحراء تحت تربية القبائل العربية. وهذه الخيول جواهر ثمينة خلقت ونشأت في بيئة وظروف صحراوية قاسية. و قد وجدت الخيول العربية في المزارع الخضراء في أوروبا او في معارض التربية الحديثة في أمريكا. و لكن تختلف هذه الخيول عن الخيول الأصيلة لدى القبائل العربية. حيت ينبغي على الخيل العربي الأصيل ان يستنشق أنفاسه الاولى من هواء الصحراء الجاف, كما ينبغي ان يقوم الحصان العربي الأصيل بخطواته الاولى على الرمال الساخنة في صحراء موطنه الأصلي. و بهذه الطريقة فقط ينشأ الحصان الأصيل الرائع الذي يحظى بالمكانة والتقدير من العالم بأكمله. و يعترف مربيون الخيول العربية في الغرب بان معتقدات و عادات القبائل العربية لقرون عديدة أدت الى رعاية هذا الحيوان الثمين والحفظ به لحد هذا اليوم و الجميع يقدرون جهودهم. لو ما احتفظوا بالخيول العربية الاصيلة ما كان ممكن نتمتع بهم اليوم.

نقدم احترامنا و سلامنا الى كل الرجال و النساء الذين قاموا بتربية هذه المعجزة الطبيعية والسلام على اشقائنا وشقيقاتنا من العرب الذين اهتموا بهذا الكنز من تراثهم الطبيعي و هو الخيل العربي الأصيل.

EXORDIUM

EXORDIUM: AN INVOCATION

The Egyptian Arabian horse is Nature's finest and most all-inclusive work of art, Nature's gesamtkunstwerk. Of all the marvelous life forms seen in Nature, none exhibits the complete, comprehensive, and ideally composed combination of form, function, sound, action, spirit and poetry. No other animal exhibits the consummate and fully integrated features of art that is characteristic of the noble, aristocratic and exotic Arabian horse. The authentic Arab horse is the epitome of form melded to function. He is the complete synthesis of all of the elements of beauty and nobility. He is a living breathing miracle. He is a prince among horses, possessing valuable features and qualities that are all his own. The Egyptian Arabian horse is a transcendental phenomenon. Speaking in his own language, he is a splendid companion, keen to please. The Egyptian horse, free in nature, is a visual spectacle unlike any other in nature.

On a breeding farm, splendor is on display when the mares and foals are turned out in the morning; it is a sight to delight all those fortunate enough to witness the spectacle. The stallions are then led out individually to their runs, and the day begins with great excitement and anticipation. It is a sight of trembling beauty.

In form, the Egyptian Arabian horse is a sinuous and lissome work of sculpture. When motionless, he is frozen dynamism, eager to explode into motion. In movement he is balletic. In spirit, the horse is noble, proud, and regal. To stand in the presence of this creature is transfixing. The thrilling trumpeting call of the aroused stallion, the terrifying sound of two stallions in battle, the gentle musical nickering of the broodmare in the evening, and the plaintiff call of the newborn foal resound in Nature's most perfect act of creative genius. To witness the earth-shaking rhythmically galloping hoof beats across a broad expanse of the plain is a primal experience, not unlike the sound of the tympani. The poetry of the Egyptian Arabian horse comes from the Bedouin themselves.

Early in the morning, while birds are still nesting, I mount my steed. Well bred was he, long bodied, outstripping the wild beasts in speed.

Swift to attack, to flee, to turn, yet firm as a rock swept down by the torrent. Bay colored, and so smooth the saddle slips from him, as the rain from a smooth stone.

Thin but full of life, fire boils within him like the snorting of a boiling kettle;He continues at full gallop when other horses are dragging their feet in the dust for weariness.

A boy would be blown from his back, and even a strong rider loses his garments. Fast is my steed as a top when a child has spun it well.

He has the flanks of a buck, the legs of an ostrich, and the gallop of a wolf. From behind, his thick tail hides the space between his thighs, and almost sweeps the ground.

When he stands before the house, his back looks like the huge grinding stone there. The blood of many leaders of herds is in him, thick as the juice of henna in combed white hair.

We returned at evening, and the eye could scarcely realize his beauty, For when gazing at one part, the eye is drawn away by the perfection of another part.

Imru' al-Qais bin Hujr al-Kindi, 6th century CE Arabian poet.

A Qasidah

In a bend in the wadi where
The grass is as tall as the Dal trees
Through which armies pass, both those with booty, and those that return empty handed
I would ride forth early
When the birds were still in their nests
And rain water was still running
In every torrent channel
On a sleek steed
A shackle for big game
Left thin by his pursuit of the herd leader's runner
On every long chase
Despite fatigue, he is ebullient, tall,
As if his withers
Despite leanness and much running
Were a large tree on a lookout hill

In galloping he vies with the wild ass
Who kicks out his legs
As his fetlock hair flies, you see
He is built like the wood of a cloth rack

He has the two flanks of an antelope
The two legs of an ostrich
And the withers of a wild ass
Standing on a lookout peak
He steps on hooves

Hard and solid
As if they were the stones of a stream
Bright green with moss.

His rump is like a sandhill
Packed down by rain,
And his withers
Like a howdah's wide saddle.

He has an eye
Like an artisan's mirror
Which she turns around her eye
To examine a veil.

He has two ears in which
You perceive good breeding
Pricked up like the eyes of a frightened oryx -doe
In the middle of her herd.
His ear bone is round
As if his reins and
Bridle were on top of
Smooth stripped palm-trunk

His tail is black
With a fleshy pliant bone
Like the date laden boughs
Of Sumayhah Spring
When he runs a double heat
And his sides are wet with sweat

You would say his breathing sounds like
The rustling of the wind as it passes a huge tree.

He turns a croup
Like a large pully
That overlooks a rump
Like a wide pack saddle.

Impatient, he chews on the tethering post
Until he seems mad
And possessed by a demon
That won't let him go
One day he pursued
A herd of white coated oryx
and another day he pursued
A wild desert ass with foal

Then while the white oryx cows
Were grazing in a thicket, Walking like maidens
In fringed white robes

We called out to each other

As we fastened his cheek strap
And my friend said

They have escaped you, so chase them

With great difficulty

We mounted our boy
On the curved back
With a strong spine

And he took off
Like an evening downpour
As the oryx emerged from a whirlwind of dust
To the leg he is fiery
To the whip like a flood
When you chide him
He takes off like an ostrich
And He reached the game
Without effort and without a second try
Whirling like a child's
Button on a string

You see the mice of the
The low soft ground
Heading for the dry hard ground
From his thundering gallop

It drove them out
From their holes
Just as a noisy evening
downpour does

He struck in succession
An oryx bull and cow
An old bull white

As a sheet of parchment

As the oryx bull of
The sand dune bellowed
He kept striking them with a Samhari spear
Reinforced with a sinew

Then old bull fell
On its white face prostrate
While another protected itself
With a horn like the tip of an awl
Imru' Al-Qais

No other creature elicits the same sense of awe and admiration in man. No other creation in the natural world moves man on the same deep emotional plane. No other animal bonds with man with such devotion and courage. The Egyptian Arabian horse is loyally devoted to his caretaker, and he forms a strong personal bond with his owner. The bond is reciprocal.

In an age as materialistic and cynical as is the present age, the Arabian horse provides a means of consolation. The Arabian horse exists in a state of perpetual becoming, and his caregivers honor this fact. It is a sublime creature, the antidote for the commonplace of life. In a world of superficiality and impermanence, it provides inspiration and a sorely needed means of hope for the future.

> "The world of to-day is sick to its thin blood for lack of elemental things, for fire before the hands, for water welling from the earth, for air, for the dear earth itself underfoot."

> "For the animal shall not be measured by man. In a world older and more complete than ours they move finished and complete, gifted

with extensions of the senses we have lost or never attained, living by voices we shall never hear. They are not brethren, they are not underlings; they are other nations, caught with ourselves in the net of life and time, fellow prisoners in the in the splendor and travail of the earth."

Henry Beston

" ...Whither thou goest I will go, whither thou lodgest, I will lodge..."

Figure 1 - *Akerf, by Victor Adam.*

Figure 2 - *Nigid, by Victor Adam.*

Figure 3 - *Hadjar, by Victor Adam.*

Figure 4 - *Omar, by Victor Adam.*

No other animal is so captivating. None are so loyal, none are so intelligent. It is the inner fire, the outer beauty, the harmony of movement and intelligence of the Arabian horse that has for centuries inspired and motivated man for to seek out and cherish these lovely creatures. The constitutional ensemble of the Arabian horse is a blending of intelligence, nobility, gentleness, curiosity, playfulness and energy, seen nowhere else in creation. The Egyptian Arabian horse is enchanting, without peer in the natural world. He embodies the spirit of freedom and gives solace to those shackled by convention and the restraints imposed by the social order. The Egyptian Arabian horse can be a companion of the first order, but he can never be dominated or commanded.

And so, in the event that someday, someone comes searching...

> The beauty and genius of a work of art may be re-conceived though its first expression be destroyed; a vanished harmony may yet again inspire the composer; but when the last individual of a race of living beings breathes no more, another heaven and earth must pass before such a one can be again.
>
> William Beebe, 1877-1962

Special thanks are extended to Dr. Karin Theime for permission to use these illustrations.

PROLOGUE

PROLOGUE

THE EGYPTIAN ARABIAN HORSE; AN OWNER'S MANUAL

Egyptian Arabian horse breeding is really very simple:

> *"To make a prairie it takes a clover and one bee—*
>
> *One clover, and a bee*
>
> *And revery.*
>
> *The revery alone will do*
>
> *If bees are few."*
>
> Emily Dickinson (1830-1866) From her book of poetry, Nature, XCVII

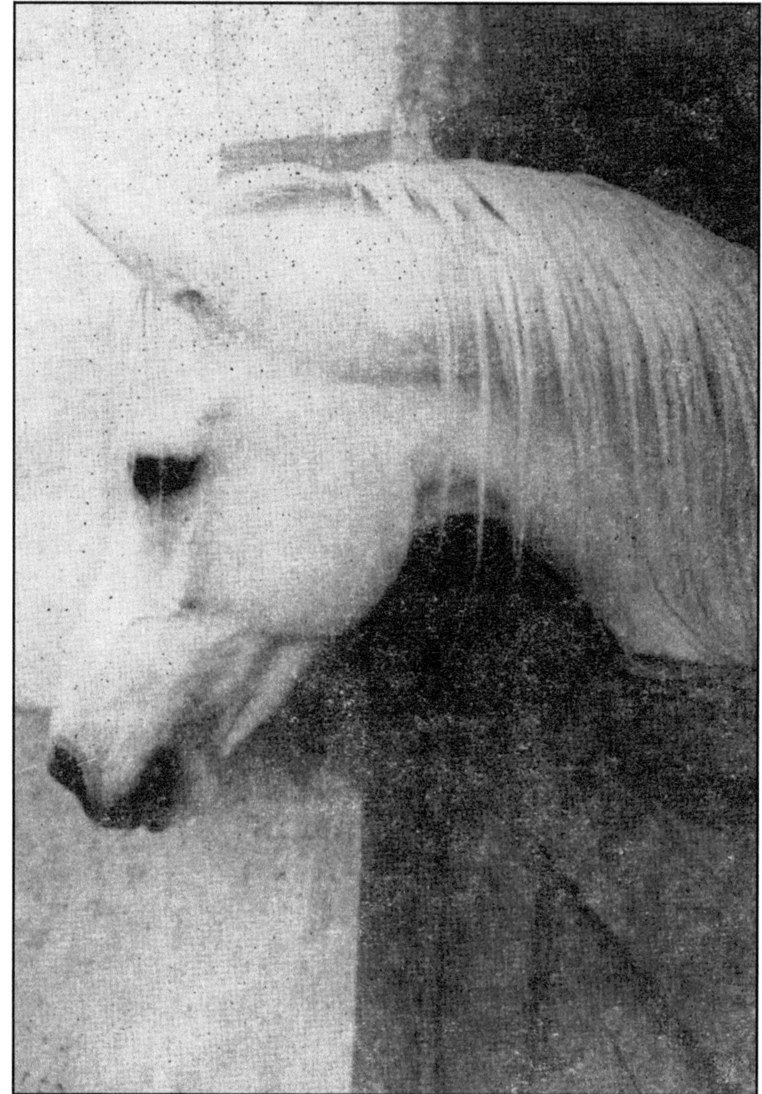

Figure 1 - *The stallion Talal (Nazeer x Zaafarana), from the EAO Studbook Volume 2, 1966.*

All very simple. One female element, the mare, and one male element, the stallion. With these two elements, a prairie full of horses can be created. The stallion inseminates the mare, and then, after some delay, a foal is born. The insemination itself takes only a few minutes. Then comes the waiting period. It's a long time to wait: eleven months. That leaves a lot of time for revery.

Most of the time, the Egyptian Arabian horse breeder exists in a state of revery, daydreaming about the future, eagerly awaiting the birth of a foal that will satisfy his ambitions, hoping for a foal that will be an improvement over the parents. Optimism prevails. There is a lot of time during which the breeder is preoccupied, lost in thought, entertained by musings about the foal to come. There is little to be done while waiting on the birth of the foal. There is some feeding and some cleaning, but basically, the breeder exists in a state of reverie; the mind wanders and is abstracted, lost in a world of the immaterial, waiting for the material to appear. He waits for the birth of a *purebred Arabian horse*. Days pass, foaling time draws near, and the mare's owner begins to become restless and anxious about the impending birth. There is so much at stake, and so many things can go wrong. At the core of the breeder's musings is the question "what is the true origin, the true genetic identity of the mare and stallion used to produce this foal?" One wonders.

This book is about the genetic identification of the matrilines of the Egyptian Arabian horses. This book asks one question: What is the actual matrilineal identity of the horses that the Egyptian Arabian horse breeder uses when he makes his apolegamic decisions? The answer to this critical question will be found in the original research published here in Chapter Eleven. In a way, this book is a search for ghosts, as it attempts to peer into the past, looking for horses long since gone.

This book also has one regrettable but unavoidable secondary purpose: to examine the perplexities surrounding the meaning of the term "*purebred Arabian horse*". The research presented here is ostensibly concerned with matrilineal connections, but the only real reason for performing such a study was to establish whether or not the horses of the EAO in Egypt were in fact *purebred*, tracing in matriline to desert sourced horses. This question need not be asked at all except for the fact that skeptics doubt the absolute genetic purity of these horses.

This book is about, among other things, words. It is about the uses and abuses of words. Words are deceptively powerful; they are often used to corrupt reason and logic. Words are often used to circumvent rigorous thought. The Greek philosopher Socrates focused on the precise and exact meaning of words, attempting to arrive at a universally valid and honest definition of certain key words. He did this because of his observation that men often commit regrettable acts, not so much because they wish to do mischief, but because of their failure to understand the proper meaning of words. Words are powerful tools in the manipulation of public opinion. Immeasurable harm can result from the use of misrepresented terms. Words are glyphs and mean different things to different people.

This book is about the questions, doubts and troubling uncertainties that the Egyptian Arabian breeder must confront.

The proper meaning of the term "*purebred Arabian horse*" is not simple, nor is it very poetic.

Broadly speaking there are now two definitions of the term "*purebred Arabian horse*".

One definition is absolute: it is defined as a horse whose ancestors all originated with the Bedouin tribes of the Arabian Desert, and which has not been contaminated by non-Arabian horse genetic material. There are no exceptions. It is the antique-type Arabian horse, the classic Arabian horse known throughout the world. For centuries, this definition was an unshakable belief held by the devotees of the Arabian horse. To many, this horse is denominated asil. There is no ambiguity.

The other definition of the *purebred Arabian horse* is both relative and ambiguous. It is an alternative fact. It is defined as a horse whose ancestors came in part from the Bedouin tribes of the Arabian Desert and in part from non-Arabian horses. A horse defined in this way contains a variable degree of non-Arabian horse genetic material. They contain many different types of genetic material, injected into the breeding pool from a wide range of non-Arabian horses. There was a time in the past when a horse of this type was considered to be a part-bred Arabian horse. But not today.

The problem with the first definition is that it is no universally accepted scientific and objective way to define the term. It is a dilemma of the deepest moment, and to the Egyptian Arabian horse breeder, it remains a problem of unsurpassed importance.

The problem with the second definition is that it constitutes an open admission that the Arabian horse of today is a mongrel, a part-breed animal consisting of genetic material from a wide range of horse types. It is a capitulation to convenience and commercial interests. It is a compromise made by those individuals who wish to expand the meaning of the term *purebred Arabian horse* in order to facilitate the international trade in a horse known by the name "Arabian". It is about the buying and selling of Nature's jewel. It suggests that the idea of the *purebred Arabian horse* is, and always was a fable. It also blocks the potential for the emergence of genuine and authentic creativity in breeding the animal. For most people, this is an acceptable position. For others, it is not.

The reader will not find a clear and definitive position on the definition of the *purebred Arabian horse* in these pages. There is not now, nor is there ever likely to be, a universally accepted definition of the term. This is in part due to the entrenchment of the present day Arabian horse breed organizations, in part due to the commercial interests of breeders and salesmen, and in part due to the absence of equine genetic data that would establish an uncontestable meaning of this term. A genetic definition would require the discovery and analysis of sufficient numbers of ancient *Equus caballus* skeletal remains from the Arabian Peninsula. This is not likely to happen. This book does not attempt to deal with the known unknowns.

This book is written specifically for people who are new to Egyptian Arab horse breeding; it is written for beginners. There has been no attempt to sanitize or simplify the subject; it is not neat and it is not orderly. It is complicated. The beginner will need to know what a classically beautiful asil Arabian looks like, and, more importantly, what it does not look like. The beginner will need to understand and appreciate the value of extended pedigrees. This book is not for professionals or academics of the Egyptian Arabian horse. Readers who wish to find authoritative research on this subject can seek the works of scholars. Likewise, experts on the subject of the Egyptian Arabian horse pedigrees are amply served elsewhere.

People yet to be born will breed Egyptian Arabian horses yet to be named. The insights of today may provide the next generation with the information needed to make informed decisions in horse breeding. In the future, the novice to Arabian horse breeding will need a reference source that presents a comprehensive and unbiased overview of the meaning of the term "*purebred Arabian horse*". This book is built on facts and the direct observations and scientific investigation of the author. It is not based on frivolous rumors or subjective assessments. The book does contain instances of hearsay and speculative opinions, but these are presented purely in the interest of providing a historical perspective for beginners. Much foolishness has been written about the Arabian horse; the beginner needs to be aware of this fact. In the Arabian Desert, the Bedu refer to Europeans who wrote such confused and contradictory opinions about the Arabian horse as "madmen". The Egyptian Arabian horse breeder Prince Mohammad Ali Tewfik was appalled by the misconceptions held by European authors who wrote about the Arabian horse. The irony of his opinion is explored later in this book is intended to be, to the last degree, objective. While the research presented here is considered by many to be tendentious, it is not intended to be so. It is just the facts.

Regarding the tone of this book, some few words are needed. Any opinions expressed in this book are purely incidental and should be disregarded by the reader. Any speculations made by the author should also be ignored. Facts alone, not opinions, serve the future of the Egyptian Arabian horse. And the facts are an avalanche of information. The problem of maintaining objectivity should also be considered.

Alexander William Kinglake discusses the problem of an author's perspective (the personal equation) in the preface to his book Eothen. "The people and things that most concern him personally, however mean and insignificant, take large proportions in his picture because they stand so near him." And so it must be.

The novice to Egyptian horse breeding is faced with a bewildering web of gossip and opinions regarding the value of particular types of Egyptian horses. There is one caveat that the novice to the field of the *purebred Arabian horse* should note. It is a mistake to judge the genetic purity of any Arabian horse based on its physical appearance. Western authorities have often suggested that the physical beauty of an Arabian horse was in some way an indicator of genetic purity or breeding potential. It is not. This is a mistake. Many Arabian horses, considered to be impure and undesirable because they did not fit the image of the beautiful Arabian horse, have contributed offspring of great significance to the breed. The reverse is also true. The Egyptian Arabian horse in its pure state is not uniform; it is multiform. There are tall Arabians, small Arabians, short Arabians, long Arabians, short-necked Arabians, and aesthetically unappealing Arabians. This observation is significant. The very idea of the standard Egyptian Arabian model misses this point entirely and misleads many beginners. There is no standard Egyptian Arabian horse. Too many genetically important Egyptian Arabian horses were never used to full effect, simply because they were not beautiful. *Purebred Arabian horses* do not all look alike. Nor should they. DNA analysis is beginning to clarify these matters.

The novice to breeding would also be well advised to disenfranchise himself from the tendency to glamorize the Egyptian Arabian horse. It is also important for the beginner in breeding to avoid anthropomorphizing the Egyptian Arabian horse. The tendency of Arab horse salesmen to glorify the horses that they sell, is, to a not inconsiderable extent, misleading. They are often peddlers of fiction. There is little about the desert and its inhabitants that is glamorous, and the romantic heady incense-filled atmosphere of the East that is propagated by Western writers and poets is, for the most part fictional. The central Arabian Desert is a scorching inhospitable furnace. In all candor, there is little about the Egyptian Arabian horses that is glamorous. Over the centuries, horsemen of the Arabian Desert and Egypt were at times depraved and cruel people. Veterinary practices were uniformly barbaric. It is a miracle that any of these horses survived the desperate conditions at all.

A young neophyte starting out breeding Arabian horses will be puzzled by the tremendous differences seen when examining Arabian horses; there are Russian Arabian horses, Egyptian Arabian horses, Polish Arabian horses, domestic Arabian horse, Pyramid Horses, Al Khamsa horses, Blue Arabians, Blue Star Arabians, Asil

Arabians, Spanish Arabians. There are sub-groups such as Kellogg, Davenport, Maynesboro, Babson, Crabbet and Selby. It is a shambolic state of affairs. What was once called "the Arabian horse" has now reached a state of fissiparous confusion, splitting into sub-groups which seem to have little or nothing in common. It is confusing. The horses are all so different from each other. The differences are so obvious that questions naturally arise, difficult questions. How can this be? Who is to be believed? How did such a seemingly simple matter as the Arabian horse come to be so vexed? It is a distressingly convoluted topic. Discussions on this topic usually lead to perplexing and apoplectic debates. This book is designed to be a guide for the perplexed.

It is a simple matter for a beginner to obtain a clear and unambiguous concept of the physical or phenotypic features of the authentic asil Bedouin-sourced desert horse of antiquity. The physical attributes of the ancient and original horse of the Arabian Desert are known and well documented. There is an abundance of artwork consisting of sculptures, bronzes, paintings, etchings, and lithographs made by artists who were eye witnesses to the *purebred Arabian horse* that they saw during the 17th, 18th and 19th centuries. Dr. W. Georg Olm's series of books Asil Araber contains excellent reproductions of many of these works. When one compares the images of horses shown in Dr. Olm's book with the Arabian horses that are seen today in the world's show rings, the differences are striking. This fact requires some reflection. It is perplexing.

There is an expression to the effect that for some people the Arabian horse must be seen to be believed. For others the Arabian horse must be believed to be seen. It is puzzling. It is intriguing. For many, it is life changing.

Future beginners new to Egyptian Arabian horse breeding will not obtain his horses by "gift, legacy or war". He will obtain his horses from salesmen. To complicate matters further, Arabian horse salesmen will tell the beginner conflicting versions of the importance of genetic purity in the Arabian horse. A beginner may not be able to sift through the contradictory opinions of modern authorities, but he can easily obtain a clear understanding of the Bedouin concept of genetic purity and the value of uncontaminated blood in desert horse breeding. The importance of purity in the Arabian horse pedigree to Bedouins in the interior deserts of Arabia is well known. This documentation is available in Dr. Georg Olm's series of books Asil Araber.

That much is simple.

In 1957 the world of Egyptian Arab horse breeding in the Western world was a quiet, simple, naïve world consisting of a small number of horses and a small number of people who, for a variety of reasons, loved them, respected them, cared for them, and used them in their proper capacity, as ridden animals. Owning an Arabian horse was an apolaustic experience. Today, all of that has changed.

For the novitiate in Egyptian Arabian horse breeding three things are needed; some horses, some land, and some knowledge of horse breeding.

Figure 2 - *Bringing in the Sheaves, photograph from the Library of Congress collection.*

Before the complete mechanization of agriculture, ranching and livestock breeding were a family tradition, passed down from generation. The horse was an essential part of the enterprise. There was a time when no rancher's son would imagine being anything other than a rancher. The knowledge and techniques of ranching and livestock breeding were passed down naturally. Boys learned about farming and ranching operations from the time they could walk. They learned directly from their fathers, brothers, and their uncles, all of whom were ranchers. They learned what a horse was to be used for and what it wasn't to be used for. They learned how to be around horses, and they learned how horses think and react. They learned that horses are not people; they learned that a horse can inflict grievous bodily injury and death to the unwary. They learned when to breed a mare and how to breed a mare. They learned when and how to discipline a stallion and when not to discipline a stallion. They learned when it was best to call it quits for the day. They learned the ways in which horses respond to the rancher's own moods and states of mind. They learned the value of a horse and they learned the value of hard work. They learned by doing, they learned by osmosis. Today, all that has changed. Man's relationship with the land has changed.

In years long past, the land was respected. Now it is just real estate. There was a time when the land itself was passed down from one generation to the next. The people were bound to the land. The land was sacred to the rancher, and it was cared for with respect and with reverence. This sense of devotion to the land was born out of the clear and profound awareness that the lives of the animals, rancher and his family depended directly on the health of the land itself; it was held to be sacred. Without the land, all life ceases. The dirt gives life; that meager three inches of soil that covers the barren and lifeless body of the planet meant everything. It was treated with respect because it sustained all living things. The land was Nature's grace to the farmer and rancher. The land and each blade of grass that grew there were regarded as miraculous. But all of that has changed now.

"Do no dishonor to the earth lest you dishonor the spirit of man"

Henry Beston

The Tuttle Farm in Dover New Hampshire was founded in 1635 by John Tuttle, an English emigrant who survived a shipwreck off the coast of Maine. His farm is still run today, on the same 240 acres, owned and operated by the 12th generation of Tuttles. Shirley Plantation in Charles City Virginia was established by Edward Hill in 1638. Today it is run by Charles Carter, a member of the 10th generation of the family. Parlange Plantation, New Roads Louisiana, has been run by the descendants of Marquis Vincent de Ternant since 1750.

The farm of Hugel et Fils in France was established in the 15th century and is still a working family farm today. The Yamamotoyama farm in Japan has been a family endeavor for three centuries. Hacienda Los Lingue and the Aculeo Stable, in San Fernando Chile, was founded in 1760, a gift to Don Aguila from the King of Spain. It is still a family owned ranch, breeding a continuous line of horses that trace to stock that originated during the Arab conquest of Andalusia in the7th century.

The fact is, the Arabian horse is no longer essential to man's survival. It is now a commodity, not unlike oil, pork bellies, coffee, corn or wheat. It is a luxury good. The future of the asil Arabian horse is now in the hands of salesmen. The desert princes, kings, and khedives are gone now, and the days in which an asil horse could be obtained only by gift, legacy or war are long past. A beginner today must buy his horse from a salesman. Dani Barbari, a highly regarded veteran Egyptian Arabian horse breeder in Cairo, once wrote a charming letter to the Arabian horses of the world. "I want to tell you that I am extremely sorry for what has happened to you. You are being turned into a commodity, something to be bought, sold, used and altered to suit the whims of a few selfish and uncaring people...people must allow you to be who you are and were meant to be...a treasure of uncommon value, a spirit with dignity, pride, loyalty, and a friend beyond riches and value."

She wrote that a long time ago.

In economic theory, two items which are considered to be commercially equivalent in all respects are said to be fully fungible. Crude oil from the North Sea is equivalent to crude oil from Texas. The two are indistinguishable. They are, in commercial terms, fully fungible. Their monetary value, as well as their composition and functionality, therefore, are equal. All horses described by the modern commercial use of the term "*purebred Arabian horse*", which is the relative definition, are considered fully fungible. One Arabian horse is, by current convention, just as good as the next. A customer cannot tell where an Arabian horse came from by looking at it, feeling, measuring it, riding it, or examining its registry papers. The salesmen know this, and they can, with peremptory confidence, manipulate potential customer's expectations accordingly. That's how capitalism works. And our world is now, pole to pole, capitalistic.

Capitalism is based on the manipulation of consumer expectations. The chicken nugget is a prime example. In general, the customer who buys a chicken nugget thinks that he is getting a nugget made of pure chicken, which in his mind means specifically the muscle component of the chicken. The American Journal of Medicine published an article in which researchers analyzed the components of three popular brands of chicken nuggets. They found that the nuggets consisted of only 40% chicken muscle. The remaining 60% was made up of non-muscle chicken parts; fat, cartilage, bone, blood vessels, nervous system elements, blood, and the internal organs of the chicken; liver, heart, lungs, kidneys, intestines, bladder, and pancreas. Technically speaking, the nugget is all chicken. It is pure chicken. But it is not pure in the way that the customer thinks that it is pure. The consumer's expectations have been manipulated. The Arabian horse of today is treated in this manner. Because of the success of

WAHO, salesmen can now claim, with complete honesty, that any horses registered with a WAHO approved registry are *purebred Arabian horses*. But they are not *purebred* in the way that most customers think they are. The term can be misleading to the uninitiated.

Arab horse breeding is an odd business. On close analysis, there really is no rational explanation why it should motivate people to go to the extremes that it does. In America, the land where the only thing that is genuinely feared is inconvenience, the Arab horse breeding business continues to hobble along. And this in spite of the fact that nothing is more inconvenient than owning and managing a breeding herd of Arab horses. It is a peculiar business, full of peculiar people. Peculiar is the very word for it. Theodore Patterson, writing to Wilfred Blunt at the Crabbet Stud in England February 28th, 1900, made the following observation; "It is a thousand pities that the interests of the Arab have been confided to such very queer people." Patterson was referring here to his dealings with the Arab horse breeders Randolph Huntington, James A.P. Ramsdell, and Colonel Spencer Borden. These three men each had peculiar turns of personality.

Today the asil *purebred Arabian horse* is on the verge of extinction. The number of horses living today that do not contain the blood of Polish part-bred Arabians is quite small. This threat has come about because the qualities that made the asil horse essential to the desert Bedouin are no longer needed for survival in the age of electronic information. This book is written to draw public attention to the plight of the Egyptian Arabian horse. The day may come when these horses simply cease to exist. Their genetic constitution is unique. The *purebred Arabian horse* may someday be needed, as it has been needed in the past, to re-infuse vigor and vitality into existing non-Arabian breeds in order to rejuvenate the mixed breeds, restore vital functions, and correct emerging structural defects. Arabian horse blood has, through crossbreeding, rehabilitated many an equine breed. The blood of the true Arab horse can restore lost vitality and correct the morphologic and metabolic abnormalities that any cross breed will eventually and inevitably develop. The problem of breed deterioration is seen most clearly in the modern English Thoroughbred horse, which has in recent years become structurally so fragile that the horses often break down or sustain leg fractures in the course of training and racing.

So why does Arabian horse breeding continue? How is it that the entire Arabian horse population does not just die out? It happened to the Auroch. The primeval ancestor to domestic cattle, the Auroch (*Bos primigenius*) was, by any definition, a truly *purebred* animal, having been present on the planet in a stable genetic state for millions of years. The oldest known Aurochs have been found in the paleontologic records in India, present there over 2 million years ago. The Auroch reached Europe about 270,000 years ago, but by the 13th century, its natural range had become restricted to central Europe. Its numbers had been decimated by *homo sapiens* who hunted the animal for food. Through cross breeding and selective breeding with the European bison, known today as the *wisent*, Neolithic man created the modern cow, *Bos Taurus*. The cow was docile, useful and easily domesticated; the Auroch, noble and resistant to domestication, remained wild and free. The last *purebred* Auroch died in the Jaktorow Forest in Poland in 1627. The Auroch is gone now, and all that we have left today is the cow. There was no preservation movement for the Auroch, no Auroch Society, no Auroch Club.

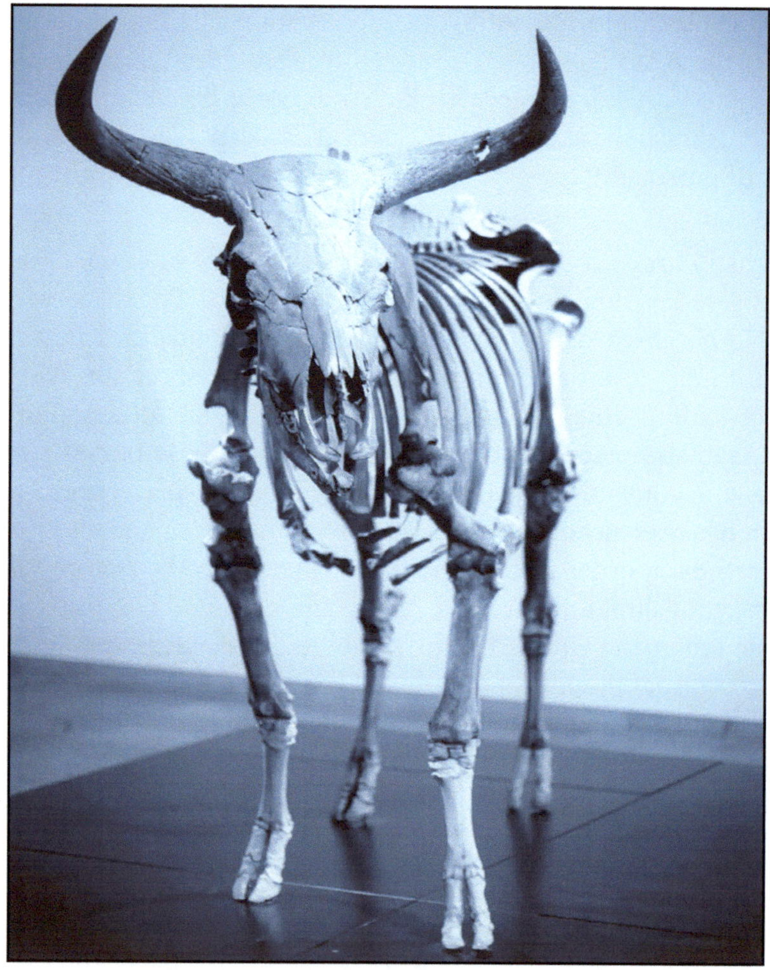

Figure 3 - *Mounted Auroch skelton, National Museum, Copenhagen.*

Figure 4 - *The Auroch depicted in cave paintings at Lascaux Caves, Montignac, France. The paleo-horse is also depicted.*

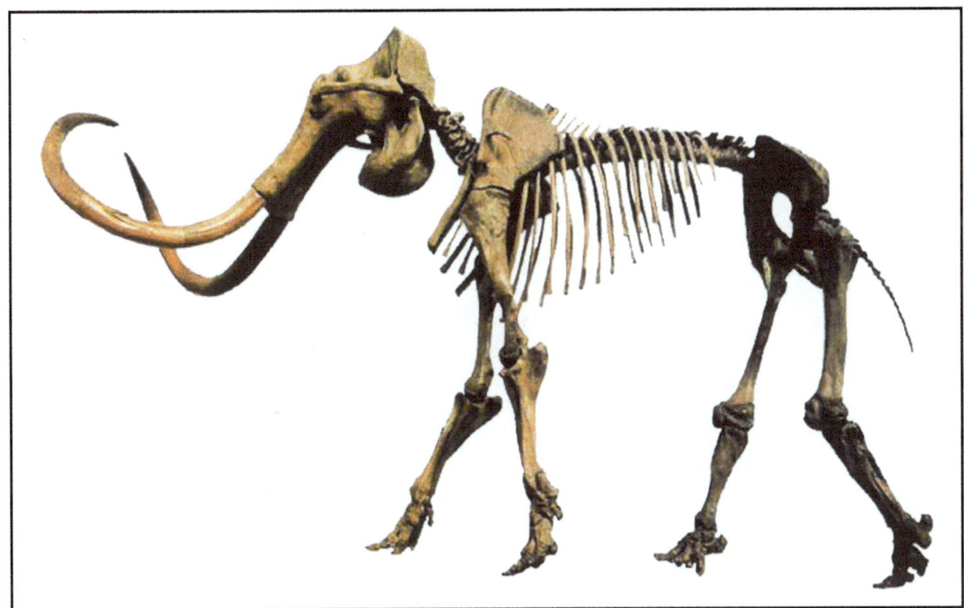

Figure 5 - *Mammuthus primigenius.*

It happened fairly recently to the Wooly Mammoth (Mammuthus primigenius). Like the Auroch, the Wooly Mammoth did not belong solely to a Paleolithic past. The Wooly Mammoth first existed in Asia, about 6 million years ago. They were hunted by Paleolithic humans, beginning about 30,00 years ago. They were killed for meat, hides, and for the prized ivory tusks, which were used for making weapons and objects of art, decoration, and ritual. They could be killed but they could not be domesticated. The last Wooly mammoth died about 4000 years ago on Wrangel Island, a small isolated bit of land on the Arctic Ocean, near eastern Siberia. They were hunted to extinction, by our direct ancestors. All that is left now are the nearly hairless cousins of the Wooly Mammoth, the much smaller and easily domesticated elephants.

In 2015, the last White Rhinoceros in the Western hemisphere died, leaving only three others in the world, all living in a wildlife preserve in Kenya. The rhinoceros has been driven to extinction by loss of habitat and the carnage of poachers who hunt them for their horns. In 2018, the last male Northern White Rhinoceros in the world died, age 45, in Kenya.

Figure 6 - *Ceratotherium simum.*

Does any of this matter? Opinions vary. There are those who feel that the purposeful and deliberate annihilation of a creature that has been present on this planet since the dawn of time constitutes an unspeakable form of barbarity, depravity, and disrespect. They are appalled at such an abomination. Others are not. To the preservationist, *et bonum quo antiquius eo melius.*

This book is, almost as a side effect, concerned with the preservation of a group of authentic animals born from Nature's crucible and now entirely at the mercy of humans. The problem with the preservation of the asil Arabian horse is that it requires three things: interest, energy, and a substantial financial commitment. To be useful to future generations, it also requires personal candor and honesty on the part of the breeder, seasoned with humility. It requires that the breeder be a conscientious and sincere servant of the breed. That is not easy. It is a constant struggle.

But then the act of creation has never been easy. The French novelist Gustave Flaubert once wrote a letter to his friend and author George Sand about the tribulations and sacrifices that he experienced as an author; *"l'Homme n'est rien l'oeuvre c'est tout."* And so it is with Egyptian Arabian horse breeding. For some, it becomes a passion. For some, a life spent breeding and caring for these creatures takes precedence above all else. No sacrifice is too great for this privilege.

The generic Arabian horse has a secure future and will survive even the fad-driven American horse buying population. The Arabian horse is in the process of slowly losing its identity, not its existence.

The same is not true for the authentic asil type horse. The threat to the genuine desert bred Arabian horse is discussed later in this book.

So why does the authentic Arabian horse of the desert continued to be bred? The reality of breeding Egyptian Arabian horses is that it is hard work. It is unpredictable work. Outcomes often do not match expectations. There is a steep learning curve. In many ways, it is accurate to say that the Egyptian Arabian horse breeder races from one calamity to the next, but each time with undiminished enthusiasm. That is really the essential element in successful breeding; enthusiasm. Enthusiasm and optimism. But then, there are rewards. For some people, there is something about the very sight, the mere presence of an Egyptian Arabian horse that touches a deep chord in the heart. Being in the presence of an Arab horse seems to wash away the dust and grime of common everyday life. Even a few quiet hours with an Egyptian Arabian horse can shield the owner from the deafening roar and chaos of the mental and emotional traffic of everyday life. Standing in a field of Arabian horses, the dull and suffocating routine of daily life vanishes. Briefly, one is at peace. Briefly, one finds refuge from the tedium, pointless monotony, and gnawing trivialities of existence. Beyond the endless chores of earning a living and paying bills, time spent with an Egyptian

Arabian horse is a powerful antidote, a restorative for the weary.

Breeding Egyptian Arabian horses is both a privilege and a responsibility. It is not difficult to maintain the initial sense of awe and excitement that the novice feels when first seeing a beautiful Arab horse. To quote the British aesthete Walter Pater "to burn always with this hard, gem-like ecstasy, is success..."

It is difficult to get to the heart of the matter. The psychological and emotional roots of this strange phenomenon must go very deep.

There is a story told about the Nez Perce Indian Chief Hinmatoowyalahtqit, known to the white man as Chief Joseph. He was born in 1841 in a camp on the Wallowa River in Oregon. His native land was stolen from his people by white American settlers who wanted the ancestral home of the tribe, the Wallowa Valley in Oregon for themselves. The U.S. government supported this action and sent troops to relocate the troublesome Indian tribe that resisted the intrusion of the settlers. Joseph and his people met the troops with armed resistance. Joseph led his warriors during the fabled and bloody Nez Perce War of 1877. Many white soldiers were killed. Chief Joseph killed American soldiers. The fierceness and bravery with which the Nez Perce fought, however, brought them the respect of their enemy, the white American soldiers. The U.S. Army under the direction of General Oliver Howard relentlessly pursued the band of fighter over 1700 miles for nearly a year. But eventually, the Indians were subdued, ground down by attrition and the relentless pursuit of the Army. The Nez Perce surrendered.

Chief Joseph handed his rifle over to the enemy at a formal surrender and peace ceremony, famously speaking to his people "My heart is sick and sad. From where the sun now stands, I will fight no more forever." He was arrested and convicted for his part in the rebellion and imprisoned for a time at the Fort Leavenworth Penitentiary. He was released on the condition that he spend the rest of his life on the Indian reservation in Lapwai in the Idaho territory, never to see his beloved Wallowa Valley again. He was forbidden by the U.S. government from returning to the Wallowa Valley. He was banished and forced to live in exile. He personally went to Washington on two occasions to plead his case before two U.S. presidents, Rutherford Hayes and William McKinley. They were amused by the appearance of this red man with his ceremonial headdress in the White House. They listened to him politely, and had their pictures taken with him; it made for a good press release. But they had no inclination to grant his request. In the Nez Perce religion, a man must be buried in the land of his birth, or his spirit will roam the world forever, restlessly seeking his final resting place. All that Joseph wanted was the right to be buried at home.

Figure 8 - *Wild Horses, a water color by Alfred Jacob Miller, an American artist who painted a series of watercolors depicting the wild horse herds of the American West. Image used with the permission of the Walters Art Museum.*

As it happened, Joseph had won the admiration of many of the American soldiers. They respected him as an honest and eloquent man. Although he fought the white man, and although he killed white soldiers, he did so reluctantly and with revulsion. He was a man committed to peace. His quiet dignity and self-possession won him many admirers. An unusually close friendship developed with one of the infantry officers who had hunted him, a man named Charles Erskine Scott Wood. Wood was impressed with the Chief's nobility, dignity, and honesty. They corresponded after the war, exchanging letters through the Indian agent. Wood sent his son Charles Wood Jr. to live with Chief Joseph, to learn the Indians ways that had so impressed the Oregonian Wood, who was by then a wealthy and influential Portland politician. After living for some years among the Nez Perce, Charles Jr. received a letter from his father, recalling him home to begin college. In the letter, he instructed the young Wood to ask Joseph what he would like to receive as a gift from Wood, in appreciation of his many kindnesses shown to the boy. On leaving, Charles Jr. posed the question to the Chief. After a long and reflective silence, Joseph said: "Tell your father to give me a horse." The boy could not believe his ears. He did not respond to the old chief. He returned to Portland, silent and puzzled. He was so dismayed by the Chief's meager request that he never told his father about the conversation. It seemed so unimportant at the time. The Chief had horses. Why did he want another horse? He thought the old Indian must have become a bit senile.

Many years passed and the boy's father and the Chief grew old. Both men died, Chief Joseph in 1904 (his doctor said, from a broken heart) and Charles Sr., wealthy and successful, in 1944. One day, after the death of his father, Charles Jr. had an epiphany. In a sudden moment of insight, he was seized by a flood of remorseful emotion. In a flash, he understood the meaning of that request made years ago by the old Chief. And he wept.

In a flash of remembrance and insight, Charles Jr. realized that he had failed to appreciate the idea of the horse as seen through the eyes of the old Chief. He had interpreted the Chief's words literally, not as they were intended.

Figure 9 - *Charles Erskine Wood, from the Oregon Historical Society Archives.*

Figure 10 - *Chief Joseph, from the collection of the Library of Congress.*

The old Indian was not asking Charles for an actual animal. He was asking him, metaphorically, for the return of his freedom, represented by the archetypal idea of the wild horse. He was asking for the freedom to return to the Wallowa Valley, the freedom to be buried there. In his religion, the dead must be buried in their ancestral homeland, or the spirits of the dead were doomed to roam the infinite spaces endlessly, restless and alone. His desire was for simple freedom, the freedom to be buried at his place of birth. And he asked in the only way that he could. He could not bring himself to say the very words to the young man. He could not bear to be turned away again. It was too painful for the old Chief. He had been disappointed too many times. So he used the best metaphor that came to him. He knew no one would fulfill him his only real desire, certainly not his wealthy friend Wood in Portland.

As a young man, Joseph had often seen the bands of wild feral plains horses, grazing in the distance or running across the hilltops. The horses came and went as they pleased. Pens did not enclose them; bridles did not restrain them. They were strong and they were fast. They were not restricted by geography or fences, nor were they subjugated by any law or authority. They were their own masters. They made their own destinies. They were free. The sight of these creatures, even their memory, stirred in Joseph a strong emotional connection with a sense of the meaning of liberty. The horse is "pure air and fire" as Shakespeare observed in Henry V, *le cheval volant, chez les narines de feu.*

This story perhaps explains why so many people have such a strong visceral connection with horses; the horse incites in us, indeed exemplifies for us, the feeling of Freedom. And who among us is free?

For Chief Joseph, incarceration was physical. In modern Western culture, incarceration is often a psychological phenomenon. The intensely personal, and at times profoundly uncomfortable, confession of T.E. Lawrence in his book The Seven Pillars of Wisdom expresses this type of imprisonment. During his time in the Arabian Desert, working closely with Arabs (Ageyl, the Bedouin) during the 1917 Revolt, the al-thawra, he was struck by the vast difference between the Bedouin psyche and the Western psyche. He admired the Bedu. "We Westerners of this complex age, monks in our bodies cells, searched for something to fill us beyond speech and sense, were, by the mere effort of the search, shut from it forever. Yet it came to children like these unthinking Ageyl, content to receive without return, even from one another. We racked ourselves with inherited remorse for the flesh-indulgence of our gross birth, striving to pay for it through a lifetime of misery; meeting happiness, life's overdraft, by a compensating hell, and striking a ledger-balance of good or evil against a day of judgement." Such are the personal demons that lead to much human misery. The Egyptian desert horse is in spirit like the Bedouin who preserved them for posterity. They exist wholly in the moment, live fully in the present. They are filled with a natural vitality; having no thought for tomorrow. They do not "fret their hour upon the stage." As such they serve Western man as an antidote to that fatal disease, modernity.

Many beginners in Egyptian Arabian horse breeding fall by the wayside. Most do.

But for better or for worse, the future of the Egyptian Arabian horse belongs to the young. This is the reason that this book has been written. Some of them will succeed, but only those that possess in sufficient measure that divine confidence that is the treasure of youth. The young are unhampered by the commercial straight jackets that constrain the older generation. They are free from idolatry to the icons of the past. They are unaffected by disenchantment, disappointment, and disillusionment. They are idealistic, not cynical. They are unhampered by doubt. They are hopeful, not bitter. They have not developed personal grudges. They do not carry the heavy load of resentment that seems to naturally accumulate year after year as horse breeders deal with the vexations of buying, breeding, raising and selling horses. Young people are free because all of their mistakes are ahead of them. They are unrealistic and completely inexperienced; these are two traits that are invaluable when beginning any great adventure. Their chances of success are greater because they have not yet learned to fear.

Nature feeds the young on a steady diet of optimism and illusion, lest they quit the journey too early, and thereby bring an end to the Egyptian Arabian horse. It takes about ten years just to get started. It takes another ten years to rectify the mistakes made in the first ten years. After this, some useful work can be done breeding Arabian horses. It is an agonizingly slow process. But it is a good life, one which challenges the beginner and calls on the breeder to use his heart, head, and hands in concert. Heart, head, and hands. The three often do not want to cooperate with each other, but their cooperation is critical to success. They often quarrel. They bicker a lot.

When all is said and done, for better or for worse, the future of the Arabian horse will be determined by that primal visceral urge which motivates all human activity, self-interest, not selfless devotion. Self-interest may be disguised by breeders; it often is. It has always been thus. This fact alone should give all those who love the Arabian horse pause to reflect. But, as economist Adam Smith wrote in The Wealth of Nations in 1776, individual self-interest is essential in the operation of capitalism, and capitalism, he wrote, serves the common good. Therefore, self-interest serves the common good. Perhaps he was right. Perhaps this is true of Egyptian Arabian horse breeding. Perhaps not. Perhaps Smith did not address the topic of enlightened self-interest. Perhaps that term is in itself an oxymoron.

The *purebred Arabian horse* is now precariously situated on the eve of its destruction. In 1917, a few days before her death, Lady Anne Blunt wrote: "It will soon be too late to save the Arabian from extinction. Let us, therefore, hear the voice of Nejd before time and civilization have beaten us and silenced it forever." Today, scientific research is beginning the address this distressing situation. And if Darwin was right, it is not too late.

Regardless, it is best for all breeders to bear in mind the axiom that "the Egyptian Arabian horse can only be destroyed by Egyptian Arabian horse breeders."

Figure 11 - *"Ignorance more frequently begets confidence than does knowledge: it is those who know little, and not those who know much, who so positively assert that this or that problem will never be solved by science." Charles Darwin, (1809-1882) photograph from the collection of the Library of Congress.*

**All things change, and we change with them.*

PREFACE

PREFACE

What is the meaning of the term *purebred Arabian horse*? There was a time when the answer to that question was simple and uncontested: a *purebred Arabian horse* was one which was derived entirely from Central Arabian Peninsular Bedouin sourced original Arab horse blood stock uncontaminated by any element of non-Arabian horse blood.

There was a time when horsemen around the world accepted without question the common knowledge that there was once long ago a type of wild feral horse that had existed for thousands of years in a state of physical isolation in the interior of the Arabian Peninsula, evolving into a state of highly specialized desert adaptation. This landrace of *Equus* was considered to be a pure race of horse, uncrossed with the blood of any type of horse from outside the confines of the peninsula.

It was presumed that the geography of the peninsula, isolated by water on three sides and on the fourth side by a desert, created a zone of isolation within which the desert horse could exist for eons without the inclusion of other *Equus* sub-types from Asia or North Africa. It was the prevailing idea that the desert horse became a creature of such remarkable characteristics precisely because of this geographic isolation. Modern genetics has examined the phenomenon of evolution in isolation, and find that specimens of unique rarity result.

It was presumed that the horse had existed there long before man came onto the peninsula, and it was presumed that with the arrival of man, the horse was captured and domesticated. The first human migration from Africa to Arabia occurred about 120,000 years ago, and the horizon of tribal organization and animal domestication was about 10,000 years ago.

Today, however, it is a vexed question.

Skepticism concerning the genetic purity of the Arabian horse breed in general as it exists today, and individual Arabian horses in particular, have sparked a series of lively and animated debates over the past 50 years. These debates have uniformly ended with inconclusive and contradictory results, due in large part to the varying and inconsistent definitions of terms that have been used in these debates to define the term *purebred Arabian horse*.

Figure 1 - *An Arabian Horse, by Victor Adam, used with the permission of Olms Archives.*

Controversy over the meaning of the term *purebred Arabian horse* has been at the center of the debate. Beyond this question, the validity and accuracy of the recorded pedigrees, once considered an unimpeachable source of the genetic composition of many specific Arabian horses have been questioned.

The two issues are closely connected.
There is simply no escape from the dilemma.

The controversy over Arabian horse pedigree accuracy has had a direct and significant effect on the decision-making process in Arabian horse breeding operations throughout the world. Animal pedigrees, like human pedigrees, are at times flawed. Accidents have been known to happen even in the best regulated families. All of this would be of only marginal interest except for the fact that modern scientific methods now exist that can ferret out these errors and establish the true family relationships of individual Arabian horses living today.

The purpose in writing this book is to present an original DNA research study of Egyptian Arabian matrilines presented in Chapter Eleven. This study provides a catalogue of the mitochondrial DNA identity of all surviving matrilines tabulated in the Egyptian Arab horse breeding records, as defined by Colin Pearson in The Arabian Horse Families of Egypt, Alexander Heriot and Co. Ltd., published in 1988. The Pearson book was chosen as the gold standard for this study because of his meticulous and thorough research methods. Pearson was motivated by a deeply felt attraction to the horses of Egypt, as well as a sense of academic zeal in connection with the questions of the origins of the horses. He was not partisan in matters of documentary conflict. He was, as his editors Kees Mol and James Fleming noted, motivated by an unwavering commitment to eliminating conjecture in his work. Where gaps and inconsistencies were found, he noted these. He did not speculate on possible interpretations.

Besides the EAO, a number of private individual breeders and institutions are included in Pearson's definition of "the Egyptian records", such as Inshass, Albadeia Stud Farm, Shams el Asil Studfarm, The Police College, Hamdan Stables, Mr. Ahmed Sherif's Studfarm, and the Sheykh Obeyd Stud Book of Lady Anne Blunt. Thus in his tabulation, horses from these sources are also denominated as Egyptian Arabians. For details see Pearson, p. lxxxiii.

The motivation behind this new genetic research is an awareness of the waning public interest in the concept of Arabian horse blood purity. The widespread prevailing public opinion today is that all living Arabian horses are genetically admixed with some degree of non-Arabian blood. Modern genetic research is beginning to provide useful data that examines this skeptical and prejudicial position.

The purpose of this book is to add a new dimension to the discussion of Egyptian Arabian horse pedigree validity and the meaning of genetic purity in Egyptian Arabian horses. The results of original mitochondrial DNA research concerning the foundation mares of the Egyptian Agricultural Organization in Cairo Egypt will be presented.

The EAO horse breeding section at El Zahraa Stud has a long-standing reputation for breeding *purebred Arabian horses* which consists of a closed herd of Arabian horses, the members of which trace in all lines of descent to the *purebred Arabian horses* of the Nejd region of Arabia. This is believed to be true in that every horse admitted into the program from its inception were evaluated by the directors of the organization and were considered in their opinion to be purely bred. As with all Arabian breeding programs, the genetic purity of the EAO bloodstock has been subjected to critical scrutiny and skepticism. The validity and accuracy of the pedigrees of some Egyptian Arabian horses has been challenged.

Prior to the year 2000, the field of Arabian horse pedigree research was based entirely on historical methods of analysis, using the written and oral record as evidence. In that year, the landmark study of Micheal Bowling was published, and an entirely new vista of potential uses of mitochondrial DNA analysis became evident.

The fallibility of this type of conventional pedigree research is well known. The furor that has often surrounded the debates over the subject of the Arabian horse pedigree has been subjected to hyperbole and exaggeration. Gaps in the data have been passed over without comment by those individuals in a position of authority and the questionable results of these debates have been solemnized and petrified by various Arabian horse registries.

Now, however, the validity of the matrilineal portion of the Arabian horse pedigrees can be investigated through scientific methods by performing mitochondrial DNA analysis. Mitochondrial DNA is a type of DNA that is transmitted by a strictly uniparental mechanism; all offspring, male or female, contain the precise and identical mitochondrial DNA of the mother. This type of investigation is beginning to provide valuable information concerning the familial identities and inter-relationships of the present day Egyptian Arabian horses.

DNA analysis and its use in pedigree investigation have been brought to bear on the questions of the modern day Arabian horse. This type of analysis is being done in an attempt to disentangle what is original in the Egyptian Arab horse of today from what is alien. This is essential. Only by removing the veil of tradition, opinion, prejudice, and conjecture can Arabian horse enthusiasts arrive at a state of understanding about this unique phenomenon, "the *purebred* Egyptian Arabian horse". This is all the more important because the genetically *purebred Arabian horse* has become extremely rare. It may even be extinct.

Scientific, as opposed to documentary research, is essential in order to arrive at an accurate appraisal of the genetic composition of the Egyptian Arab horse of today. The science of mitochondrial DNA analysis provides a method for taking the first step in deconstructing the baroque layers of fancy and misinformation that have been deposited in successive layers for centuries over the question of Arabian horse pedigrees.

Most significantly for the horse breeder, mitochondrial DNA technology provides a method of inferentially determining the mitochondrial DNA identities of horses that lived long ago, even without direct examination of their tissues or blood samples. The ability to determine the mitochondrial DNA sequence of Ghazieh I of Abbas Pasha, for example, has certain obvious applications.

Figure 2 - *Ibn al-Haythem, featured on a Qatari stamp.*

The scientific method provides a useful counterpoise to the work of pedigree specialists. Science trusts no statements without verification, tests all propositions and theories thoroughly and rigorously, and keeps no secrets. Science is never arbitrary. It is not affected by personal interest or bias. Science is never prejudicial.

Scientific investigation does not reveal which horse is superior or valuable, or what type of horse is better or worse. It takes no position on the question of what type of horse should be considered ideal. Science can only reveal what "is", and it makes no judgments.

Ironically the Western scientific method itself originated with an Arab, Ibn al-Haythem, who was born in Basra Iraq in 965 CE and died in Cairo in 1040 CE. Ibn al-Haythem was a talented polymath who is credited with writing some of the earliest treatises on optics, mathematics, and formal physics, working at a time when London was a village. By applying the methods of observation, hypothesis formation, experimentation, and verification, he laid the groundwork for the flowering of the modern scientific method that developed during the European Renaissance.

The structure of mitochondrial DNA is unique within a mare line. This structure constitutes a distinct genetic identification code which identifies all members of a matriline. It is a genetic sequence that serves to identify the matriline to which any individual belongs. Mitochondrial DNA analysis provides no information pertaining to the genetic identity of any other elements in the pedigree.

It reveals nothing about the genetics of the sire lines and has no bearing on the matrilines of the sires. It provides no information about the "purity" of a horse. It does not directly address the identity of the *purebred Arabian horse*. It is only applicable to the investigation of matrilineal connections or relationships. Unless interrupted by the insertion of an incorrect female into the matriline, the mitochondrial DNA sequence is consistent and identical across thousands of generations. Thus mitochondrial DNA research has become useful in "detective" work, discovering facts that were previously unknown and unknowable.

Mitochondrial DNA analysis has been applied to pedigree investigations of living Egyptian Arabian horses. Egyptian Arabian horse breeders have taken a lively interest in this new technology in that it would allow them to determine if their horses are in fact descended from the matriline that is indicated on their pedigrees as issued by the Arabian Horse Association or other registries.

For over 80 years, the validity of any Arabian horse pedigree in the U.S. was entirely dependent on the honesty and attention to detail of the individuals providing the written documents that were submitted to register foals with the Arabian

Horse Registry of America, now the Arabian Horse Association. Incorrect pedigrees, whether the result of fraudulent intent or inattention to details do exist. The level of dishonesty in the Arabian horse business can be gauged by a perusal of the Arabian Horse Association monthly list of Censures and Suspensions posted on their website. It lists the names of people who have been suspended from AHA membership due to unethical behavior, and it lists horses that have been de-registered due to the discovery of falsified pedigrees.

Prior to 1976, the Arabian Horse Registry of America based its registration of foals on a physical description of the foal and a statement certifying the purported dam and the purported sire. This information was submitted by the breeder of the foal. Errors, both intentional and unintentional, were made. But the AHRA had no tools with which to clarify errors such as coat color inconsistencies, or to verify the data as it was presented to them. Registration was conducted on the honor system. There was no mechanism to prevent a breeder from intentionally falsifying the pedigree when registering a foal. Registration errors have been made for as long as horses have been registered.

To combat this problem, the AHRA introduced the first biological test to assess the validity of the purported parentage of the foal presented for registration in 1976. This was the blood typing system, and while it could not prove the parentage of any foal, it was capable of detecting stallions and mares that could not possibly have participated in the given breeding. The accuracy of this system and its ability to detect fraud or error in the foal registration application were limited.

In 1991 the AHRA instituted the use of autosomal DNA analysis. Biological samples obtained from hair roots became the mandatory method of verifying the purported parentage of a foal presented for registration. This method is based on the determination of microsatellites, small DNA sequences found on nuclear chromosomes that serve as identifying markers. This type of DNA analysis, like blood typing, works by exclusion. It cannot prove that a given sire or a given dam is, in fact, the sire or dam of a particular foal, but it can detect a stallion or a mare that could not possibly be correct in the proposed breeding of the foal. The accuracy of this method is about 99%.

Unlike nuclear DNA, mitochondrial DNA can be recovered and analyzed from extremely small samples of biological material. The reliability, accuracy, and reproducibility of mitochondrial DNA analysis is well established. It is accepted in forensic pathology and judicial proceedings. The Federal Bureau of Investigation began using mitochondrial DNA in 1996, and now has a laboratory, the DNA Analysis Unit II, devoted exclusively to this technology. The FBI maintains a mitochondrial DNA database section in its National Missing Persons DNA Database. It also maintains the Scientific Working Group of DNA Analysis Methods mtDNA Population Database. The technology is used in forensic investigations by the Department of Homeland Security and the Drug Enforcement Administration.

Mitochondrial DNA is admitted into evidence in courts of law, and its use has been vital in obtaining convictions that otherwise would have been impossible to obtain. It has also been used to free individuals wrongfully convicted of a crime. In 2007, the Supreme Court of Georgia heard an appeal concerning a case in which a man was charged with the murder of a woman JAL in 1999. She had been strangulated until dead, and then her body was run over by a car. Evidence pointed to the accused. Blood stains on the carpet of the accused man's car were recovered by forensic investigators and analyzed in the FBI lab. The samples were found to be too deteriorated for adequate identification. However, a strand of hair was obtained from a shock absorber from his car. The hair root bulb was analyzed by the FBI mitochondrial DNA unit, and it matched the mitochondrial DNA from the murder victim. On the basis of this finding, the accused was found guilty by a jury of malice murder and felony murder and he was sentenced to life imprisonment. The Supreme Court upheld the decisions of the lower courts.

Figure 3 - *Grand Duchess Anastasia Nikolaevna.*

This type of research was used to identify the remains of Anastasia Nikolaevna, daughter of The Russian Tsar Nicholas II and Alexandria Fyodorovna. The Russian Royal family was executed by the Bolshevik secret police in 1918, and the remains were buried in a mass grave, not discovered until 1991. Neither Anastasia nor her brother could be identified from forensic evidence, fueling long-standing rumors that one or both children had escaped the massacre. A second grave containing the remains of two children was discovered in 2007, and mitochondrial DNA analysis done in 2009 confirmed that the bodies were the remains of Princess Anastasia and her brother.

Figure 4 - *King Louis XVI and Marie Antoinette in the Basilica of Saint-Denis.*

The genetic results from the Russian samples were also useful in disproving the claims of the dramatic and flamboyant Anna Anderson (1896-1984), a Polish woman with a long and disturbing psychiatric history who for an entire lifetime insisted that she was Anastasia, Grand Duchess of Russia. Her flair for attracting publicity and her exotic accent kept her in the public eye until her death. The details of her claims to being Anastasia bordered on the incoherent.

The same technology was used to resolve the controversy over the identity the mortal remains of the French crown prince Charles Louis, the son of King Louis XVI and Marie Antoinette. History relates that the eight-year-old boy died in a Paris prison after the execution of his illustrious parents, and his heart was removed by the family physician and preserved in a crystal urn in the Basilique St.-Denis in Paris. Rabid Royalists of the time created the rumor that the urn contained the heart of an unknown child; they claimed that the boy had in fact been rescued by his supporters, had survived into adulthood and had sired offspring. This rumor persisted for 200 years, and a number of "Charles Louis descendants" had subsequently emerged. In an attempt to disprove the endless string of pretenders to the title, mitochondrial DNA analysis of the boy's heart, recovered from the urn, was carried out and was compared to the mitochondrial DNA of a hair sample of his famous mother. The match was identical: the heart did, in fact, belong to the son of Marie Antoinette.

Figure 5 - *Richard III (1452-1485) killed at the Battle of Bosworth Field at age 32.The location of his grave was unknown to historians. The remains of a man whose bones fit the description if Richard III were recently discovered under a car park near the scene of the battle.*

The skeletal remains of the English King Richard III were identified using the same mitochondrial DNA technique. Richard's mitochondrial DNA sequence was obtained from bone samples taken from the body in question. The next problem was in finding a living matrilineal descendant of Richard from whom biological tissue could be obtained. Knowing Richard's mitochondrial DNA sequence provided the genetic identity of his mother, Cecily Neville, and consequently the DNA sequence of her two daughters, Anne of York and Margaret, Duchess of Burgundy. These two females established the point of origin for a genealogic search that eventually located a 17th generation matrilineal descendant, a woman who had recently died. Her only living offspring was a son, but his mitochondrial DNA sequence would be identical to his mother's. His tissue sample analysis proved conclusively that the body thought to be the earthly remains of Richard III had been correctly identified.

Figure 6 - *Joseph Smith.*

In a related matter, the remains of Richard's two nephews have been preserved in an urn at Westminster Abbey since their death in the 17th century. Dramatized in the Shakespearean tragedy, the nephews were thought to have been murdered on orders of their uncle; the urn in the Abbey is inscribed "stifled with pillows by the order of their perfidious uncle Richard the Usurper". Despite the availability of mitochondrial DNA tests which would establish the accuracy of this bit of history, the British government has blocked several attempts to have the contents of the urn submitted for DNA testing.

This technology has been used to investigate the validity of some of the foundational assertions made in the Book of Mormon of the Church of the Latter Day Saints.

The Book of Mormon was written by Joseph Smith in 1823 who claimed that he had found the original text inscribed on golden plates hidden by ancient prophets in the vicinity of Palmyra, New York, in the U.S. The Book states that the American Indians were descended from the "ten lost tribes of Israel". The Book claims that four groups of dispossessed Israelites crossed the Atlantic Ocean about 600 BCE and found the North American continent uninhabited. The four groups are called the Nephites, the Jaredites, the Lamanites, and the Mulekites. These people were the first to settle in the continent. A lengthy history of these four groups followed, marked by wars between the tribes, and their subsequent loss of sanctity. Over time, the book states, the tribes lost all memory and artifacts of their previous existence. According to the legend, they became savages; they were the Native American Indians found living in North America by European explorers.

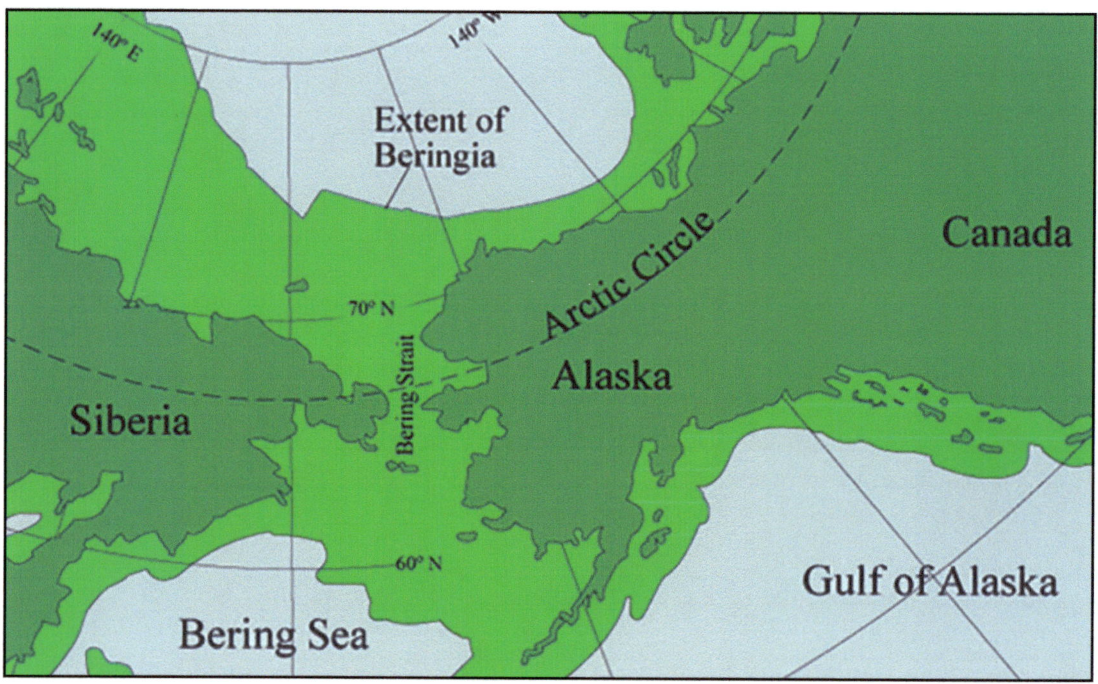

Figure 7 - *Beringia is the name given by paleontologists to the dry land bridge that connected Asia with North America. The light green area indicates the extent of dry land at its maximum. Beringia ceased to be about 10,000 years ago when sea levels rose and the Bering Strait was formed.*

Mitochondrial DNA analysis of modern day American Indians has shown that they are genetically derived from Asiatic sources, not Israeli or Middle Eastern populations. This finding is in agreement with the generally accepted conclusion that the American Indians of both North and South American came to the Americas by crossing over the Bering Land Bridge from Asia about 15,000 years ago.

Mitochondrial DNA analysis has been used to establish the source of the human settlers who established the Minoan civilization during the Neolithic period, about 7000 BCE. For many years, scholars thought that the civilization originated with humans who came to the island of Crete from North Africa.

Figure 8 - *Crete, photograph by NASA.*

Figure 9 - *Minoan fresco from the museum at Knossos, Crete, 1600 BCE.*

In 2013, mitochondrial DNA sequencing and haplotype determination was conducted by paleo-geneticists from the University of Washington using tissue samples taken from 37 skeletal remains from a burial site of the earliest Minoans. The results were compared to known haplotypes of ancient humans from Asia, North Africa, and Europe. The surprising results indicated that the first Minoans came to the island from southern Europe. They formed part of a large-scale migration of Neolithic humans from what is present-day southern France and northern Spain, a migration of people that also settled Anatolia. The fine quality of the Minoan art and the centrality of bull acrobatics suggests the association with ancient southern Spain.

Figure 10 - *NASA satellite photo illustrating the proximity of Siberia and Alaska.*

Mitochondrial DNA research has been used to assess the genetic composition of animals such as the American bison (*Bison bison*). Recent analysis has shown the presence of domestic cattle mitochondrial DNA in most of the existing bison herds in North America. The ability of the bison to cross breed with domestic cattle (*Bos taurus*) would appear to violate the genetic law that individuals from differing species cannot produce offspring, and certainly not fertile offspring. The possibility of different genera successfully interbreeding is even more unexpected. However, this is exactly what has taken place.

Figure 11 - *American Bison, "Bison bison".*

For over 100 years, Western cattlemen have conducted hybridization experiments, intentionally cross-breeding bison bulls with female cattle. The first generation females are fertile and appear to be fully and genuinely bison, but they contain the mitochondrial DNA of their cattle mothers. They reproduce generation after generation, even in the wild. These females have the phenotypic appearance of being bison, and when bred back to hybrid bison bulls, they produced offspring that appear to be bison but carry the cow female's mitochondrial DNA. The progeny appears to be bison. However, all matrilineal descendants contain *Bos taurus* mitochondrial DNA.

This tell-tale genetic marker is the only way that the hybrid Bison can be identified. Of the 500,000 bison being preserved today in North America, the only "*purebred*" herd known to be free of cattle mitochondrial DNA consists of a small isolated group of bison in the Henry Mountain herd in Utah. This herd was created using bison originally captured in Yellowstone Park. Preservationists are continuing to use mitochondrial DNA to identify and isolate the few remaining *purebred* bison, conserving them in a pure state for use by future generations of breeders and scientists.

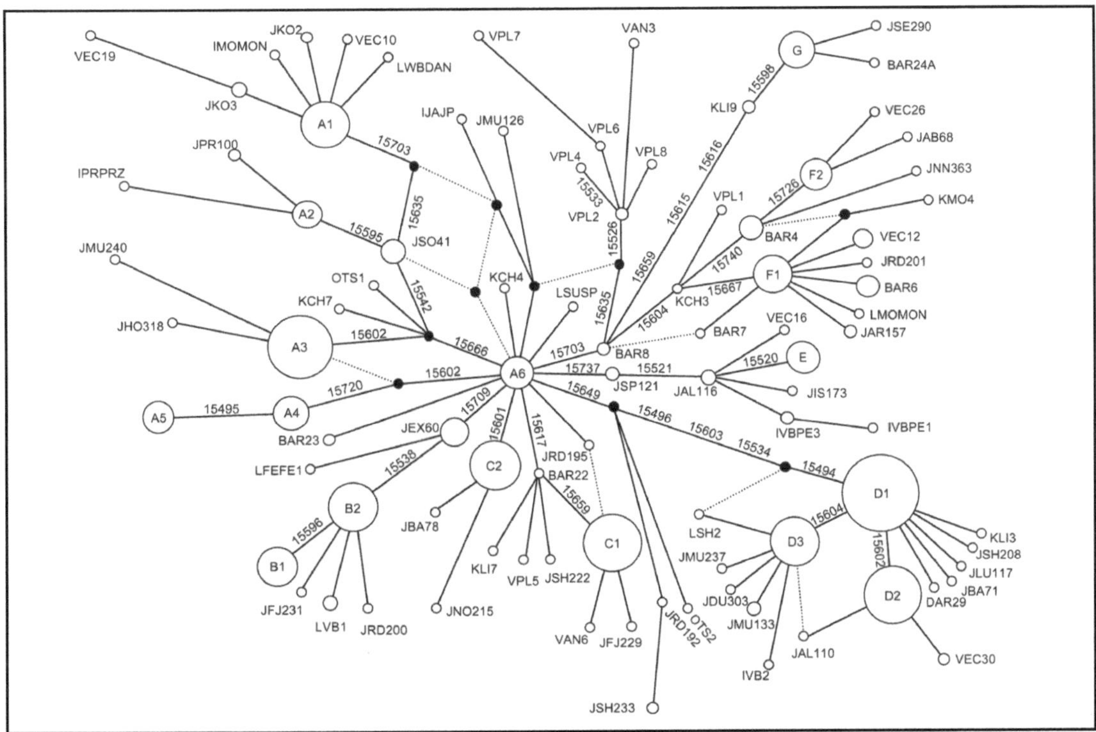

Figure 12 - *Origin of the domestic horse.*

Researchers at the University of Cambridge have used equine mitochondrial DNA analysis, published in the *Proceedings of the National Academy of Sciences*, to investigate the question of whether modern domestic horses were derived from a single source or from multiple sources. The diagram shown above, Figure 12, represents their meta-analysis of over 600 individual equine mitochondrial DNA haplotypes from their own studies and the studies of others. They revealed that a small core of 17 haplotypes accounted for a large percentage of modern horses. The 17 phylogenetic clusters form star-like patterns of proliferation. This suggests that they were formed in the most remote reaches of the ancient past.

Mitochondrial DNA analysis continues to be used for unusual and unexpected investigations. The 2013 scandal concerning the mislabeling of ground meat in Europe illustrates the usefulness of mitochondrial DNA in an unexpected field of study. Random screening of uncooked fresh ground meat labeled "beef" revealed the presence of horse meat in a large number of samples taken from leading food distribution chains and supermarkets. An investigation by the Irish government, epicenter of the contamination, used mitochondrial DNA analysis to trace the source of the horse meat to unlicensed meat traders and agents. The presence of horse meat in European food raised questions about the possible presence of veterinary drugs such as steroids, antibiotic, and phenylbutazone in the food as well. Concerns about contamination of ground meat contaminated with the virus responsible for Equine Encephalitis, a disease of horses and humans, were raised. The investigation also disclosed the presence of pig meat in a number of randomly collected samples. This raised concerns in the Muslim community. The Irish Minister of Agriculture, Food, and the Marine Simon Coveney introduced sweeping reforms in the Irish meat industry including DNA analysis of meat products, combating food fraud, issuing horse passports, and strengthening laws related to package labeling and certificates of origin. Offending abattoirs identified in continental Europe are under investigation by Europol and other EU member states.

Figure 13 - *Mountain Zebras.*

Figure 14 - *Bontebok from Namibia.*

But the problems did not stop there. Wild meat testing in South Africa revealed widespread mislabeling of wild game meat. Meat labeled as kudu, springbok, ostrich, horse, impala, kangaroo and many others were mislabeled. A major concern was expressed over the fact that samples labeled "zebra meat" were actually from Mountain Zebra, *Equus zebra*, an animal red-listed by the International Union for Conservation of Nature as being near extinction. The investigations widened, and meat fragments from *Damaliscus pygargus*, the Bontebok, a "near threatened" species, were found in mislabeled meat sold for human consumption. The problem was traced to the unlicensed game markets at which unidentifiable dressed animal carcasses are brought in by the hunters and managers from wild game farms. Researchers have recommended the institution of mitochondrial DNA analysis of wild game sold in supermarkets. This method represents a genetic "bar code" that conclusively identifies the genus and, more importantly, the species, of any meat being sold. The mitochondrial DNA sequences for any species are available in DNA "libraries" located in a number of research centers.

PEDIGREES, MITOCHONDRIAL DNA AND THE EGYPTIAN ARABIAN HORSE

Mitochondrial DNA study has been applied to the study of horse populations as well. The study of mitochondrial DNA identification of the horses of EAO/Egypt presented in this book is one such example. It is not a dry academic exercise in tedious theorizing but is instead a valuable tool which can be immediately applied to the practical problem of successful selective breeding.

The purpose in writing this book is to present the results of an original study conducted on the matrilines of the EAO, which provides information that discloses the true matrilineal relationships among the existing female lines of the Egyptian Agricultural Organization.

Egyptian Arabian horse breeding has become highly competitive and increasingly expensive in the 21st century, and this has led to the increased use of selective breeding strategies. Selective breeding is the intentional mating of a particular stallion to a particular mare, both chosen in an attempt to produce offspring that exhibit particular desired characteristics. However, the success of selective breeding is dependent on the assumption that the pedigrees of the horses involved are correct. The work of Bowling et.al. published on 2000 was the first to raise a red flag with regard to the validity of EAO matrilines in general. His findings confirmed the genetic identity of Ghazieh I and the Yamama descendants bred by Prince Mohammad Ali.

The possession of an accurate pedigree is indispensable to the success of planned Egyptian Arabian horse breeding. For most of Western history, random horse breeding was the rule. Generic *equids* were a commodity. They had commercial utility and commercial value. Random breeding was based on the need to produce large numbers of horses for military and agricultural. For most of human history, horses were bred locally from readily available stock.

In Egyptian Arabian horse breeding, the pedigree is critical to understanding the breeding potential of the horse. Possessing a correct pedigree is vital. Record keeping at the RAS/EAO has been thorough and complete, having been spared the devastation and destruction experienced by the Arabian horse breeding studs of central Europe during World War I and II.

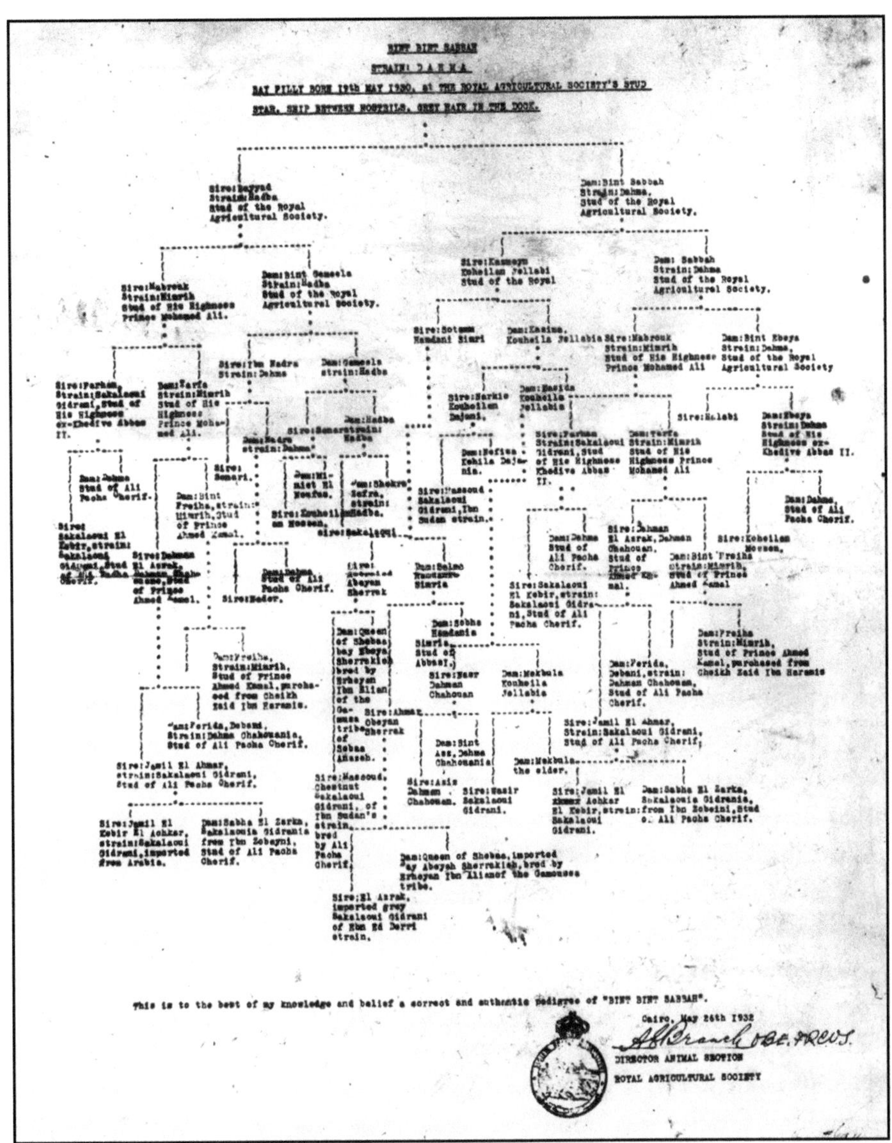

BINT BINT SABBAH

STRAIN: D A H M A

BAY FILLY BORN 19th MAY 1930, at THE ROYAL AGRICULTURAL SOCIETY'S STUD

STAR, SNIP BETWEEN NOSTRILS, GREY HAIR IN THE DOCK.

This is to the best of my knowledge and belief a correct and authentic pedigree of "BINT BINT SABBAH".

Cairo, May 26th 1932

A.E. Branch O.B.E. F.R.C.V.S.

DIRECTOR ANIMAL SECTION

ROYAL AGRICULTURAL SOCIETY

Figure 15 - *The pedigree of Bint Bint Sabbah a 1930 filly bred by the Royal Agricultural Society, with the seal and signature of Dr. A.E. Branch, director of the animal section of the Royal Agricultural Society. This pedigree accompanied Bint Bint Sabah when she was sold to Henry Babson in the U.S. in 1932. This documentation format was used by the Royal Agricultural Society while Dr. Branch was director. Throughout its long history, EAO management made every effort to maintain complete and accurate pedigrees of its horses. Modern DNA analysis has revealed errors. Image used with the permission of Bruce Johnson.*

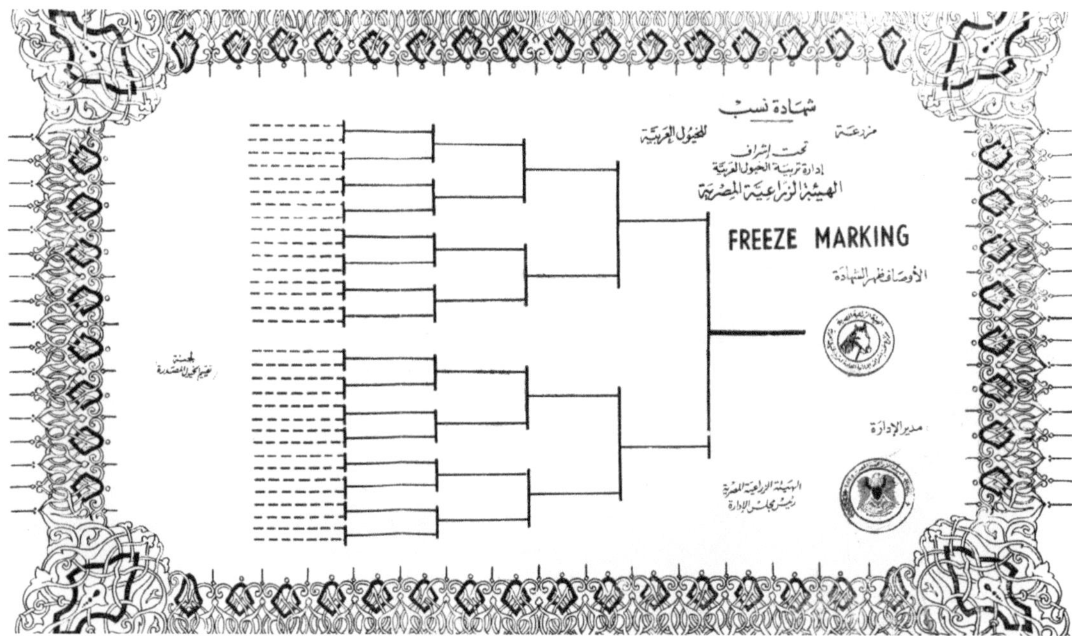

Figure 16 - *Freeze Marking: This sample worksheet from the 1974 EAO Studbook Volume IV outlines the earliest attempts to use technology to verify the identity of horses.*

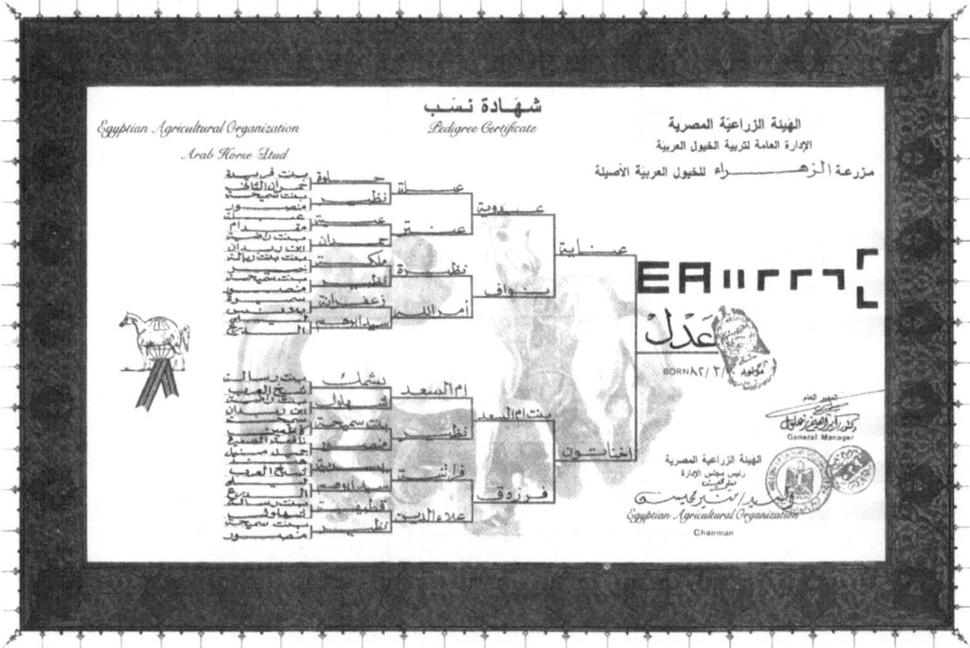

Figure 17 - *Pedigree of the EAO stallion Adl, with the matriline shown across the top of the pedigree, tracing to Bint Farida, and eventually leading back to El Dahma, one of the most influential foundation mare of Egyptian breeding. EAO horse breeders acknowledge the importance of the female line graphically in this way, in contrast to the Western convention of placing the male line at the top of the pedigree and the female lines by placing the female line across the bottom.*

The modern day Egyptian Arabian horse market is much smaller, more focused, and much more competitive. Financial rewards have increased for successful breeder and financial losses have escalated for unsuccessful breeders. The modern market is global, and the bloodstock has become highly mobile. Mares and stallions can be bought, transported and bred anywhere in the world with ease. Transported semen, ova, and embryos have further expanded the availability of bloodstock from which breeders can choose when making selective breeding decisions.

The requirements of the Egyptian Arabian marketplace have become more stringent. Current economic conditions demand that breeding programs strive for commercial viability. Breeders can no longer breed randomly and without intention. They must be able to produce particular types of progeny that possess particular types of characteristics that make them attractive to the horse buying public. A breeder must develop a rational system of mating mares and stallions. Very few breeding programs can survive if their produce cannot be sold at acceptable prices.

Selection of mares and stallions to be used in a selective breeding enterprise is based on one of three criteria; success in the show ring, selection by "championships", by "pedigree", or selection by "type".

1) Using show performance criteria, "breeding by championships" either at halter or in performance, a champion stallion may be bred to a champion mare with the goal of reproducing the parent's qualities in the offspring. The mare and stallion undergo progeny testing, in which they are bred together to produce three test foals. The foals are then evaluated. If they do not meet the breeder's expectations, they are discarded and the stallion/mare mating is abandoned. If they do meet expectations, then the stallion/mare mating is continued. This was the policy used by the RAS/EAO throughout its history. Possessing a correct pedigree is less essential when using this system of selective breeding because the presence of an incorrect or pedigree in either sire or dam will eventually become apparent. This method of breeding is aimed at the short term. The offspring resulting from breeding incorrectly identified animals will eventually deviate from type.

2) Selective breeding by pedigree follows pragmatically determined genetic theories that are based on the notion that certain genetic lines of individuals, male or female, are prepotent and tend to produce offspring which predictably and reliably manifest certain qualities. These qualities may relate to the conformation of the horse, anatomic proportions, the "typiness" or beauty of the horse, or the horse's athletic ability. Using this method, individuals are sought out for breeding purposes based in part on their pedigrees. Using this system of selective breeding, any inaccuracy in the

pedigree is problematic.

3) Breeding by visual appearance. In this method, the breeder evaluates phenotype only, which is the appearance of the horse. He determines what phenotype he prefers and selects mare and stallion which possess the features that the breeder would like to see combined in the anticipated offspring. In this case, an inaccurate pedigree will cause the offspring to deviate from parental type. Revelations of contamination may not become manifest for many generations.

Clarity of purpose and intent is essential to breeding success, but no breeder can expect to achieve the anticipated results if the genetic constitution of the raw material he is working with is incorrectly recorded in the pedigree.

But mitochondrial DNA has another use in Egyptian horse breeding besides the detection of incorrect pedigrees. It also serves as a guide in understanding matrilineality.

Figure 18 - *An Arabian Horse, by Achilles Constant Theodore Emile Prisse.*

It is difficult to exaggerate the importance of the female line in Arab horse breeding. Breeders have struggled with this issue for centuries. In 1854 the French Arab horse breeder Duc de Guiche wrote "certain families (strains), though of good descent, cannot be mated together because they will produce only mediocre or outright bad offspring. Their blood seems to be possessed of opposite tendencies."

In Egyptian Arabian horse breeding, pedigree analysis has occupied a special place in selective breeding. This is due primarily to the matricentric nature of breeding, a practice that has been appreciated and followed by historically important breeders, stretching back in time to the deserts of Arabia and to the Bedouin, the original breeders of the Arab horse, Matricliny refers to the fact that mare lines, or tail female lines, tend to reproducibly and reliably transmit their genetically determined characteristics from generation to generation. The intensity, or extent, of this transmission among the mare lines, has also been observed and reported. In some female lines, the family lineaments are quite clear. In some lines, this phenomenon is less significant.

Patrilineal influences tend to be of secondary importance in the long term results of Egyptian Arabian horse breeding. The influence of the stallion may in some cases be striking, but the effects on progeny tend to be significant for only one or two generations. After this, the effects tend to be less predictable.

While there is no scientific or genetic explanation or support for the phenomenon of matriclinous importance, Egyptian Arabian horse matrilines have been found by empirical observation to possess the predisposition to predictably transmit certain of their traits to future generations of offspring. Some mare lines have been found to reproducibly transmit increased body size to subsequent generations, while other matrilines are noted for their ability to transmit correct leg conformation. Some lines may transmit enhanced athletic ability while others may produce offspring that exhibit the quality of "typiness", or close adherence to the breed standard for symmetry, balance, and beauty. In addition to passing down desirable traits, some matrilines have been noted that predictably pass on undesirable traits as well. Poor conformation, gait disturbances, neurologic disorders, infertility and deranged maternal instincts are among the traits that breeders wish to avoid. Matriclinical tendencies of the mare lines are given close scrutiny by the Arab horse breeder whose goal is long-term predictability in the type of foals produced.

Mare lines have been identified that have the ability to reinforce certain existing desirable traits while others have been found to be useful in preventing conformational faults that exist in potential parents from appearing in offspring. Some matrilines can predictably and reliably improve faulty leg conformation in the offspring that certain stallions tend to produce. This phenomenon has been anecdotally observed by Arab horse breeders for centuries and has provided them with an important tool that can be used in planning breedings. This type of breeding is limited by the fact that it is only effective if the ancestral elements in the pedigrees are correct.

Figure 19 - *Horses of Schubra, by unknown artist, courtesy of Olms Archives. Schubra was once the site of the stables of Mohammed Ali the Great.*

Modern day Egyptian Arabian horse female lines trace to foundation mares that lived over 150 years ago. The physical and behavioral characteristics of their descendants have been meticulously documented in pedigrees, photographs, and written descriptions. A veteran Egyptian Arabian horse breeder will be personally familiar with the individuals that make up the previous ten or more generations of any horse that they intend to breed. Breeders apply the principles of selective breeding by studying the specific desirable and undesirable traits of each of the many descendants of each matriline and, by extension, arrive at an understanding of which features are unique to each matriline. These characteristics include adherence to breed standards, esthetic appeal, height, weight, fertility, coat color, speed and performance attributes, and temperament.

Breeders seek out horses from particular female lines to incorporate into their breeding programs in an attempt to produce certain desirable traits in the offspring or to eliminate particular undesirable traits in the offspring. This strategy is based on the assumption that the acknowledged pedigrees of the horses are correct. If a pedigree is inaccurate, especially the matrilineal portion, then this breeding strategy will be ineffective. Selective breeding based on invalid matrilineal pedigrees will fail to produce the characteristics in the offspring that would be expected of that matriline.

The pedigrees of the foundation matriarchs of the EAO are in large part based, technically speaking, on hearsay, since the information was set down by persons with no first-hand experience of the matters involved, having obtained their information from other persons who did have first-hand direct personal experience with these matters. Caution, therefore, should always be observed when dealing with pedigree documents.

Mitochondrial DNA analysis offers a method to separate fact from fiction: it provides a means of uncovering falsified or erroneous matrilineal pedigree information that has been introduced into the registered pedigree of any living Arabian horse. The matrilineal portion of a pedigree may be false due to inadvertent errors in record keeping or may be the result of intentional falsification of the identity of a foal's mother.

The validity and accuracy of the tail female element of Egyptian horse pedigrees has until recently been a matter of uncritical acceptance of the pedigrees as published by the EAO. However, Egyptian breeders are an inquisitive, argumentative and skeptical lot, (not to mention competitive), and the produce of female EAO lines has been the object of close scrutiny and analysis as the stock has reproduced generation after generation. Questions have been raised; doubts have been circulated.

The role of matrilineality became apparent to the English Thoroughbred horse industry at a very early stage. From its inception, the General Stud Book, the English Thoroughbred registry, has cataloged horses according to the distaff or female side of the pedigree.

The matriclinous tendencies in English Thoroughbred horse breeding have been well studied. In the late 19th century, an Australian accountant by the name of Bruce Lowe became involved in the business of Thoroughbred breeding and racing. Lowe noticed that certain female Thoroughbred lines tended to produce an inordinately high number of winners in the major races. This led him to analyze the Thoroughbred pedigrees and mare lines, tracing and documenting the information contained in the General Stud Book. He found that there were 43 root mares listed in the General Stud Book. Lowe then compiled a database of the winners, tabulated according to the tail female connection, of the Epsom Derby, Epsom Oaks, and the St. Leger Stakes. He then ranked the mare lines from 1 to 43: family number one being "the running family" (the family with the most wins), and family number 43 being the family with the least wins. This tabulation was made available to horse breeders who used the data to make informed and accurate decisions concerning the selection of mares to be used in planned breeding.

Figure 20 - *An English Thoroughbred horse.*

Lowe's work was revolutionary in the sense that it challenged the cult of "stallion worship" that permeated the Thoroughbred industry at the time. Lowe believed that all Thoroughbred champion racing stallions of the day owed their fame and breeding prosperity to the dams from which they had been bred. His Figure system was the first attempt to rationally and systematically investigate the age-old observation that "blood will tell".

Figure 21 - *A Caravan Halt, courtesy of Olms Archives.*

Lowe's findings and tabulations were published posthumously in London by William Allison in 1895 as "Breeding Racehorses by the Figure System". This compilation has influenced the Thoroughbred industry for many years, and the "running families" have indeed produced such 20th century champions as Northern Dancer and Man O' War.

During the 1950's the work on matrilineality in Thoroughbred selective breeding was advanced by the work of Captain Kaziemierz Bobinski and Count Stephan Zamoyski whose interpretation of the pedigrees resulted in the identification of a total of 74 Thoroughbred female lines. Additional lines came from previously undocumented mare lines that had been preserved in America, Argentina, Australia, and Poland. For example, the Bobinski Families A - A37 descended from a mare registered in the American Stud Book who could not be traced with certainty back to lines found in the GSB. Their work was published as Family Tables of Racehorses, now commonly referred to as the Bobinski Tables. The Bobinski classification established a numerical matriline system that is used today in sales and auction catalogues to aid buyers in selecting potential breeding stock. In 1988 Toru Shirai of the Japanese Bloodstock Agency bought the copyright to the Bobinski Tables and expanded the tables to include all horses born since the 1950s.

The first use of mitochondrial DNA analysis in Arabian horse breeding was the A.T. Bowling paper published in 1998, in which the long debated pedigree of the bay mare Domow (1913), registered as the daughter of the chestnut stallion Abu Zeyd and the chestnut mare Wadduda (1899), was analyzed using mitochondrial DNA sequencing of samples taken from living descendants of the mare. The results proved that Wadduda was, in fact, the dam of Domow, establishing that Abu Zeyd could not have been the sire. The identity of Domow's sire remains unknown.

The next contribution to the field came with the research of A.T. Bowling, who in 2000 published "A pedigree-based study of mitochondrial D-loop DNA sequence variation among Arabian horses". The Bowling study included horses from both Egyptian and direct Arabian Desert sources and was designed to assess the plausibility, validity, and accuracy of the strain designations that usually accompany Arab horse names.

> In general, it has been assumed in the West that a horse sharing a strain and a substrain designation are likely to be derived historically from a common maternal ancestor and that different substrains of a particular main strain reflect a female line connection more distant in time.
>
> (Bowling, Animal Genetics, 2000)

The results of mitochondrial DNA analysis failed to show a genetic connection between strain names and mitochondrial DNA haplotypes. Strain names were not reflective of matrilineal descent. In some cases, mares descended from the same strain designation were found to have different mitochondrial DNA haplotypes, and some mares of separate strain designations were found to have identical haplotypes. It was concluded that whatever the strain names mean, they do not refer to the parentage or matrilineality of the horses involved. Strain names were in no way connected to genetic factors.

Figure 22 - *Colin Pearson.*

An even more significant finding resulted from the Bowling study when it was noticed that the study horses descended from Ghazieh I had the same haplotype as the study horses descended from *Maaroufa, said to be from the Jellabi strain. By comparing the conventional pedigrees with the results of mitochondrial DNA testing of living descendants of these lines, the Bowling study demonstrated that the Egyptian Arabian mare *Maaroufa was derived in matrilineal line of descent from the Seglawi mare Ghazieh I line of Abbas Pasha, not from the historical Jellabi line as *Maaroufa's importation documents claimed. *Maaroufa and Ghazieh I shared a common female ancestress. However, the Boling study only sampled two EAO matrilines, and the need for a complete analysis of the EAO matrilines was evident.

The discovery of a significant pedigree error in Egyptian horse breeding has raised questions about the possibility that other errors may exist in the Egyptian Arabian horse pedigrees. If investigative mitochondrial DNA research confirmed or disproved the possibility that Egyptian Arabian horse pedigrees contained errors, future interest in breeding Egyptian Arabian horses would be impacted significantly.

Similar research methods have been applied to the Polish Arabian horse population. Mitochondrial DNA research conducted at Gdansk University in Poland has produced data indicating that a previously unacknowledged mare in Poland named Malikarda produced some of the early Polish Arabian stock that have been historically attributed to the mare Milordka, a highly regarded and influential root mare in Polish Arabian horse breeding. The similarity between the names Malikarda and Milordka appear to be the source of this confusion. This pedigree error continues to be repeated in registered Arabian horses being bred today. Neither Milordka nor Malikarda were *purebred Arabian horses*. They had no pedigrees.

The Bowling research highlighted a powerful new scientific method that would prove invaluable in the future investigation of the identities of matrilines. This technique involves the comparison of the historical family lines with mitochondrial DNA data from living descendants of these lines. Disparities between the historical pedigrees and the results of mitochondrial DNA analysis of living Arabian horses point to inaccuracies in the recorded pedigrees. As Bowling stated in the 2000 paper in *Animal Genetics*, "The power of derived mitochondrial haplotypes to answer maternity questions for horses from which no biological material is directly available was demonstrated."

Building on this concept, a scientific study, presented in this book, was formulated and conducted. Its purpose was to determine the mitochondrial DNA identification code of all of the existing female lines of Egyptian and EAO origin. This research served two purposes: first, it established the correct family relationships between the foundation mares of the EAO; secondly, it provided a method for detecting the presence or absence of pedigree errors in the matrilineal limb of the pedigrees of living Egyptian Arabian horses. This research was felt to be especially necessary due to the fact that several of these lines are nearly extinct. Many of the original female lines of Egyptian horse breeding stock are already extinct within Egypt/EAO breeding, among them Kuhayla Jellabi, Gamila Manial, Koheila Mimrieh, and Badria, the latter from the Inshass Stables.

Using this technique, it is possible to determine the mitochondrial DNA identities, known as haplotypes, of the original foundation mares of the EAO. The haplotypes can be reconstructed by inference from the analysis of mitochondrial DNA sequences of living descendants of these foundation mares. The correct identification of the founding matrilines will provide information useful to breeders of these horses, and may play a role in the preservation of some of the nearly extinct dam lines.

Figure 23 - *A Horse of Abbas Pasha, from Arabian Exodus by Margaret Greely.*

The problem of mare owners providing incorrect sire and dam information to horse registries when registering a foal is not an insignificant issue. Even an organization as venerable and serious-minded as the EAO is not immune from the pitfalls of flawed record keeping. When Colin Pearson cataloged the entire contents of the EAO Arabian horse pedigree records, he found so many obvious inconsistencies that he was compelled to add a section of explanatory notes to the end of his book The Arabian Horse Families of Egypt in which 46 inconsistencies were recorded. For example, in EAO Vol. II, Sameh (1945) was listed as the son of Samira (1935). Sameh was in fact out of Sameera (1934)

The discovery and scientific confirmation of an error in the long accepted pedigree of a horse of Egyptian origin such as *Maaroufa inescapably brings to the forefront echos of an ongoing smoldering debate, now over 50 years in duration, regarding the accuracy of the published pedigrees of many Arabian horses. A schism in the community of Arabian horse breeders has developed over several decades with regard to the question of the genetic purity of the Arabian horse and the validity of published pedigrees.

The Egyptian Arabian horse is an ancient and primeval horse breed with a record of written pedigrees reaching back over two hundred years. The *purebred Arabian horse* had traditionally been defined as one which "traces in all lines of descent to the horses bred by the desert Bedouin of Arabia". However, due to the events and developments in the world of Arabian horse registration and authentication during the latter half of the 20th century, the majority of Arabian breeders, as well as the general Arabian horse public, now accept the fact that most Arabian horses today are not descended entirely from 100% pure desert stock, having been genetically altered and adulterated over the course of time as the result of the introduction of unacknowledged foreign and non-Arabian equine genetic elements into the breed.

Figure 24 - *The Pyramids of Egypt at sunset. Image used with the permission of the Library of Congress collection.*

The majority opinion of the pedigree relativists is expressed in the philosophy of the World Arabian Horse Association, which takes the position that since the majority of horses known by the generic term "Arabian" have clearly demonstrable instances of non-Arabian blood in their pedigrees, the most reasonable course of action is to abandon the search for certainty and proof of purity and accuracy in the pedigree of the Arabian and concentrate rather on creating a worldwide commercially vital global enterprise in which the preponderant mass of generic Arabian horses could be endorsed under the arbitrary umbrella entitled the *purebred Arabian horse*.

> There is no need to concern oneself with the question of which horses in what historical period entered today's population of Arabian horses and how they got there. That is bound to remain an unsolved riddle since there simply isn't enough reliable historical material; and the farther back one goes in the ancestry, the more the individuals fade into the mists of the past. No group of breeders, neither the "Blue" breeders of the USA nor the "Asil" breeders anywhere else, can vouch for the proven origin of their horses . . .
>
> (Dr. Hans Nagel, WAHO president)

On the other hand, there are Egyptian Arabian horse breeders who believe that, in spite of the convincing evidence that most living horses known by the generic term "Arabian" have documented instances of non-Arabian genetic components, a small percentage of living Arabian horses may truly be 100% genetically *purebred.* Purist breeders adhere to the conviction that a wild species of genetically unique horse existed on the Arabian Peninsula before the advent of man, evolving in a state of geographic isolation. This, they believe, was the *purebred* Arabian horse. Purists believe that actual unadulterated descendants of the horse of the Arabian Desert, free from genetic equine contamination, may still exist. They believe that the pedigrees of some Arabian horses have been accurately preserved.

Purists and the preservationists are motivated not by a fey and fanciful delusion of wish fulfillment but by a firm adherence to the time-honored Bedouin dictum that any non-Arabian blood in the pedigree of an Arab horse will eventually lead to deterioration in the quality of its offspring. They take this dictum to be absolutely true.

The purists are supported by men with considerable academic and authoritative credentials. Dr. W. Georg Olms wrote in Asil Araber III "A function from which WAHO is barred…the Asil Club, Al Khamsa Inc., and the Pyramid Society have assumed; they separately register breeds which are to be regarded as asil, according to Bedouin tradition. Their horses trace exclusively to Bedouin breeding on the Arabian Peninsula... they must not be lost to the total stock."

Pedigree cynics, skeptics, and relativists consider these groups of individuals to be suffering from what has been called "document delusion". In his book Hanan, Dr. Nagel describes Arabian horse pedigree purists as "idiosyncratic pedigree fetishists".

This point of view suggests, of course, that all generic Arabian horses have some element of non-Arabian blood, a possibility that the purists would find discouraging and that the skeptics would embrace wholeheartedly.

Dr. Nagel, writing from a position as both a broadly experience Arabian horse breeder and as a widely respected equine authority, recapitulated the relativist opinion in his 2013 book The Arabian Horse:

> All the living Arabians in the Middle East have now to be considered as cross-breds between different populations of earlier times. A lot of them returned through re-emigration from the North to the South or due to trade and traffic.

It is a thorny problem. Most of the world's Arabian horse breeders find the topic quixotic and puzzling. The position of the relativists is now generally considered to be an obvious fact, so obvious that debate over the issue seems in many parts of the world to be so untenable, so devoid of facts that the issue has come to be regarded as pointless. For most Arabian horse breeders, it hardly merits discussion. Except for a few pockets of resistance in the U.S. and Germany, the question of absolute genetic purity is no longer even considered.

But it gets worse. Once the threshold of absolute genetic purity is breached by the presence of just one non-Arabian individual in a pedigree, the question becomes more difficult to manage. Is there a difference between the breeding potential of an Arabian horse that is 99% genetically pure as compared to a horse that contains only

50% genetically pure Arabian blood? Is there any place in Arabian horse breeding for the concept of the "purest of the impure"? Modern advances in DNA analysis have a role to play here too.

Figure 25 - *A Scene in the Desert, by Eugene Fromentin, used with the permission of the Hermitage Museum, St. Petersburg, Russia.*

The written pedigree record, far from becoming obsolete in the age of relativism, remains vitally important to the genetic investigator. Planning, conducting, and interpreting the results of DNA studies require a full understanding of the history of the breed and some degree of familiarity with the documentary and textual record pertaining to the pedigree of the Egyptian Arabian horse. The ability of a DNA analyst to arrive at dependable and reproducible conclusions regarding the accuracy of the historic record requires a thorough knowledge of the contents of that record. The geneticist cannot work in a state of isolation from the object of study but must sift through the details of the pedigree history as the work of reaching verifiable facts is pursued. The horse breeder cannot effectively carry out the process of breeding without having an understanding of the scientific data and facts emerging from Arabian horse DNA pedigree verification research.

The problem of pedigree research, investigation, and clarification is complex. Thus far, simple answers have been elusive. The course of inquiry taken in this book is neither simple nor conclusive. Any attempt to know the whole of a subject requires the consideration of the opinions of all persons who possess the widest possible range of orientation and motivation. Every modality of past research must be included.

In the course of compiling this book, many place names, personal names, horse names, and terms in the Arabic language have been presented in English. Since there is no universal standard for the transliteration of Arabic words into English, it has been the practice throughout the book to render the words transliterated by the same English counterparts that were contained in the research source material. There has been no attempt to standardize these terms, and the reader will find the same place names, for example, in variable forms in English in different portions of the text. The various spellings that were found in the source materials used in the research for this book have been reproduced as they were found. Appended horse name information (for example* or #) is not used in this book.

Figure 26 - *Image credited to Mari Art.*

Throughout this book, the reader will encounter several related terms which have caused a great deal of confusion over the years. The term 'matriline' and 'tail female' are used with the same highly specific meaning; the actual genetically proven female to female line of descent. The term 'strain' is used only in its historical context; it has no scientific or specific meaning in terms of the research presented in this book. The term "historical family" is used in a very specific way; to indicate the recorded matrilineal line of descent of all horses from a given matriline as tabulated by Colin Pearson in The Arabian Horse Families of Egypt.

**Better untaught than ill-taught.*

INTRODUCTION

INTRODUCTION

THE ARAB HORSES OF EGYPT, THE ROYAL AGRICULTURAL SOCIETY, AND THE EGYPTIAN AGRICULTURAL ORGANIZATION

Egyptian Arabian horses are a specific type of *Equus caballus*. Originating with the Bedouin of the central Arabian Deserts, they have been preserved for posterity through a variety of channels. One of these preservationists is the Royal Agricultural Society/Egyptian Agricultural Organization (RAS/EAO) Arabian horse breeding program (Egyptian Agricultural Organization, El-Zahraa Stud, Ahmed Esmat St., Ein Shams, Cairo, Egypt), an Egyptian state-run breeding farm which maintains a closed horse breeding population consisting of *purebred Arabian horses*.

These horses are unique in that each individual chosen for inclusion into the breeding herd was selected by a group of highly qualified program directors who, based on the best available information, believed each horse to be descended entirely from Arabian Desert Bedouin sources, free from the admixture of any non-Arabian blood. While today the idea that a *purebred Arabian horse* of this description ever existed is for many horse breeders little more than a fairy tale, the matter was for several centuries of intense seriousness to horsemen of Arabia and beyond. While the importance of genetic purity and authenticity has, on occasion, been subject to hyperbole, it is difficult to dismiss the importance of the truly *purebred Arabian horse* with the stroke of a pen.

H.E. Fouad Pasha Abaza, Director General of the Royal Agricultural Society, described the program in the following manner in 1947:

> The horses bred by the Royal Agricultural Society are the descendants of the authentic Arabian horses imported into Egypt from the Arabian Peninsula by the late Abbas Pasha I, Viceroy of Egypt, and afterwards bred by the late Ali Pasha Cherif.

The forerunner of the RAS/EAO, the Egyptian government-mandated Horse Commission, was founded in 1892 under the direction of Omar Tousson. Charged with improving the quality of cattle and horses in Egypt in general, the Commission found that it was unable to pursue its goal of breeding the pure blooded desert horse, and was forced instead to use English Thoroughbred stallions to service the Egyptian mares that were available. This situation was brought about by two factors. First, the Ottoman government in Constantinople would not permit the importation of asil (*purebred*) stallions into Egypt and second, all asil stallions that were then in Egypt were in the hands of private breeders, primarily members of the Royal families.

The Egyptian government devoted considerable attention to matters of livestock management since the Army and Police required a regular supply of mounts. In 1908, the Breeding Section of the Royal Agricultural Society of Egypt was founded, taking over the operations of the Horse Commission. From 1908 to 1914, it continued the policy of producing Anglo-Arabs in order to supply the working horses needed in the agrarian economy of Egypt.

In 1912, a group of concerned individuals met for discussions about the possibility of creating a government-sponsored *purebred Arabian horse* breeding program in Egypt. Under the patronage of H.H. Abbas Pasha II, the Khedive of Egypt, this movement was called the International Society for the Preservation of the Arab horse. These were all men of distinction, well educated, veteran horsemen, and deeply committed to the idea of preserving the *purebred Arabian horse* for posterity. Members included H.H. Prince Mohammed Aly, Prince Youssef Kemal, Russian Prince Alexandre Sherbatoff, and Prince Kemal el Dine Hussein. This group of breeders was concerned about the continued use of English Thoroughbred in the RAS program and felt that there was a need for an alternative: a breeding program whose aim was to preserve the type of 100% *purebred* Nejd Bedouin breeding stock. The Society's project was stalled due to the unrest leading up to the Great War.

Figure 1 - *A Purebred Arabian Horse, Palestinian Government Horse Stud, Acre, early 20th century, photo from the Library of Congress.*

Egypt was not alone in its commitment to the preservation of the rapidly declining stock of *purebred Arabian horses*. The crisis in obtaining high-quality desert Bedouin horses was appreciated by many Arabs who valued the *purebred Arabian horse* which was their birthright from antiquity. In the early 20th century the Palestinian Government established a state-run *purebred Arabian horse* breeding farm in Acre. Unlike the experiment being carried out in Cairo, the Acre program did not flourish.

In 1914, on the eve of World War I, with the hegemony of Constantinople beginning to wane, the decision was made by the directors of the RAS to end the experiment with the use of English Thoroughbred sires, as the resulting stock were considered by the Egyptian breeders as well as the Egyptian public to be both vicious and ugly. At this point, the officials governing the RAS began to assemble a group of *purebred Arabian horses* of pure Abbas Pasha breeding which were then present in Egypt. The purpose was to maintain a nucleus of *purebred Arabian horses*. The Society rejected the idea of collecting horses from Arabia itself due to the prevailing suspicion that most of the peninsular stock had been contaminated by cross breeding with foreign animals that had been introduced by the successive hordes of invading European and Ottoman armies.

The selection of mares was given special attention by the Society. Of the many foundation mares brought together by the RAS since its origin, only a few mare lines still exist in tail-female lines of descent. Four of these lines are represented today by only a few living female representatives.

The fact that the *purebred Arabian horse* was becoming quite rare both in Arabia and Egypt was apparent to all of these early preservationists. Four factors responsible for this crisis were:

1. The unification of the Arabian Peninsula by the Saud family had led to an end to inter-tribal warfare. The state of pacification made the horse obsolete. It had become an expendable luxury for the Bedouin.

2. Mechanization and the automobile made the horse redundant as a means of transportation. Modern methods of warfare had made the horse too vulnerable to be of any practical use in times of war.

3. The expense of keeping horses in the desert played an important role. The Arabs had very few stallions since the mare was more practical for riding. Colts were usually killed at birth unless they came from mare lines that were particularly valuable. Gelding was unknown in the desert.

4. Equine infectious disease was a constant threat in Egypt; because of regular and recurrent cycles of African Horse Sickness. The number of Arabian horses in Egypt has throughout history has been very low.

The RAS gathered most of its foundation stock from the few remaining sources of desert-bred horses with known pedigrees or from reliable sources, all originating from the Bedouin tribes of Arabia. The Society established standards in selecting its foundation stock. The horses must be "sound animals of the best blood and good pluck and free from any vicious tendency of temper, or other hereditary defect". Many of these horses were in the hands of members of the Egyptian royal family, and most of them were from the stock of Abbas Pasha I, who was considered to be an unimpeachable source of pure desert horse breeding. In 1919, the Society bought 18 stallions and 2 fillies from the Blunt Arabian bloodstock breeding program in England. The Society's original stables were at Behtim, but in 1930 the facility was moved to larger and more modern stables at Kafr Farouk.

The Egyptian Kings maintained a separate private Arabian horse stud or khassa, beginning with King Fuad and continuing with King Farouk. This stud was known as Inshass and consisted of some of the same desert blood lines used by the RAS. Unlike the RAS, the King's khassa also contained a number of desert horses. Their genetic composition was considered to be stainless, free from the inclusion of non-Arabian blood. Not all of the horses given to the khassa as tribute were kept. The King's stud manager carefully evaluated each horse, keeping only those horses that were considered to be of superior quality and to be of unsullied desert breeding.

Figure 2 - *The Stallion Stables at the EAO, from the EAO Studbook Vol. II, 1977.*

Figure 3 - *A Bedouin hunting party on the mud flats between Mudawara and Guwaira.Wadi Rum and the mountains of Edom are in the background. Photo from The Raswan Archives.*

By 1936, the breeding program of the RAS was making great strides in building up the breeding stock. The stud consisted of 39 stallions and 30 mares. The Society was developing an international reputation and was beginning to sell 3 or 4 yearlings every year to buyers America and Europe. The introduction of modern veterinary techniques and improved stabling and feeding conditions led to an expansion of the number of horses being produced.

With the passage of time, a new and unanticipated dilemma was encountered. The RAS directors came to realize that the desert horse of known and certifiable pedigree could no longer be found among the Bedouin desert tribes, and an appreciation of the value of this nearly extinct animal began to grow.

In Egypt, the directors of the RAS were becoming increasingly concerned about the highly inbred nature of the Egyptian horse breeding experiment. Without new outcross blood, they feared that the RAS stock would become "all pedigree and no horse". Egyptian breeders would have to go to the desert of Arabia to secure more stock. In that year Dr. Ahmed Mabrouk, at the request of the RAS Directors, was sent on a journey throughout the Middle East, seeking asil horses among the Bedouin tribes of Hedjaz, Nejd, Iraq, and Syria.

Dr. Mabrouk observed:

> If inbreeding is allowed to go on without the infusion of new blood of the pure Arab horse, which has been famous for centuries, it will disappear in a limited number of generations, giving place to an inferior breed of animals which will lack the stamina, courage and handsome appearance of the original breed.

The first stop on Dr. Mabrouk's journey was in Tayef and the stables of H.H. Feisul. Breeding had reached its nadir there, and Feisul would not vouch for the purity of the horses since most of them were imported from Syria and Iraq. H.H. was confident of the purity only of the few authentic Nejd horses remaining in his private stables, and these were very old.

In Riyadh, Dr. Mabrouk was welcomed by His Majesty King Abdel Aziz Saoud. His Majesty deplored the decline in breeding of the ancient and authentic Nejd horse, and attributed this to the peaceful life of the Bedouin, a peace that His Majesty himself had imposed on the Nedjeans through force of arms. The inter-tribal conflicts, said the King, had made unification of the country impossible. The Bedouin, now comfortable and at peace, had no reason to expend the money and effort needed to breed horses. The King also remarked on the excessive loss of military mounts during the battles for unification of the kingdom. He personally counted 450 horses dead after one skirmish. Mabrouk inspected the King's stables at Kharg. He found the horses to be small, and there were very few stallions. The stallions, he said, were selected on the basis of pedigree only, and appeared to him to be unattractive and unsound, with poor conformation of the legs.

In Bahrein, Mabrouk found a small number of horses owned by the ruling family, but he found that they did not possess the individuality of character that he considered essential to the true desert horse.

The horses in Iraq were concentrated in Baghdad and Bassora. These were centers devoted entirely to the export of race horses. Mabrouk was doubtful that any of them were of pure Nejd breeding. They were very tall, long legged, and many had very large heads. These factors suggested the presence of non-Arabian blood in their ancestry.

The widespread decline in horse breeding, he felt, was attributable to the use of the automobile for transportation and the increasing worldwide competition from Arabian horse breeders in Argentina, Brazil, Australia and South Africa. Globalization in the marketplace was having a significant effect in the Middle East.

Figure 4 - *Hunting by auto on the peninsula rapidly led to the extinction of the Arabian gazelle, (Gazella arabicus,) the Arabian oryx (Oryx leucorhyx) and the native Ostrich, (Struthio camelus syriacus). Photo from the collection of the Library of Congress.*

Figure 5 - *Arab Raiding parties became mechanized around the turn of the century, leading to the obsolescence of the war mare. Horse breeding on the peninsula never recovered. From* The Raswan Index.

Figure 6 - *The now extinct Arabian Ostrich, from the book* Kitab al Hayawan, *by Al Jahiz, born in Basra in 767 CE.*

Figure 7 - *The black tents of the Bedouin. By the time Dr. Mabrouk made his foray into Arabia, the migratory patterns of the Bedu had changed, and many of them had become settled in towns and villages. Sights like this were becoming increasingly rare. Photograph used with the permission of Saudi Aramco.*

Mabrouk, unable to determine a reliable and valid pedigree of any of the horses that he saw, could recommend only one of them to the management at the RAS without reservation. This was the stallion Kroush, a Kuhaylan Nawwaq purchased in Beirut, who was used at stud at the RAS in Cairo for about four years. He left no lasting impression on the .program. Kroush was from the stables of Saad el Din Chatilla and was bought by the RAS in 1936. He died four years later. At the RAS he sired Bushra 1940, Tamie 1937, and Madiha in 1938.

> The first difficulty which Dr. Mabrouk encountered was the absence of stud records. The rapidly dwindling interest in horse breeding amongst the Bedouin is responsible for this. Although some fine mares were found, their origin was always in doubt. Even the Amir Feisul would not give a certificate of authenticity to any imported animals, however, fine they may be. He recognized only the Nedj horses bred

> by himself, which were very few . . . stallions were scarce and horses were seen dying of plague . . . during the past few years, there has been a big change in the Bedouin mode of life and habits. Whereas in days past the Bedouin tribes were always at war with one another, carrying out raids and always on the move from place to place, today peace prevails. The tribes have, to a large extent, settled down in a definite district. The consequence is that the possession of a horse is not imperative as it was before, when it was essential that the Bedouin should have a fast-moving animal for the sake of his safety and that of his family. Owing to the difficulty of meeting the cost of feeding . . . only the Kings . . . can afford to keep horses.
>
> The horses bred by the Royal Agricultural Society are the descendants of the authentic Arabian horses imported into Egypt from the Arabian Peninsula by the late Abbas Pasha I, Viceroy of Egypt . . . and has been carried on ever since by members of his family.
>
> Breeders and lovers of the authentic Arabian horses abroad are well aware of the fact that most of the famous studs of Arab horses in Europe were devastated during each of the two Great Wars, and Egypt will, therefore, remain an important centre for breeding and preserving the authentic Arab horse.
>
> (H. E. Fouad Pasha Abaza, Director General, RAS/EAO Stud Book Vol. 1, 1947)

By the 1940s, the Arab horse breeding program in Cairo was at a crossroads, and the directors sensed the need for strong and authoritative leadership. Mohamed Taher Pasha, uncle of King Farouk and director of the Royal Agricultural Society, selected an exceptional horseman from the Hungarian Babolna Stud, General Tibor von Pettko Szandtner, to direct the Egyptian stud. His skillful use of the stallions Nazeer and Sid Abouhom produced a decade of steadily improving quality at the stud.

Szandter was born in 1886 at a Hungarian stud farm managed by of his father. Following in his father's footsteps, he became commandant of the Babolna Arabian stud farm where he managed over one thousand *purebred* and part-bred Arabians. During the Second World War, he was instrumental in saving many of the horses from the invading German army.

As hostilities escalated, he escaped to Sweden where he managed a friend's farm that bred Hungarian Arabian horses. He worked there until 1949 when he was chosen to take over the management of the Egyptian stud at Ein Shams in Cairo. He was director of the RAS during the tumultuous and dangerous years of the Revolution of 1952 and the subsequent years of hardship for the horses in Egypt. He served there for eleven years and retired at age 74 due to declining health. He died the next year in the care of an admirer, Prince Ruprect of Bavaria.

Figure 8 - *General Tibor von Pettko-Szandtner.*

Figure 9 - *Cheval Arabe, by Horace Vernet, from the British Museum.*

Szandtner's years as head of the Egyptian stud produced exceptional stock. His skillful use of the stallions Nazeer, Sheikh El Arab, and Sid Abouhom led to a decade of steadily improving quality at the stud. This was a Golden Age of breeding at the Cairo stud.

THE REVOLUTION

In 1952, the government of the Egyptian King Farouk was overthrown by the Free Officer's Movement, a group of volunteers from the Egyptian military ranks headed by Abdel Gamal Nasser.

The vast majority of the commoners of Egypt lived in a state of poverty, decay, and hopelessness. The country had no effective government, no reliable food supplies, minimal electricity, and erratic transportation. Cairo lacked even the most basic sanitation services.

Meanwhile, the royal family lived in a parallel universe, wealthy beyond reason, and fully Westernized. The Royals copied American styles and fashion, and the women adorned themselves according to the fashions of Hollywood. They absorbed all of the worst influences coming from America, notably the worship of money and the cult of celebrity. They accumulated the trappings of the Western elite, building palace after palace, all carried out amid the squalor and despair of the common Egyptian. The people regarded the behavior of the Royal family to be disrespectful of Islam.

In 1950, the average income of the Egyptian worker was USD 500. The country was seething with dissatisfaction over its monarchy's indifference to their suffering, and the stage was set for action.

The Free Officers began to informally coalesce into a cohesive group following the Egyptian loss of the 1948 war with Israel. The war was especially personal for Nasser as he led a military unit involved in some of the fiercest fighting with the Israeli forces. King Farouk was personally involved in corrupt arms scandals and overt interference with the officers of the Egyptian Army during the war with Israel. The officers blamed the loss to the Israelis on lack of support from the King, and their discontent led them to accuse him publicly of irresponsible behavior, mismanagement of the government and public finances, and his disdain for the welfare of the poor and the oppressed. They demanded that the King abdicate by the 4th of Zul Qa'ada 1371.

At 7:30 AM on that day, the Egyptian people awoke to radio broadcasts announcing the overthrow of the King and the establishment of a Revolutionary Government. The voice that the people heard over the radio that morning was junior Free Officer Anwar Sadat.

It was a bloodless coup.

The Army, under the direction of General Muhammad Naguib, removed Farouk as King. Although there were many causes of the uprising, the main cause was the global movement away from monarchy and toward collectivism. In Egypt as elsewhere, there was public outrage over the decades of financial and political corruption of the royal family. The new military government regarded displays of personal wealth and collections of Arabian horses as only one of the most potent symbols of the egregious excesses and extravagant lifestyle of the royal family and their apparent indifference to the suffering of the common man.

These men were socialists at a time in history when the ideologic conflicts between communism, socialism, and democracy consumed the post-World War II world. Socialism in Egypt took the form of governmental ownership of the means of production of goods. The production of goods that directly benefited the common man was given priority.

Figure 10 - *King Farouk, before the Revolution.*

The members of the extended Royal family were specifically targeted and all fled the country. Those Royals who had the foresight to stockpile their wealth in European banks were able to live comfortably after their expulsion from Egypt; those who did not see the Revolution coming did not fare as well.

The movement sought to abolish the constitutional monarchy, and bring about an end to the power of the aristocratic families of Egypt. Their goals were to establish a republic in Egypt and end British involvement in Egyptian affairs; Britain had virtually run the country since 1882. The Revolutionary Council sought to support the cause of Arab nationalism throughout the region as Pan-Arab sentiments were running high.

Sadat's Declaration of Revolution made the following accusations:

1. The King accepted bribes to ensure the defeat of the Egyptian Army in 1948.

2. Traitors have been commanding the Army.

3. The government of the King is capricious and unstable.

4. The Commander of the current Army is either ignorant or corrupt.

Figure 11 - *The Free Officers group in its formative years. This group, formed in obscurity, contained Egypt's future leaders.*

After the announcement by the Free Officers, the King's assets were immediately frozen. Farouk left Egypt from the port in Alexandria on board his yacht Mahroussa.

Even in exile, Farouk continued to live an opulent lifestyle with his rapacious and gluttonous ways. His connections with international bankers had allowed him to retain a great deal of his wealth. This was in stark contrast to the poverty in which his relative Prince Mohammad Ali Tewfik lived in Geneva.

Figure 12 - *Prince Mohammad Ali Tewfik was abroad when the Revolution occurred. As heir apparent to King Farouk he was a man of considerable interest to the Revolutionary authorities. He returned home to Egypt to discover his fate.*

An acquaintance described Farouk as a "stomach with a head". The ex-King died in Rome at age 45, suddenly stricken following a typically enormous meal. The wheel of fate had come full circle for a man described by his Oxford tutor as "untruthful, lazy, capricious, vain, and irresponsible".

The horse stock of the Inshass stud of King Farouk, as well as those of the Royal Agricultural Society, was nearly destroyed during the turmoil of the Egyptian Revolution of 1952.

Egyptian Gazette 8 September 1952

PRINCE MOHAMED ALY TO LEAVE EGYPT SOON

Resigns from Mohamed Aly Club

H.R.H. Prince Mohamed Aly on Saturday resigned his presidency and membership of the Mohamed Aly Club. This gives confirmation to the widespread rumours that the Prince wishes to leave the country and settle abroad.

"Al Ahram" states that Prince Mohamed Aly, who since his return from Europe, has been seldom seen in his usual clubs and has refused to meet anyone except those whose official position demanded it, hopes to leave Egypt within a few days.

His hurried return to Egypt, "Al Ahram" says, was dictated by the present political situation. Having conferred with the national leaders, he now wishes to complete his rest-cure in Europe and settle there definitely, resigning all rights to the title of Heir Presumptive.

There is some doubt whether the laws governing the transfer of money from Egypt to a foreign country will apply in this case. The matter is reported to be at present under study.

The Prince possesses 7,100 feddans, according to "Al Ahram." Of these 2,000 are a Wakf left by his mother, 2,000 are arable land and 3,600, situated in the Fayum, are infertile land which the Prince has been reclaiming.

His palace at Roda with its huge garden was constituted a Wakf by the Prince, with the intention that the palace should serve as museum of antiquities and the garden as a public park.

PRINCE MOHAMED ALY, the heir to the throne, who is reported to be considering leaving Egypt to establish his residence abroad.

Figure 13 - *Pmat 1952. Prince Mohammad Ali Tewfik returns to Cairo from a trip abroad.*

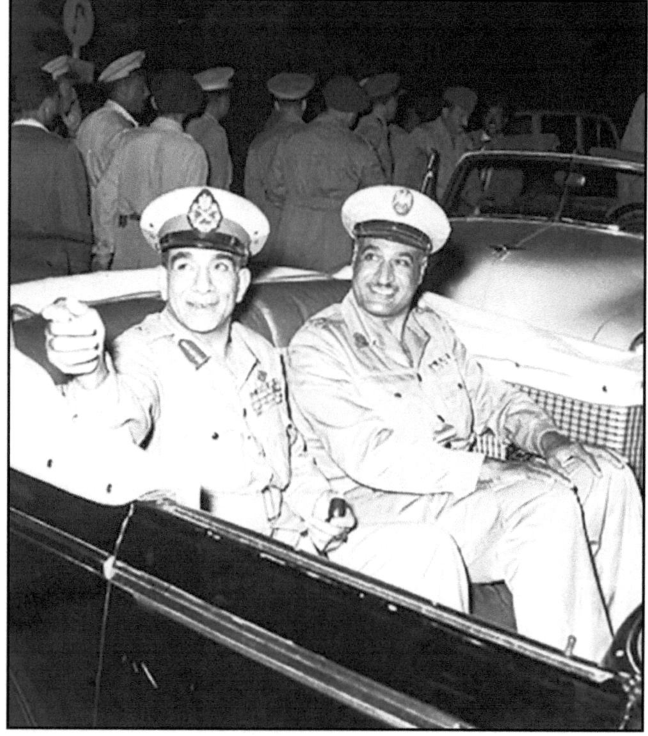

Figure 14 - *Muhammad Naugib and Gamal Abdel Nasser.*

Figure 15 - *Gamal Adbel Nasser's tumultuous welcome in Alexandria as he entered the city after the coup.*

The horses of the King were initially destined for liquidation. The herd of Arabian horses at the King's private stud called Inshass was seized and reconstituted under the newly created Egyptian Agricultural Organization. Some of the Inshass herd horses were incorporated into the EAO. Most of the horses in the Inshass stud were disposed of through sales, culling, and transfers prior to the merger of Inshass stock with the RAS stock. The name of the newly minted Arabian horse farm was changed from Kafr Farouk to El Zahraa. Since absorbing a portion of the Inshass horses at the time of the 1952 Revolution, the EAO has bred and promoted this closed herd of horses without the introduction of additional genetic material.

The only exception was the addition of the stallion Sharkasi who was obtained by the EAO in 1955 on the death of his owner T.B. Trouncer. However, Sharkasi and all of his offspring were discarded due to their poor quality as individuals.

The Revolution was a dangerous time for horses in Egypt, and the threat continued long after the 1952 Revolution itself. The new government considered the Egyptian

Arabian horses to be useless, wasteful, and a painful reminder of the excesses of the Farouk regime. The Egyptian people had suffered extreme privations under Fouad and Farouk, and emotions ran high.

Anyone owning more than a few horses was identified by the government as being excessively wealthy and was targeted to have their estate and horses sequestered and sold at auction. The money that was seized was funneled into public works projects. Some horses were bought by merchants to be used as drafts animals, and some were directed to the Cairo Zoo to be butchered for use as meat.

Von Szandtner was principally in charge of the process of choosing those Inshass horses that would be added to the old RAS pool, and deciding which were to be disposed of. He was rigorous and disciplined in his approach, rejecting all but the finest specimens.

Sayed Marei, a long time Arabian horse breeder before the Revolution, was a staunch defender of the horses, urging the new government to preserve them. He was also a member of the new establishment and became Minister of Agriculture in the new government. His influence in preserving the horses was considerable.

The government for its part was dealing with chronic poverty of the people of Egypt. The Revolutionary government was attempting to bring the essentials of food, transportation, electricity and health care to a populace that was suffering greatly. The new government was composed of a group of firmly committed socialists, and they were intent on the sequestration of the property of any wealthy Egyptian, whether they were royalty or not.

An American woman was responsible for saving several of the Inshass horses from being killed by the Revolutionary Government. American Sara Loken had been posted in Egypt along with her husband. She was an avid horsewoman and developed a deep love of the horses that she came to know at the RAS. The Inshass stallion Hamdan was seized in 1952 by the new government and sequestered for butchering, to be used to feed the carnivores at the Cairo Zoo. Through sheer dogged determination, Loken was able to demonstrate the worth of the horse to the zoo officials. They relented and Hamdan was spared. He was bought by Ahmed Hamza who established a small private breeding program built around the aged stallion, called Hamdan Stables.

The crisis for the horses did not end in 1952. Internal political extremism practiced by the Muslim Brotherhood had become a direct threat to the stability of the new Revolutionary government. The assassination of Nasser had been planned by the Brotherhood. Laws were passed which made its activities illegal and it was suppressed. For the government in crisis, the fate of the Arabian horses in the government stud seemed to be unimportant. Allocating money to feed horses seemed indecent while millions of Egyptians were destitute.

The new Egyptian Republic came under attack in 1956 when the Tripartite Aggression of the combined forces of Britain, France, and Israel invaded Egypt. The loss of life was costly for the Egyptians, but they were able to repel the invasion, and, for the first time in its history, achieved complete and independent control of the Suez Canal. The British would not leave the Canal Zone quietly, and their removal followed repeated military attacks by the Revolutionary Command Council. In 1956, the last British soldier evacuated the Canal Zone.

1956 also marked a turning point in the political structure of the Egyptian government. Nasser announced the creation of a new Constitution, setting up a democratic presidential system. Women were given equal political and voting rights. Nasser was the first president.

Figure 16 - *Nasser appearing before his supporters in Mansoura, in 1966.*

Further hardship was imposed on the Arabian horses when the EAO stables were relocated in 1960 from the pastures of Ein Shams to the desert of Heliopolis.Dr. Mohammed Marsafi was director by then, and his task was made difficult by continuing governmental strictures. The hardships imposed by the government continued into the 1960s. Horses remained a low governmental priority.

In 1966, the prominent Arabian horse breeder Ahmed Hamza, founder of Hamdan Stables in Cairo, was caught up in yet another wave of governmental confiscations and sequestrations. The Pasha was targeted for his "conspicuous consumption" and was confined to his residence. His wealth and all of his possessions were sequestered and placed under government control. The stables and horses were confiscated as well. They were poorly cared for and suffered marked privations.

Under President Nasser, the socialistic dogma became more rabid, and there was increasing pressure to hunt down and to root out men of wealth. The government wanted to take every opportunity to make examples of the men of wealth and to vilify instances of anti-socialist activity. In 1967, Nasser's animosity toward Western corrupting influences boiled over in his May Day speech. Hostile anti-foreign sentiment forced many Westerners to leave the country, and the attitude of the Egyptian public discouraged outside buyers from attempting to obtain horses from the EAO. Nasser's tone was violent and his policies were harsh and punitive. He died from a heart condition in 1970. 5 million mourners attended the funeral.

The Socialist aim of empowering the common man continued with renewed vigor. The Nasser government instituted agricultural reform and initiated extensive industrialization programs. Urbanization and infrastructure development continued.

It would be many years before these policies would change. Nasser's fear of the spread of Communism, Western aggression, and regional religious extremism fueled a rigid and restrictive political environment in Egypt. Nasser's concern for the small group of horses housed on the edge of the desert was minimal. The budget which he allotted for the care and feeding of the animals was inadequate for their needs.

Conditions at El Zahraa slowly began to change. The Egyptian economy improved, rabid political ideologies were modified, and the condition of the horses began to improve. By 1988 the EAO breeding program was so successful that it began serving as a registry agency for 39 local private Egyptian Arabian breeding farms. By 2014, the EAO supervised over 600 private studs in Egypt. The EAO herd now consists of over 350 horses. The EAO has published 8 volumes of stud records. As a

WAHO member and registering body for the nation of Egypt, the EAO is required to issue registration documents accepting any WAHO approved registered horse sold to any Egyptian buyer. In doing this, the EAO is, by adhering to the WAHO definition, conferring the status of *purebred Arabian horse* to any horse entering the country whose new Egyptian owner wishes for it to be registered by the Egyptian authorities. If the horse has unknown parentage, known non-Arabian parents, or carries the genes for fatal diseases in offspring, the EAO is required to accept the horse into the Egyptian National Registry. This fact is especially perplexing since the EAO is, by its own description, "the official channel which preserved the purity of Arabians..."

Figure 17 - *Yearlings in the paddocks at the EAO in Cairo, Egypt in 1984. Photo used with the permission of Polly Knoll.*

Figure 18 - Tagweed, *by Gad Allah ex Tee, El Zahraa Stud, EAO, Cairo, Egypt, 2011. Photo from the author's collection.*

Figure 19 - *Defaf, 2004 stallion by Ouf and out of Sarhana, El Zahraa Stud, Egyptian Agricultural Organization, 2011. Photo from the author's collection.*

Figure 20 - *El Zahraa, Cairo, Egypt, 2011. Photo from the author's collection.*

CHARACTERISTICS OF THE AUTHENTIC EGYPTIAN ARABIAN HORSE

The Egyptian Arab horse, also known in centuries past as the Oriental or Eastern horse, is unique and distinct among modern day Arabian horses, easily recognizable and identifiable when they are compared to horses which today are called by the generic term "Arabian". To compare the authentic horses depicted in this chapter to the Arabians of the international show ring of today, the reader is referred to the chapter on the Arabian horse bubble.

> He is a very perfect animal; he is not large here and small there. There is a balance and harmony throughout his frame not seen in any other horse. He is the quintessence of all good qualities in a compact form... the build of the Arab is perfect. It is essentially that of utility.
>
> Homer Davenport, 1867-1912 - American importer of desert sourced Arab horses

> The Arabs do not consider that the height (of their horses) has any relative value; it is proportions that make the value, and we shall see that they preferred the smaller horses.
>
> Mohamed Ali

> The original horses (like the zebra today) are always of square or nearly square built. Only man shaped it into a rectangle (see our modern harness and competition horses); thus it is small wonder that the most of the breeds of horses which are still more or less natural produce short back, that is square, horses, the former "prototype of a riding horse," for which the riding instructions were developed until 1789. Short and broad loins and small size seem to have actually been quality stamps of the purity of the race of these horses.
>
> Sadko G. Solinski, Lohhof, Center for Traditional Pleasure Riding.

Solinski (1937-2005) was a Swiss equestrian expert, well known in Europe for his new style of riding technique, designed to be a pleasure both to the rider and the horse. His books about his technique were published by Olms Press in Germany. Solinski's method was based on the sentiment "Escoutes toun chivau".

Symmetrical and balanced bodies are regarded by humans as the primary determinant of beauty. This is true when humans evaluate other humans for the purpose of mating and is true when horsemen evaluate the asil horse to assess its quality.

Symmetry and balance are appealing to the eye, but they are in fact features determined by Nature for functional soundness and enhanced survivability in a habitat filled with dangers and existential challenges. In Nature, only those horses that can bolt from a dead stop, accelerate rapidly, and maintain a gallop for hours at a time will survive. Flight is the natural horse's defense mechanism. Nature selects those horses to survive that possess functional superiority. Human conceptions of beauty had nothing to do with it.

To understand the physical and phenotypic features that constitute the ideal Arabian horse, mathematical laws and parameters prove useful. Carl Raswan had an intuitive sense that this must be true.

> "Nature still designs and creates in her own way with the aid of laws which we have incorporated into our sciences of space and numbers and have applied to physical research (and thus to animal husbandry too!) We have found that no human hand or mind can alter these perfect rules--rules which work "mysteriously" also within the strains or blood-lines of a pedigree (and thus in what this pedigree presents to us: an Arabian horse. We may apply these unseen, but controlling (and therefore reliable) powers and use them either for benefit or loss to ourselves."

These laws, he felt, would lead the breeder to the harmonious and balanced "Original". His comments prefigured the advent of DNA analysis.

The physical proportions and anatomy of man and animals have occupied the interest of scientists for centuries. The Fibonacci sequence (0, 1, 1, 2, 3, 5, 8, 13,..) has been central to this research. Leonardo Da Vinci (1452-1519) described the mathematical relationships of the various parts of the ideal human body in his work with the Vitruvian man.Taking the human head as the fundamental unit of measurement, he found that the ideal man was eight heads or eight units, tall. The number eight is in the Fibonacci sequence. But his notes written above and below the image of the man revealed much more. The arm span of the Vitruvian man, he found, is eight units. The width between the shoulders was two units. The length of the forearm from the elbow to the outstretched fingers was two units. (a cubit). These are all Fibonacci numbers. DaVinci was not interested in functional matters; he viewed the Vitruvian man as a key to understanding the cosmography of the microcosm. He did not recognize the proportions as part of the Fibonacci sequence, being much more interested in the cosmologic proportion phi (0.61803551...) This proportion was known as the golden ratio and was used by artists and architects during the Italian Renaissance.

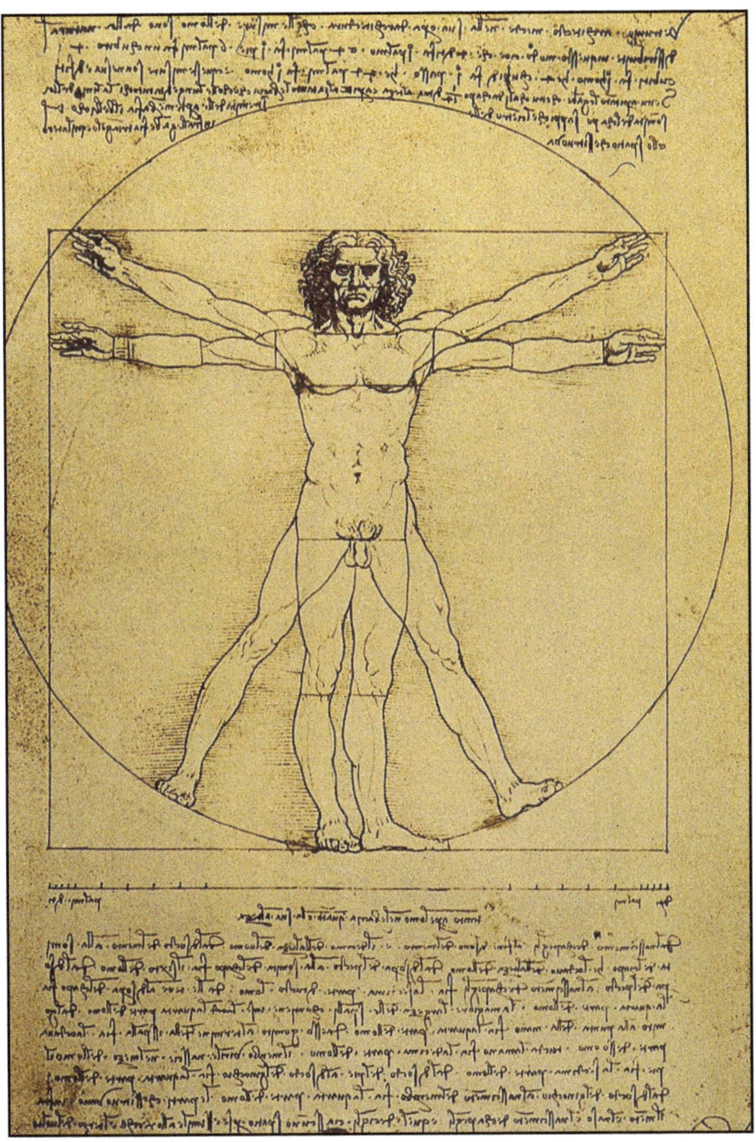

Figure 21 - *Vitruvian Man. This iconic image was drawn by Leonardo Da Vinci based on notes by the ancient Roman architect Vitruvius in his book De Architectura. The original drawing is kept at the Gallerie dell'Accademia in Venice. Using the square and circle, Da Vinci was attempting to discover the proportions in the ideal human figure as they relate to the proportions and harmony of the cosmos. He implies in his notes that the symmetry of man's ideal physical form corresponded with the symmetry of the universe.*

Figure 22 - *Leonardo Da Vinci.*

The Fibonacci sequence is a numbering system found throughout the natural world. From the logarithmic spiral of the shells of the nautilus and the snail to the arrangement of stems around a branch and the leaves around the stems, all are Fibonacci numbers. Even the packaging of seeds in the head of the sunflower, the aster, and the columbine are Fibonacci spirals and numbers of seeds is always a Fibonacci number. The sequence is not a mystical secret of the universe; it simply happens to describe the best way to make a plant or tree by which maximum sun exposure can be achieved.

Similar mathematical relationships of the structure of the ideal asil horse are also in evidence as members of the Fibonacci sequence. Using the length of the head as the starting point, considered as one unit of measurement, then the length and height of the ideal Arabian horse is not simply square and equal, but are both five units in measurement. The fore cannon bone is one unit; the forearm (the radius/ulna) is two units. The hind cannon bone is two units, and the distance from the "ankle" to the "knee" (tibia) is two units.

Figure 23 - *This type of research can be applied to the purebred Arabian horse as well. The analysis of the physical composition of the ideal horse, in this case, a purebred Egyptian Arabian horse, begins with the measurement of the head. This measurement is considered the cornerstone of the animal's overall proportions, and is regarded for this purpose as one "unit".*

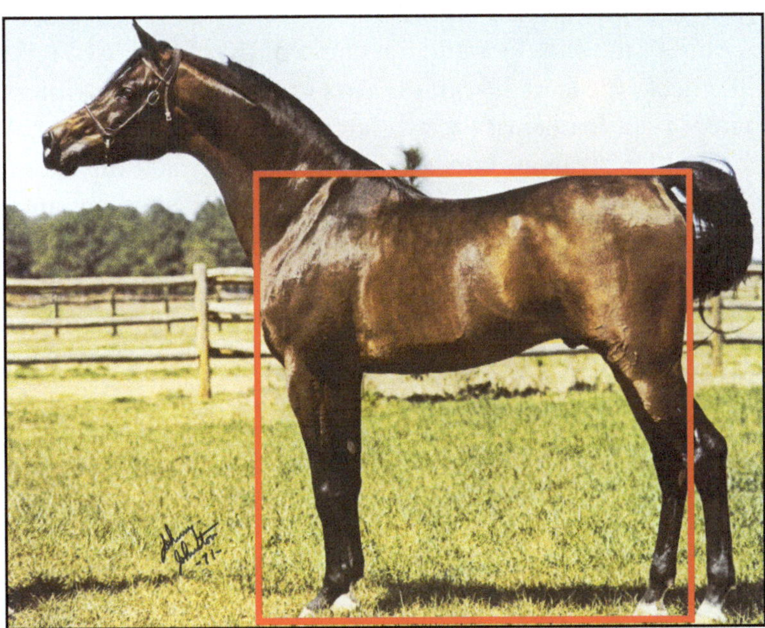

Figure 24 - *An ideally proportioned asil Arabian horse in which the "squareness" is demonstrated. The landmarks for assessing overall body configuration are the chest, the croup, and the rump. The height is precisely the same as the length. Moreover, the height and length are five units of measure, the number five being part of the Fibonacci sequence. The animal used in this illustration is the purebred Egyptian stallion Ibn Hafiza, born in 1959, bred by the Egyptian Agricultural Organization in Cairo, Egypt. The photograph is by Johnny Johnston.*

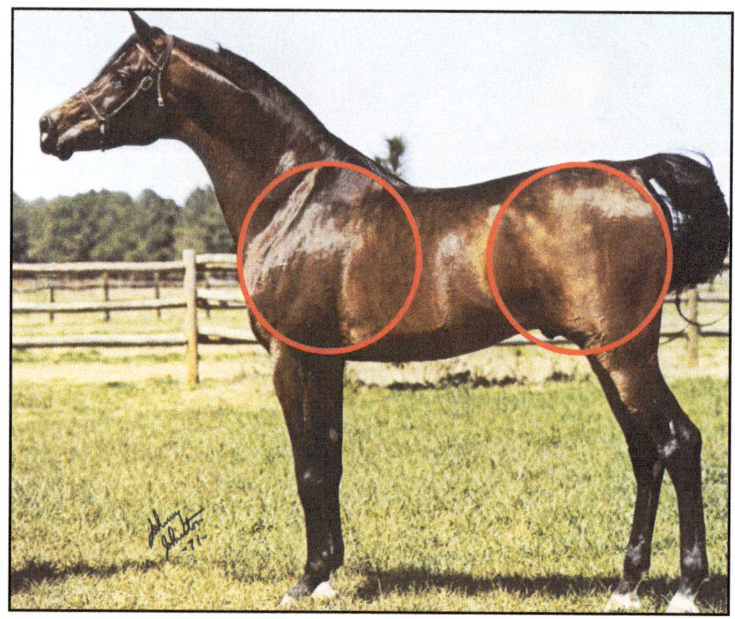

Figure 25 - *This image depicts the same asil horse with measurements to illustrate the balance of the horse. The front circle defines the proportions of the chest. The rear circle defines the anatomy of the hip and croup. The diameter of the circles is the same, for the ideal Egyptian Arabian horse is balanced, both fore and aft. The diameter of the circles is two units, the number two being one of the Fibonacci sequence.*

The circle defining the shoulder and the circle of the hip are two units in diameter. The length of the neck, measured from the ears to the end of the withers is two units. The neck is an important feature of the asil horse, being neither too short nor too long. Thus the ideal Egyptian Arabian horse is anatomically based on the numbers in the Fibonacci sequence (1, 2, 5, 8, 13, and so on). This comes as no surprise since the configuration of all plants and animals are based on this sequence. One has only to look.

The attractiveness, as well as the functionality, of the Egyptian Arabian horses, derives from its symmetry and balance. It is a 'square' horse, in that it is just as tall as it is long.

These proportions are useful in evaluating the quality of a horse. A horse that deviates from these proportions will have impaired function. It will not be able to propel itself properly, and will eventually experience permanent deterioration of the joints and spine. And, what's more, such a horse is beautiful to behold.

The authentic Egyptian Arabian horse is characterized by its modest size (14-2 to 15-2 hands high) and its sculpted gazelle-like head, with large dark sloe-type eyes set low in the skull. The anatomic proportions and beauty of the head are the hallmarks of the breed. The eyelids are dark black and well-folded, and the eye itself is large, lustrous, limpid and expressive. The eyelashes are long and fine hairs that are densely arranged along the eyelid. The eyelashes slant laterally, a feature not seen in other breeds. The iris is large in diameter and darkly pigmented. The sclera as a consequence is minimally visible. The gaze of the horse reveals intelligence and gentleness, inquisitiveness and a detached insouciance. The size of the head is appropriate to the body, neither too large nor too small, and is triangular or wedge-like in shape, with great width between the eyes, tapering to a small delicate muzzle. The nostrils are large and thin-walled, flaring and trumpet-like when the animal is excited. In profile, the head exhibits a concavity.

The ears are small and held in an attentive manner, tipped slightly inward, alert and responsive to sounds in the environment. The distance between the base of the ears is moderate, being no more and no less than the width of a human hand. The sense of hearing is very sharp. The horse exhibits sudden attentiveness and restlessness on hearing an unusual sound, even though it may be far away. The head of the Egyptian Arabian horse at maturity is described as "dry". The facial bones become prominent and angular, creating a sculpted appearance. There is very little subcutaneous tissue to fill out or soften the features, and this lends the head the appearance of having been chiseled by an artist from solid marble. The jibbah, the broad, flat and noble forehead, is prominent and imparts an aesthetic element of assala.

The mitbah, or throat latch, is refined and arched, forming a gently curved section that attaches the head to the neck. The neck is swan-like and gracefully set, rising vertically from a well sloped shoulder. The throat and trachea are free and loose under the neck when at rest, and are not bound in or restricted when the horse is in motion. The distance between the mandibles is great, creating a large ganache (from the French for "jowl"), the space containing the soft tissues and vital structures that pass from the head to the neck. The flexibility of the cervical vertebra is noteworthy.

The space between the mandibles is known as the waride. The distance should be as large as a man's palm. The wide waride provides the space needed for the trachea and great vessels to comfortably pass through. This permits greater breathing capacity with exertion.

The skin of the horse, paper thin, is the color of al kohl, an antimony-based Bedouin cosmetic, the Koheilan black. The hair of the coat is very short and gleams with a dazzling metallic brightness. This brilliance is seen especially in the burnished coat of the chestnut horse. The mane and forelock are long and silky, never overgrown or coarse.

The shoulder of the ideal Egyptian is well laid back. The angle of the shoulder is measured from the top of the wither to the point of the shoulder, which is roughly the course of the scapular spine. The angle of the shoulder should be about 45 degrees. This is essential to allow the horse increased foreleg extension which results in an increased stride length. The sloping shoulder also serves to minimize the effect of ground concussion from being transmitted upward from the hooves. A well laid back shoulder allows the shoulder joint to be open and swing freely in its action, producing a relaxed and comfortable gait.

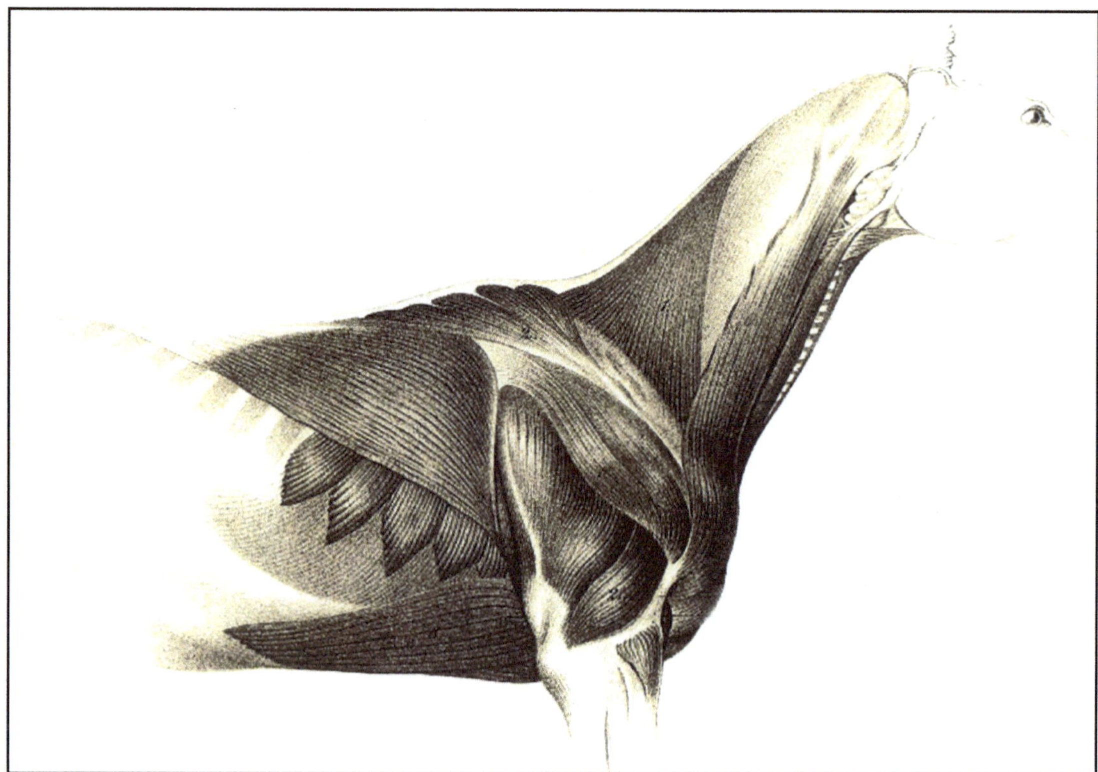

Figure 26 - *The musculature of the shoulder and neck of the horse.*

The withers, comprised of the spinous processes of the 8th through 12th vertebra, are high and well demarcated. The thorax is barrel shaped in cross section, not flattened or oval shaped, not "slab sided", the ribs have a marked degree of curvature and are said to be "well sprung". A filled-out thorax provides respiratory reserve that gives the animal respiratory endurance over long distances. The expanded thorax is associated with maximal width of the breast, which is the distance between the forelegs.

The back is short and exhibits a slight concavity, suggesting springiness. The Egyptian horse is short coupled and full in the haunches. The croup is long and flat. The peak of the croup is slightly higher than the withers. A horizontal croup allows the horse to have a long flowing stride. The tail is set high on the croup. The dock of the tail at its base is ideally very thick. The hair of the tail is thick, long, flowing and, when in motion, falls freely and delicately, like a woman's bridal veil. There is a silken quality to the hair. When the horse is in motion the tail is raised, creating a flag-like appearance as the hair streams out horizontally in the wind. Most authorities consider this feature to be an essential element of the asil horse.

The muscles forming the insides of the thighs should be full and well filled out, nearly touching in the center in such a way that they seem to be one unified whole, as if grown together.

Figure 27 - *Arab Cheval, by Carle Vernet.*

The horse has a constitution of iron, with hard bones and elastic but firm tendons. The bones are dense, with an ivory-like quality. The hooves are hard and flinty, small, round, well formed and symmetrical.

The structure of the Egyptian horse's cannon bone is unusual and distinctive. It is short in relation to overall leg length. The forearm is long in relation to the entire leg. A short cannon bone prevents excessive torque and instability of the carpals. This gives the Egyptian horse excellent stability and agility when navigating over rough or uneven ground.

Tradition holds that the ideal cannon bone is flat. This is not the case. When the cannon bone is palpated in a living animal, the complex set of tendons and ligaments creates the impression of flatness. Together they are shaped like a sword, with the sharp edge of the "sword" to the rear. The blunt edge of the "sword" is the leading edge of the cannon bone. When palpated, the cannon bone feels "flat". The cannon bone is anterior. And the posterior component is the superficial digital flexor tendon.

The conformation of the pasterns represents an ideal combination of length and slope. The pasterns are of average length and form a 45-degree angle with the ground, avoiding the disadvantages of the short upright pastern which transmits ground shock up the leg. A properly configured pastern also eliminates the predisposition to injury that a long and excessively sloping pastern creates.

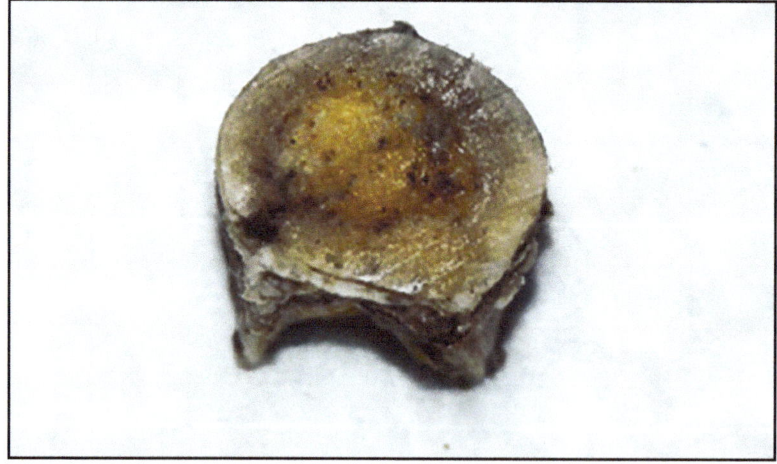

Figure 28 - *A section of a cannon bone freshly collected at necropsy from a 12 year old Egyptian Arabian stallion of pure EAO breeding. All tendons and surrounding tissue have been removed. The bone is round with an indentation on the posterior aspect which accommodates the suspensory ligament. The bone is almost entirely cancellous bone, hard and compacted with very little trabecular bone (marrow) in the center. The bone is thus capable of withstanding tremendous concussive force. The bony prominences noted at edge of the groove are the splint bones, a holdover from the era when the paleo-horse was three toed.*

Figure 29 - *The same specimen showing the posterior groove through which runs the suspensory ligament.*

Figure 30 - *A frontal view of the same specimen.*

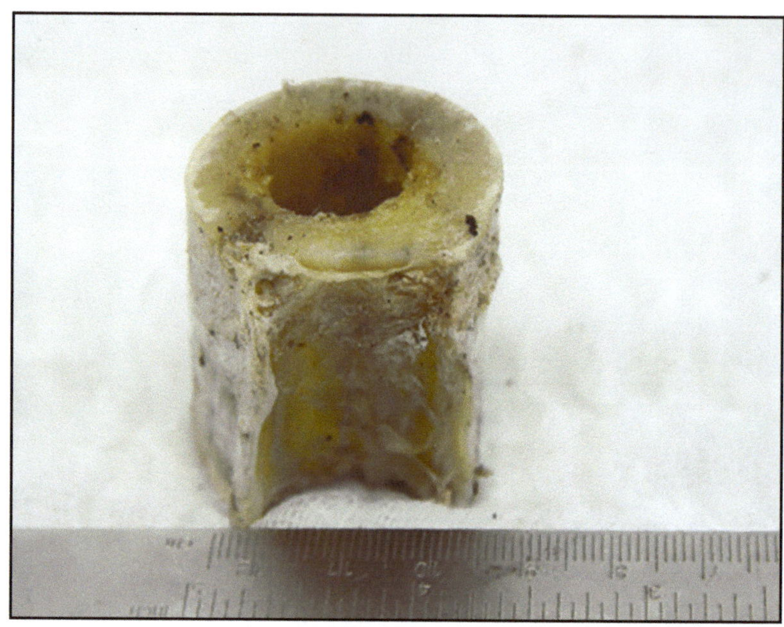

Figure 31 - *The same specimen with the bone marrow removed. The bone itself is 3.5 cm in diameter and the wall of the bone is 0.6 cm in thickness.*

When shown in hand the Egyptian horse reveals himself to be a regal creature, confident, majestic, distinguished and dignified. He is comfortable with himself and is eager to perform for those who watch his display. When observed at liberty the horse is animated, energetic and graceful in movement with a spirited and proud carriage. There is a subtle fluidity in the movements of the horse, his head held high and his eyes blazing with the fire of primordial creation. There is present in the animal a throbbing vitality that suggests to the observer both danger and excitement. In motion the horse is agile, moving with power and enthusiasm, while the tail of the horse is held erect, at times curved forward nearly touching the croup. The nostrils flare, and the horse snorts and blows, as he steps lightly, with such powerful impulsion from the haunches that he seems suspended in the air, no longer touching the earth. He is proud and dignified, and he is as confident as a peacock. The spontaneous exuberance of the horse at liberty is a marvel of power and gracefulness; turns and halts are executed with apparent effortlessness. The tremendous speed and accuracy with which the horse executes maneuvers is the result of the strength and dexterity of the hindquarters, and it is because of the power of the haunches that the forehand appears so light.

Wilfred Blunt described the Arab's action this way:

> There should be little high knee action, but the whole limb should

> be thrown forward and the hoof "dwell" a second in the air before it is put down. This, with corresponding action behind, like that of a deer trotting through fern, is most important in a sire, and a great test of quality.

Under saddle or being driven, the Egyptian is a natural athlete, swift and sure-footed, and amenable to instruction. His step is confident whether on smooth ground or rough. He is an easy horse to sit on. Having an aristocratic nature, he will cooperate with his rider but will not submit to coercion or force, and resents abuse. He will hold a grudge. As an athlete, the Egyptian horse excels in the three characteristics required of a riding horse: speed, strength, and stamina. As a hunting mount, he is a match for the ostrich and the gazelle.

The gallop is the Arabian's most natural gait. The ground covering stretch of the pace is less elongated when compared to the English thoroughbred, but it is accomplished with an ease and fluidity that make it seem effortless to the observer. The forelegs are thrust forward forcefully and with precision, and the powerful propulsion from the hindquarters creates the impression that the horse floats above the ground. The Egyptian Arabian horse exhibits the "stag neck" when at full and unrestrained gallop, with the head drawn up and rearward.

The Egyptian Arabian horse is known by his nobility of character; he is calm, good natured and free from vices. He is energetic and spirited, and full of life. He is innately fond of humans, with whom he is affectionate and familiar. They have an inherent sense of trust in people; they do not fear man unless taught to do so. Once taught by abuse to fear man, the horse is permanently imprinted with aversion and he will not forget ill treatment.

Unlike some horse breeds in which the character is violent and unpredictable, the Egyptian is amiable. He is alert to his surroundings. He seldom becomes vicious unless provoked. The behavioral characteristics of the Egyptian Arab consist of a vibrancy and vitality mixed with curiosity and high spirits. When needed, the horse will exhibit courage and resolution when speed and endurance are required. He is a patient animal. Docile, gentle, and eager to learn, this horse is both pliant and tractable under saddle. He is a forgiving creature, willing to overlook his handler's mistakes. He will not, however, overlook his handler's insensitivity, and is positively incensed by intentional cruelty. The harsh training practices used to beat a horse into submission do not work with this horse. An aggrieved horse becomes a danger to humans.

He is a clever animal, interested in everything, and is easy to teach and willing to cooperate when in training. He is curious about anything new. When he comes across something in his path that he does not recognize or understand, he stops and intently investigates, eyes wide and ears alert. When he has satisfied himself about the nature of the object, he will then proceed calmly, untroubled. He responds to voice commands promptly. He will not, however, submit to impertinence and cruelty on the part of his handler. He is a loyal partner but a formidable adversary. It is difficult to exaggerate the animal's high degree of sensitivity.

In both form and substance, the Egyptian horse is a creation in which the qualities of harmony and balance are manifest. He can be as fiery as the desert sun or as placid as a desert spring. The desert horseman understood this. When treated with harshness and violence, the horse responds with a sense of indignation. He can be turned into an obstreperous creature if improperly managed and handled. The Egyptian horse has an infallible memory with regard to places that it has been before and the treatment that it has received.

> It is terrible to see how merciless and gruesome the handling of these animals has become in Europe. In the Orient, such conditions are unheard of. The mildness with which the noble horse is handled, according to the rules laid down by the Coran, and the care with which it is looked after, are richly rewarded by the horse itself.
>
> (Berlin Almanac, 1828, in Asil Araber)

The Egyptian horse is an animal of native and feral vigor, noted for its freedom from asthenia, and known for good health. Its fertility exceeds that of the average equine. The Arab horse grows and develops more slowly than other breeds, not reaching maturity until age six. Because of its slow maturation, it lives long, and what is more, it retains its vitality and spirited essence long into old age. Its general demeanor is one of nobility and refinement, its hallmarks being balance and harmony of form, beauty, and intelligence. In conformation the Egyptian is harmonious, with the various parts of the body well-proportioned, standing in relationship to each other with appropriate balance. Symmetry is the ruling principle.

The most vivid impression made on the observer, however, is not the perfection of the pieces of the animal, but the wild and uninhibited animation of the Egyptian stallion when he is displayed or aroused by the presence of a mare. Like a peacock spreading his plumage, the stallion begins an electrifying and dramatic display of vitality and dominance, with his powerful frame springing into action and every muscle brought to life, every sinew straining with exuberant energy. His veins fill and his eyes are enlivened with a fire-like intensity. It is a thrilling display of feral nature, free and uninhibited.

Tossing his head fiercely, he prances and performs the capriole with a fiery grace and accuracy that is impressive. He throws his tail high into the air, over his back, and arches his swan-like neck into a rainbow of dynamic perfection. Intent on impressing his audience, he flares his thin-walled nostrils, vigorously snorting and neighing. His ears sharpen and become focused, while his eyelids retract to emphasize the dark and bulging eyes. He wheels about, his front hooves high in the air, pawing. The effect is visceral. It strikes a primordial chord, and the emotional impact is not easily forgotten.

All of the movements taught to horses in the *haute ecole* of classical dressage can be seen as Egyptian horses live and move freely in their pastures, interacting with each other in the family groups in a natural and spontaneous manner. They spontaneously exhibit all gaits, both collected gaits and extended gaits. One can witness the *Les Saults d'ecole*, the airs above the ground. It is a wonder to see spontaneous courbettes, caprioles, levades, piaffes, passages, pirouettes and smooth effortless changes from a left hand lead to a right hand lead. These are all natural and elegant untaught movements.

The Egyptian Arabian horses each exhibit their own unique personalities. The range in personalities is wide, and there is great variability in temperament. Some horses are quiet and docile, and some are fiery and constantly in motion. Some have an innate close connection with humans and some are aloof and distant. Some are showy and extroverted, while others are more quiet and reflective introverts. There are highly a few highly intelligent horses, while some are less so. Some of them are ill tempered, but most of them are tractable and cooperative when put to a task. A few of the horses are extremely charismatic. A few of them are unforgettable. But each one is unique and requires unique handling and management.

The key feature of the Egyptian Arabian horse that distinguishes it from all other horses known by the generic name "Arabian", however, is also the feature most difficult to describe; it is the quality of beauty. The Bedouin, the Egyptians, and Europeans all have been affected by the striking physical beauty of the creature as well as its inner intangible beauty. But beauty is a subjective quality and difficult to define. Tastes vary from person to person and the capacity for aesthetic sensitivity is not the same for everyone.

Beauty is the name given to that quality that is found in a concrete or abstract entity that summons from the unconscious of the observer an emotion that is at once delightful and deeply affecting. Since the emotional arousal that is called by the name "beauty" arises from the hidden recesses of the human psyche, language cannot comprehensively describe its qualities. Language can merely hint at its true nature. Its effects, however, are familiar to everyone.

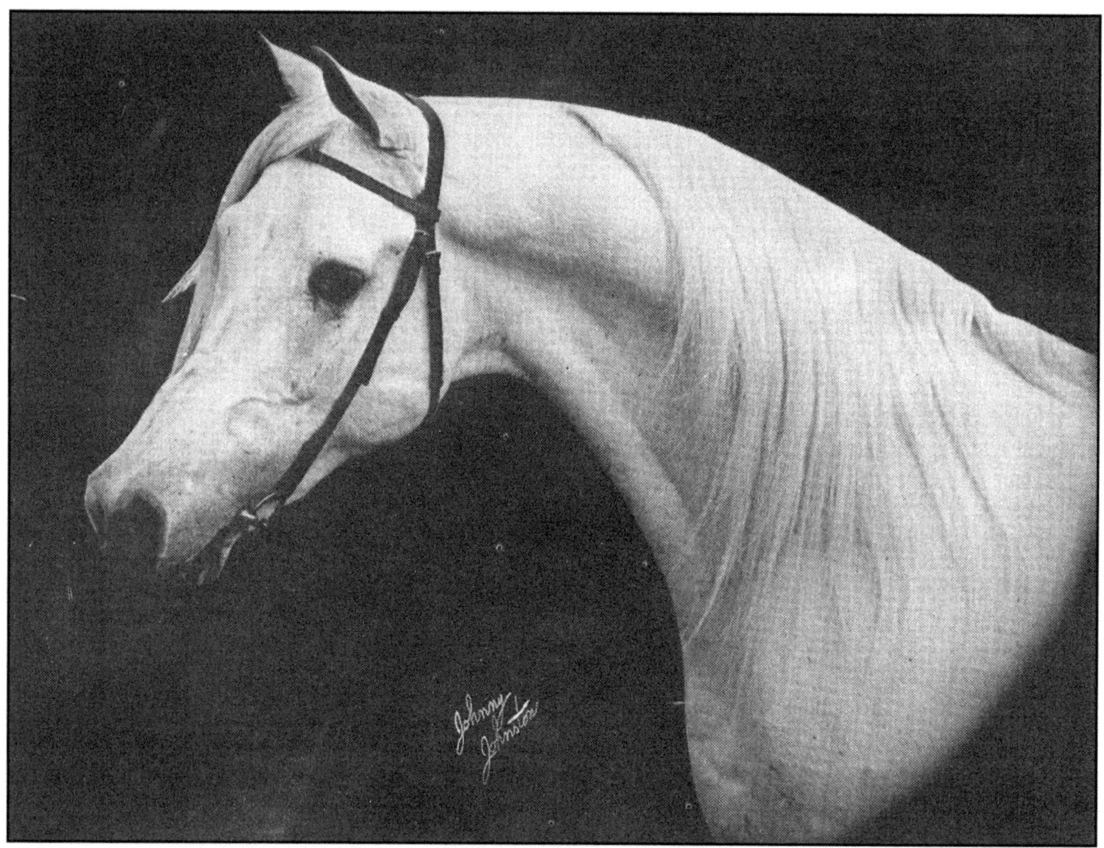

Figure 32 - *Morafic. Photo used with the permission of Johnnie Johnston.*

Somerset Maugham observed that:

> People talk of beauty lightly, and having no feeling for words, they use that one carelessly, so that it loses its force; and the thing it stands for, sharing its name with a hundred trivial objects, is deprived of dignity. They call beautiful a dress, a dog, a sermon; and when they are face to face with Beauty cannot recognize it.

The difficulty in conveying the subtleties of complicated or refined feeling makes any discussion of the beauty of the Egyptian Arabian horse perilous. Feelings are themselves inconsistent and at times self-contradictory. When put into words, feelings reveal much more about the inner tensions and psychological peculiarities of the speaker than about the object itself. Certainly *le coeur a ses raisons que la raison ne connait point.*

There is also the definition of beauty based on utility; beauty is the organization of the parts into the whole that is characterized by economy, clarity, cleverness, simplicity, and the ability to exist and perform a natural function with the maximum efficiency and the minimum of wasted energy and the absence of unnecessary ornamentation, artificiality, and complications. This definition favors elegance, balance, symmetry, order, and above all, proper utility.

The Egyptian Arabian horse fits into this model of beauty as well.

Most observers agree that the Egyptian Arabian horse is a creature of unsurpassed beauty.

Spare is her head and lean, her ears set close together;

Her forelock is a net, her forehead a lamp lighted,

Illumining the tribe, her neck curved like a palm branch,

her withers clean and sharp. Upon her chest and throttle an amulet hangs of gold

Her forelegs are twin lances.

Her hoofs fly forward faster even than flies the whirlwind.

Her tail held aloft, yet the hairs sweep the gravel.

Red bay he: his loincloth chafing the ribs of him

Shifts as a rainstream smoothing stones in a river bed.

Hard is he: he snorteth loud at the pride of him,

fierce as a full pot boiling, bubbling beneath the lid . . .

Lean his flanks, gazelle like, legs as the ostrich's;

He like a strong wolf trotteth; lithe as a fox cub he.

Stout his frame; behind him look, you shall note of him

Full filled the hind leg gap, tail with no twist in it.

Polished, hard his quarters, smooth as the pounding stone

Used for a bridegroom's spices, grind slab of colocynth.

Wilfred Scawen Blunt, Poetical Works, 1914

A COMPARISON OF THE CRANIAL FEATURES OF THE EGYPTIAN ARABIAN HORSE WITH THE COMMON GRADE HORSE

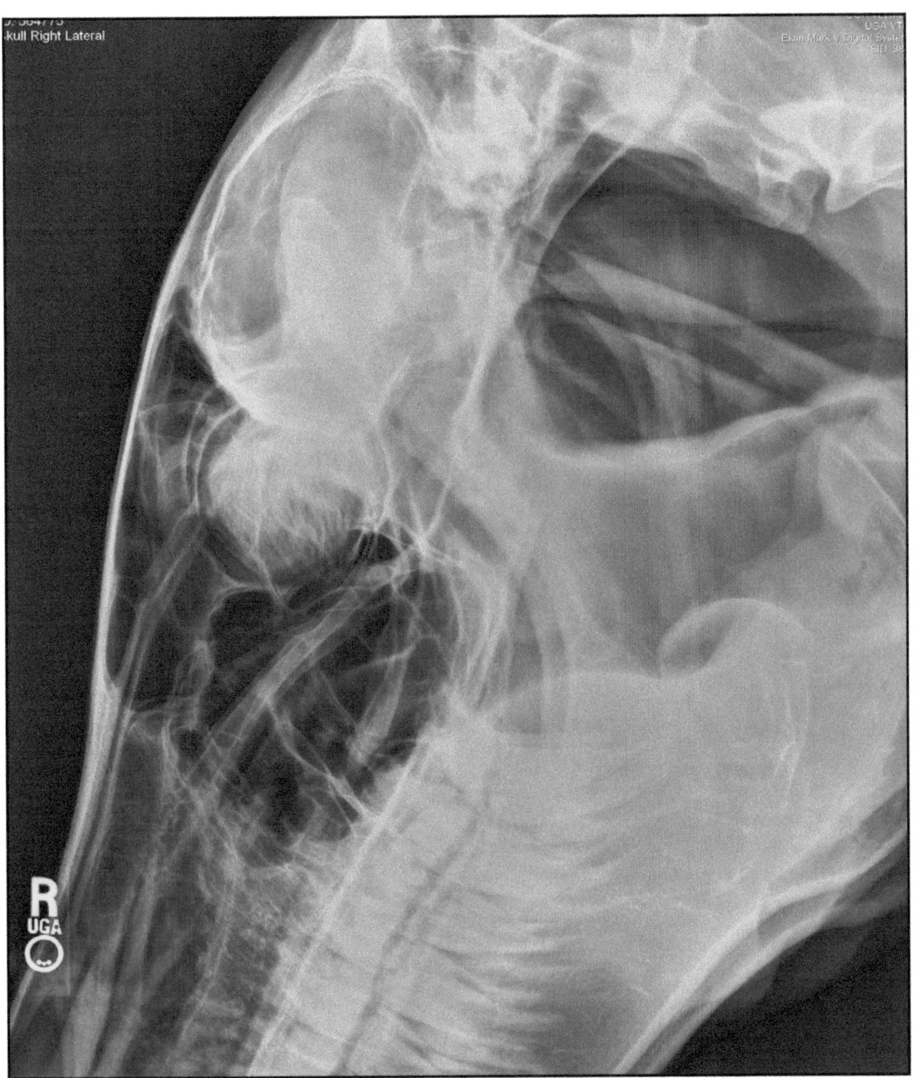

Figure 33 - *A radiograph of the skull of a 900 pound mare of pure EAO breeding. Compared to the skull of an unpedigreed 1200 pound American grade horse, there is little variation in basic structure. When adjusted for body weight, the Egyptian skull is quite small and short. Adjusted for body weight, the cranium and thus the brain size of the Egyptian mare is considerably larger than the grade horse.*

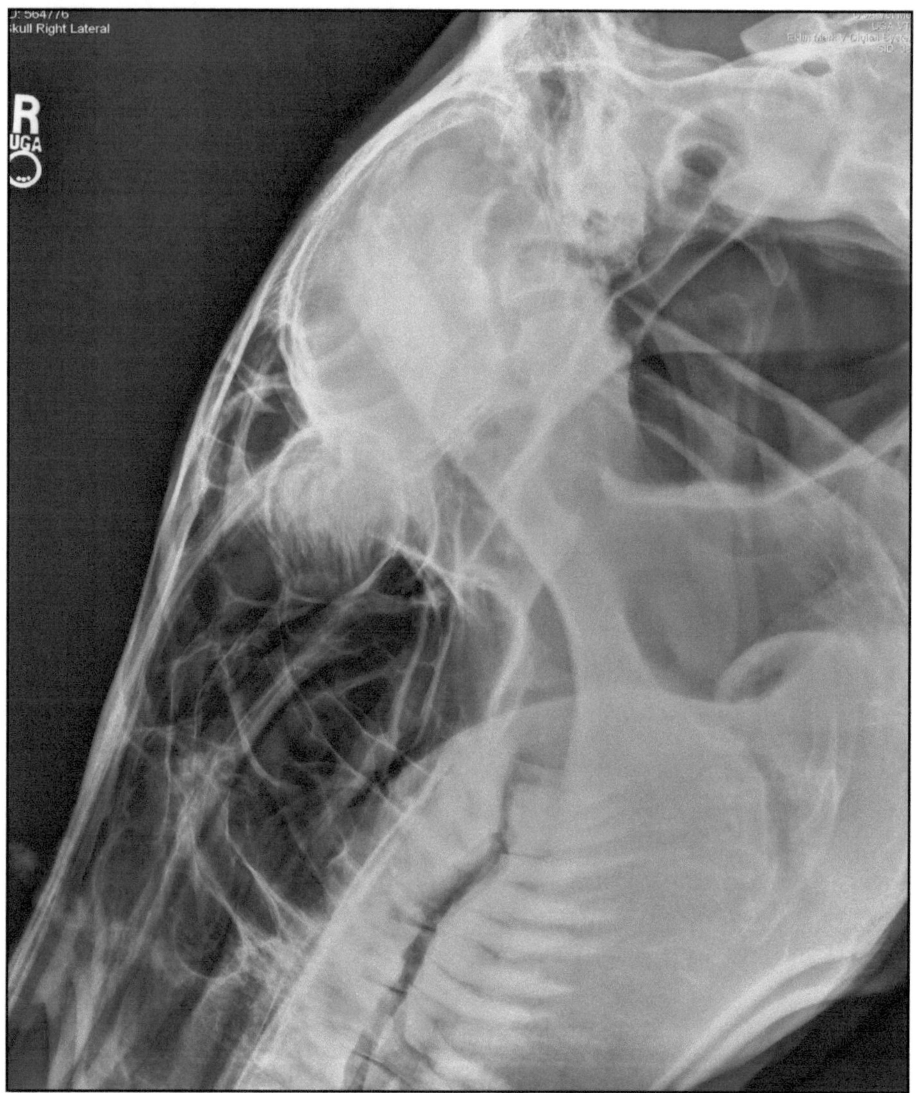

Figure 34 - *A radiograph of the skull of an unpedigreed 1200 pound common grade horse.*

Figure 35 - *The skull of an unpedigreed grade gelding. Contrary to popular belief, the structure and volume of the nasal passages are not significantly different from the Egyptian horse. The canine teeth of the male horse are noted.*

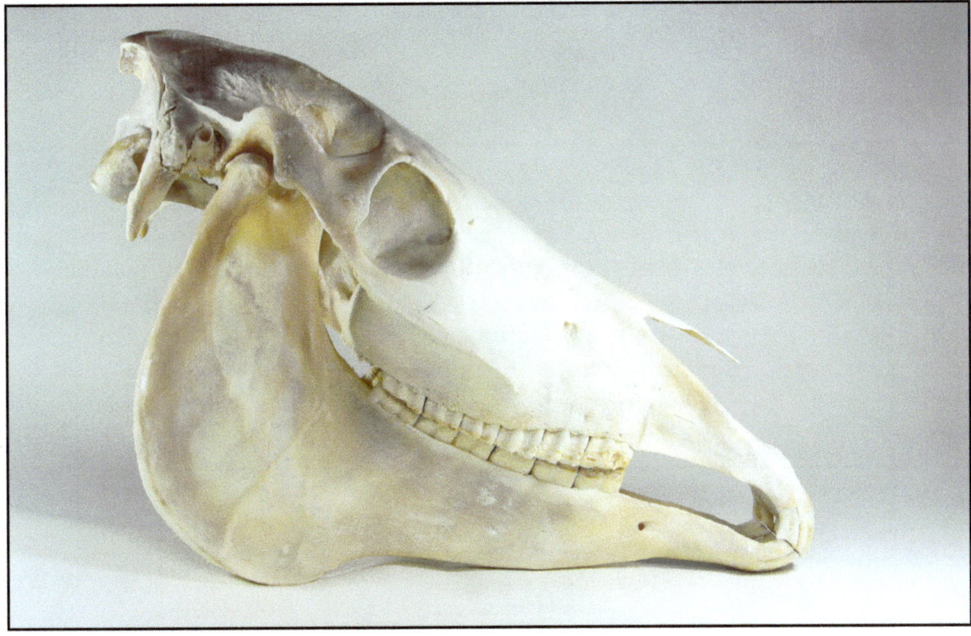

Figure 36 - *Skull of a horse of pure EAO breeding. The skull of a mare, ten years old, of pure EAO descent. Apart from the fineness of bone, there are no clear osteological differences to justify considering the Egyptian horse as a separate species from other* ***Equus caballus****.*

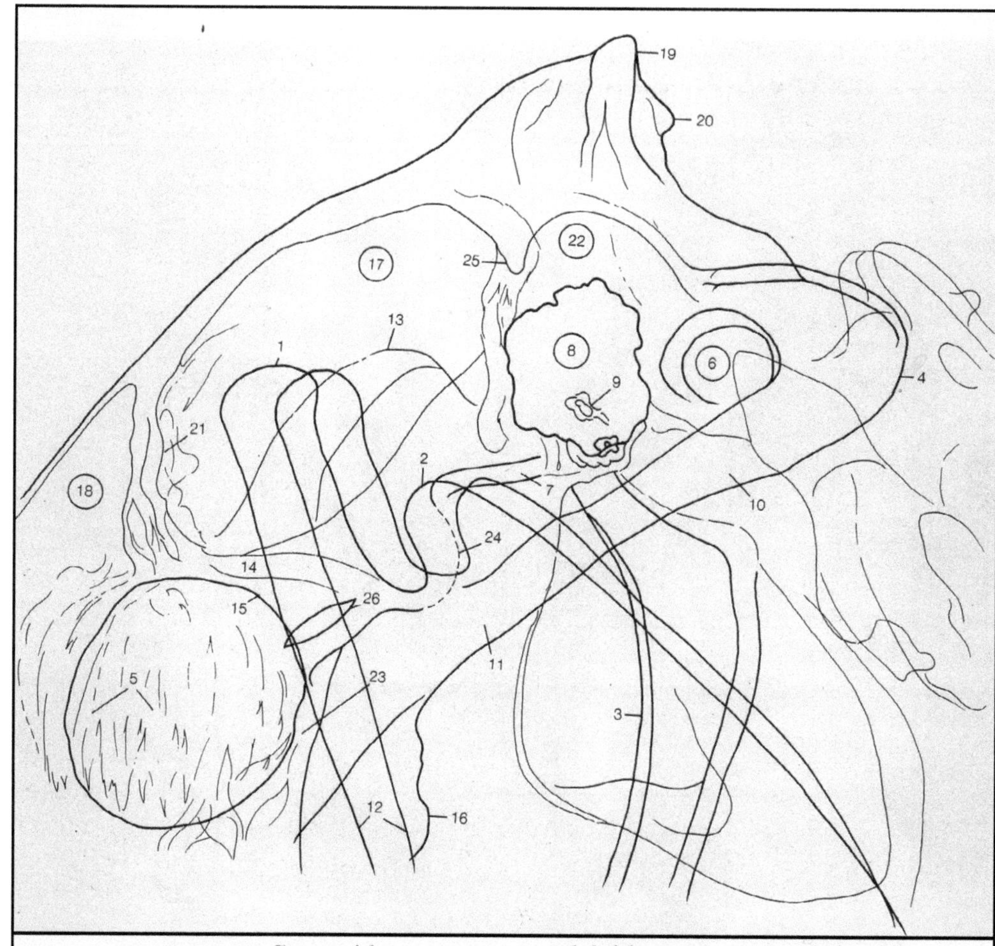

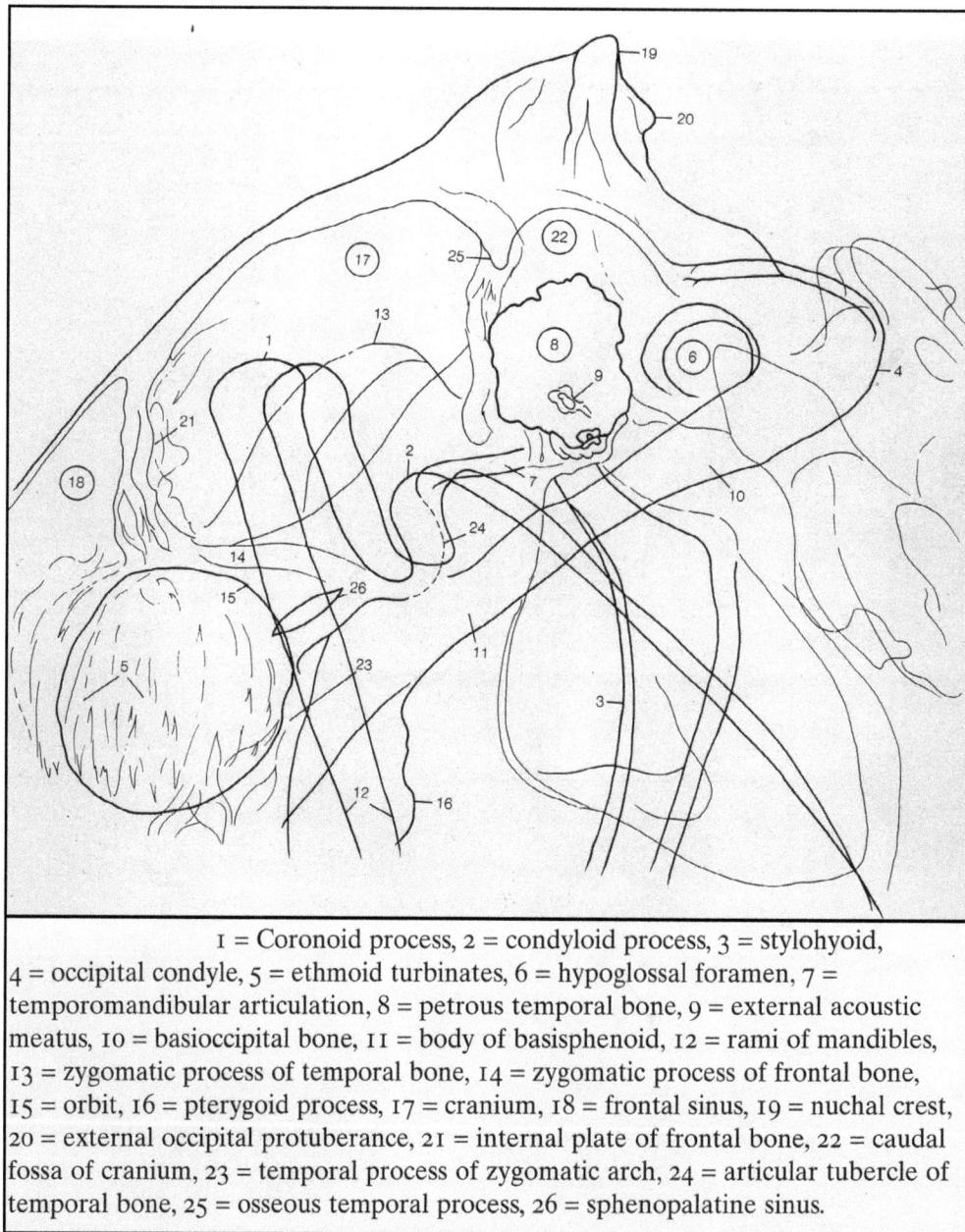

1 = Coronoid process, 2 = condyloid process, 3 = stylohyoid, 4 = occipital condyle, 5 = ethmoid turbinates, 6 = hypoglossal foramen, 7 = temporomandibular articulation, 8 = petrous temporal bone, 9 = external acoustic meatus, 10 = basioccipital bone, 11 = body of basisphenoid, 12 = rami of mandibles, 13 = zygomatic process of temporal bone, 14 = zygomatic process of frontal bone, 15 = orbit, 16 = pterygoid process, 17 = cranium, 18 = frontal sinus, 19 = nuchal crest, 20 = external occipital protuberance, 21 = internal plate of frontal bone, 22 = caudal fossa of cranium, 23 = temporal process of zygomatic arch, 24 = articular tubercle of temporal bone, 25 = osseous temporal process, 26 = sphenopalatine sinus.

Figure 37 - *A diagram and key to the basic structures of the equine skull. From* *<u>Clinical Radiology of the Horse</u>, Blackwell Sciences.*

A PHOTO GALLERY OF EGYPTIAN ARABIAN HORSES

Figure 38 - *Ibn Hafiza, 1959 stallion, by Sameh and out of Hafiza, bred by the EAO. Photo used with the permission of Polly Knoll.*

Figure 39 - *Fakher El Din, 1960 stallion, bred by the EAO, by Nazeer and out of Moniet El Nefous. Photo used with the permission of Polly Knoll.*

Figure 40 - *Nabiel, 1970 stallion, by Sakr and out of Magidaa. Photograph used with the permission of Polly Knoll.*

Figure 41 - *Morafic, EAO stallion, born in 1956, by Nazeer x Mabrouka. Photograph used with permission of Polly Knoll.*

Figure 42 - *Tuhotmos, 1962 bay stallion, bred by the EAO, by El Sareei and out of Moniet El Nefous. Photograph used with the permission of Polly Knoll.*

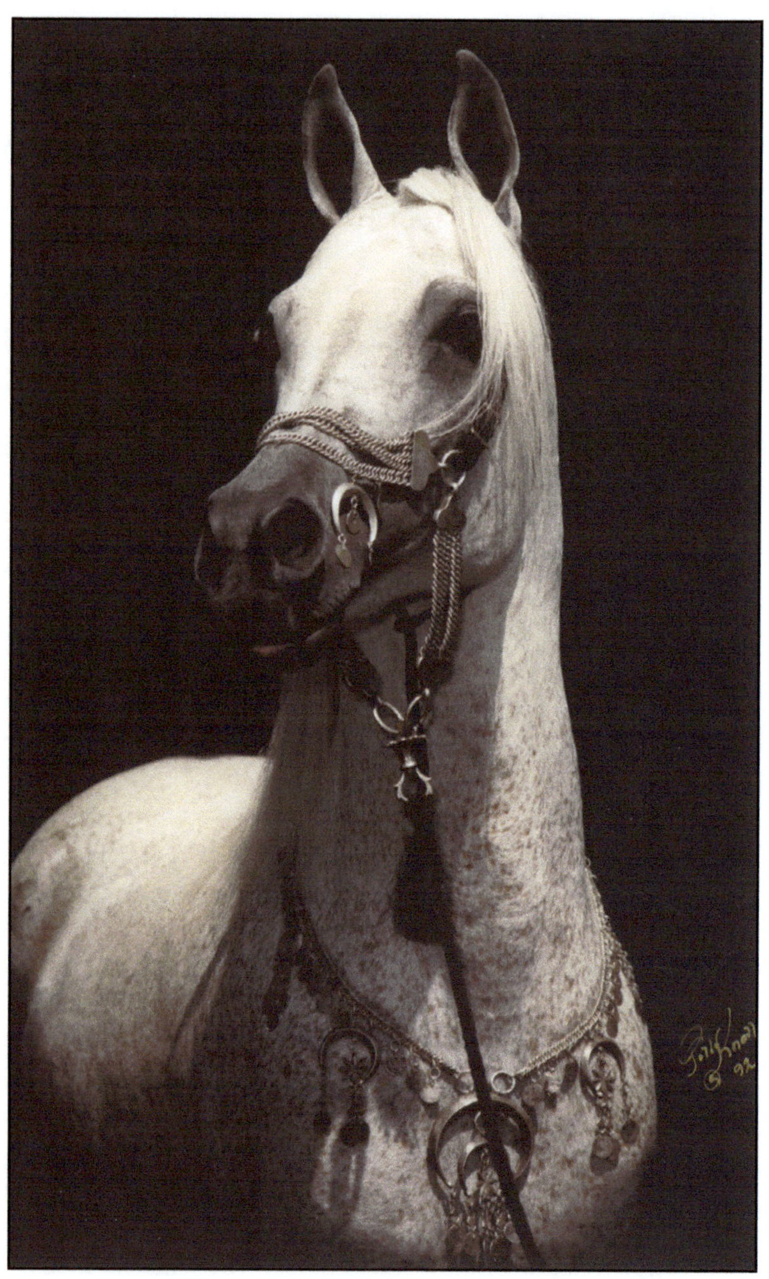

Figure 43 - *GAF Waseem, 1973 grey stallion, bred by the EAO, sired by Waseem and out of Bint Bukra. Photograph used with the permission of Polly Knoll.*

Figure 44 - *Soufian, 1968 chestnut stallion bred by the EAO, by Alaa El Din and out of Moniet El Nefous. Photograph used with the permission of Polly Knoll.*

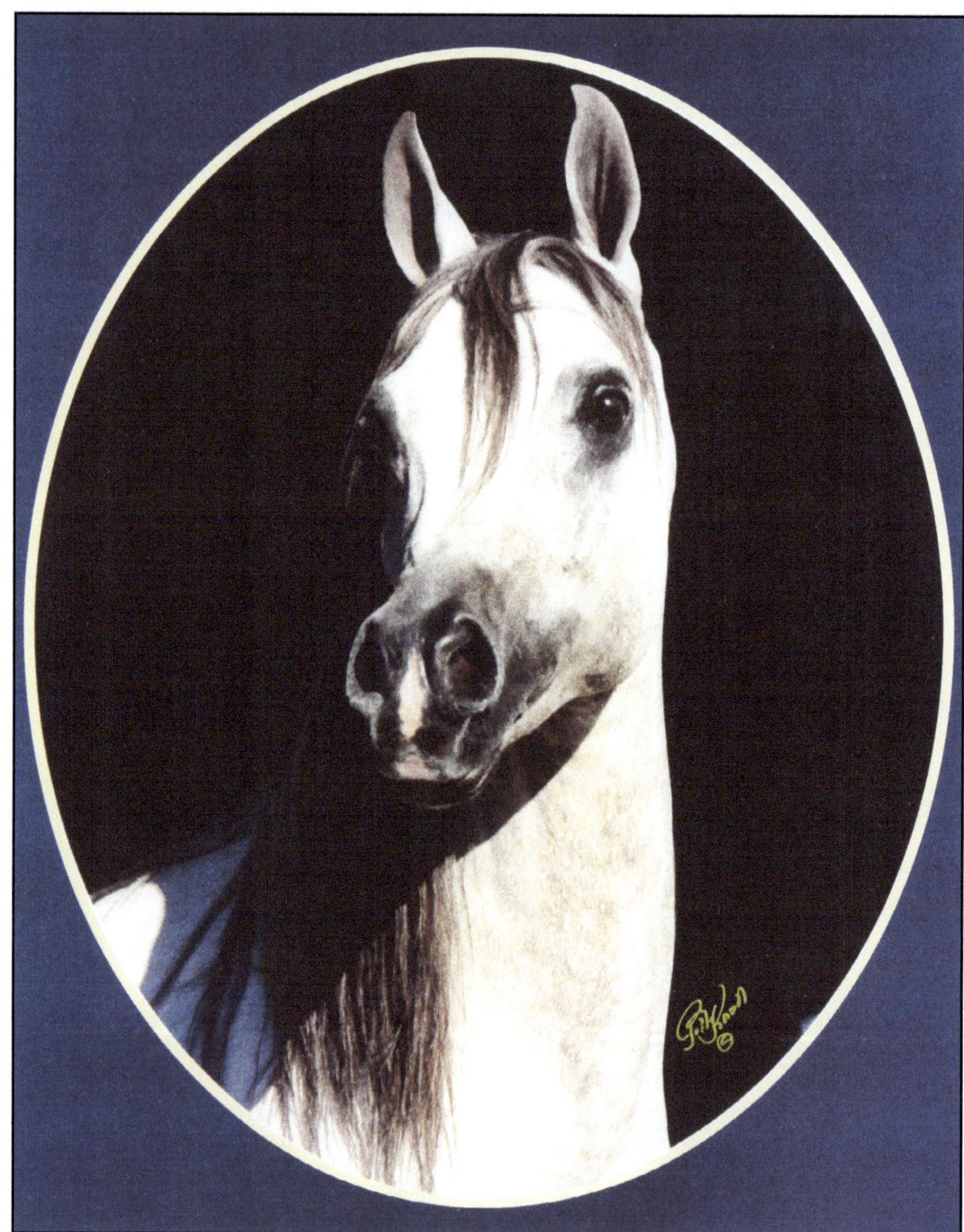

Figure 45 - *Farazdac, 1966 grey stallion, bred by the EAO, by Alaa El Din and out of Farasha. Photograph used with the permission of Polly Knoll.*

Figure 46 - *Sheikh Al Badi, by Morafic and out of Bint Maisa El Saghira. Photograph used with the permission of Polly Knoll.*

***All horses appear to be equivalent to those who are not aware of the facts.**

CHAPTER ONE

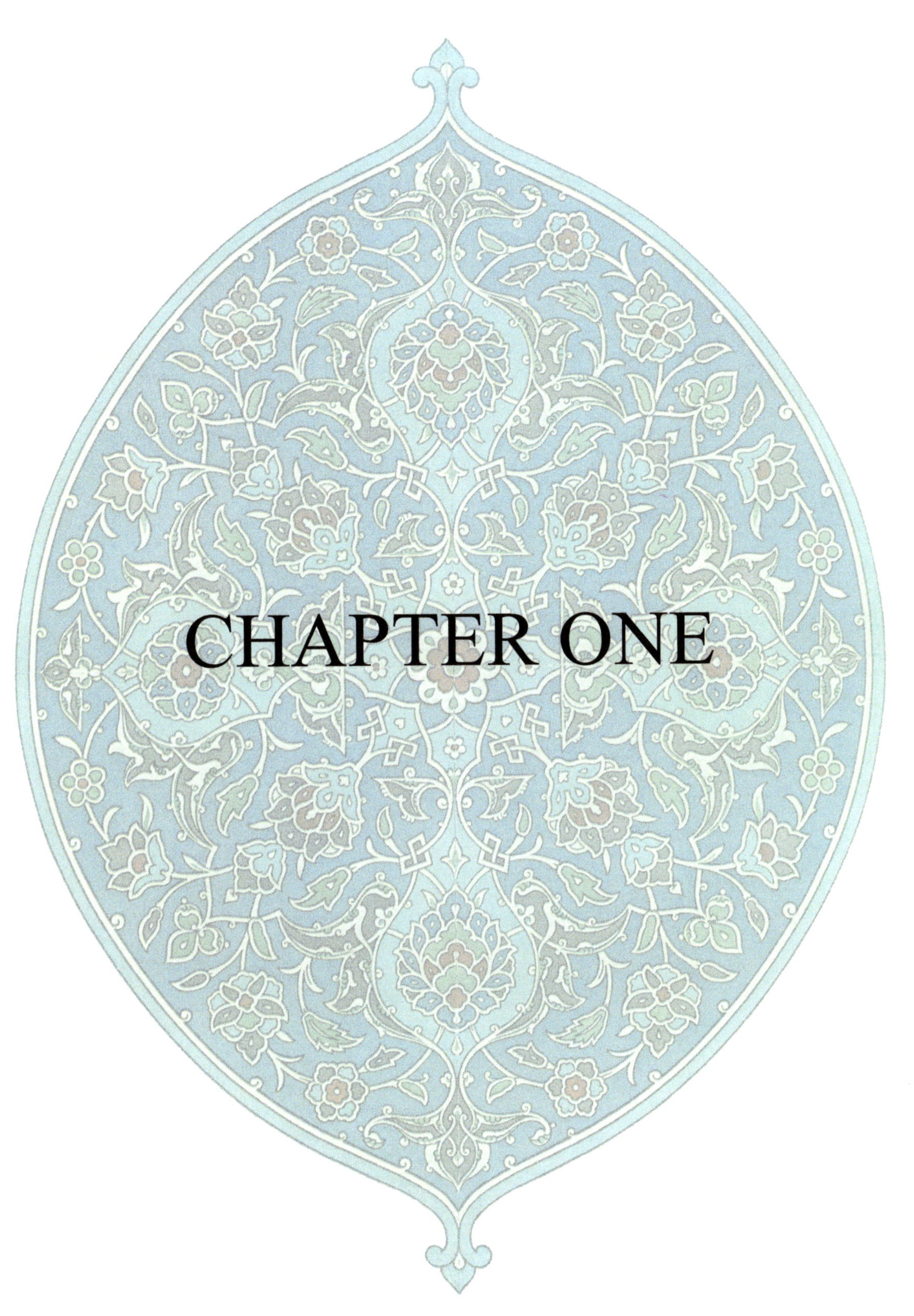

CHAPTER ONE

THE QUESTION OF THE GENETIC PURITY OF THE EGYPTIAN ARABIAN HORSE; A DIALECTICAL REVIEW

يا مآمن للرجال يا مآمن المية في الغربال

The interest that has been shown in earlier chapters regarding both the origins of the Egyptian Arabian horse and its genetic purity has been neither the result of idle curiosity nor academic interest, but has been the result of the longstanding desire on the part of Arabian horse breeders and pedigree researchers to understand more about these two issues. This data is needed in order to help breeders make better and more informed decisions about their choice of breeding stock.

In making breeding decisions, Arab horse breeders have one goal: to produce a particular type of horse of the highest possible quality with the highest degree of predictive accuracy possible. The type of horse that a breeder wants to produce varies from breeder to breeder. But the goal of producing quality stock does not vary. This is the one constant.

Figure 1 - *The Coming Storm, by Abraham Cooper, (1787-1868). Cooper was born in London and in his youth he displayed a remarkable talent for watercolors of animals. His career mainly consisted of producing artworks for English sports journals. His works are now largely forgotten, with the exception of the highly regarded Battle of Bosworth Field, in which he displayed his signature mastery of equine anatomy and movement which characterized his best work.*

It is difficult to sell horses. It has always been so, and probably always will be so. The reason is fairly obvious; it is difficult to sell horses because it is difficult to consistently and predictably produce horses of superior quality. Customers want superior stock, not average quality stock. There is no ready market for the average horse. There is no market at all for horses of inferior quality.

The consumer often wants to purchase a top quality horse, because he often intends to become a breeder himself. The likelihood of success for a young breeder starting out in the business is extremely low. If the horse buyer plans to become a breeder himself, he learns very early in the process that he needs to begin with the best quality live stock that he can afford.

Or, the new horse owner may just like the competitive spirit of showing Arabian horses, and he wants to buy a horse that can win in shows. He still wants to buy the best horse possible.

Or, he may just be a horse trader who intends to arbitrage the horse; that is, buy the horse at a place (a remote rural farm) where the price is low and sell the horse at a place where the price is high (a large metropolitan area or the Middle East).

Or, he may just be a collector of fine art and antiquities, and if he becomes interested in Arabian horses he will naturally be attracted to the highest quality, most beautiful of the horses that are available to him.

Most Arabian horse buyers have limited funds to expend on horses. Buyers must make choices in the horse buying process. No buyer can always find or afford exactly what he wants, exactly when he wants it.

There are always ready buyers for the higher quality horses, and there are virtually no buyers for lower quality animals. The largest group of horses, those of average quality, sell with difficulty in times of favorable market conditions and do not sell in bad market conditions at all, at any price. Many horses that a breeder produces must be given away, or "re-homed". Some horses fare even worse. The Arabian horse market, like all markets, is cyclical, and in difficult economic times, difficult choices must be made by horse owners.

The reason that Egyptian Arabian horse breeders often have so much trouble achieving their goals is that horse breeding, like all livestock breeding, tends toward the production of mediocrity, which is a kind of genetic entropy. The produce of any breeding program, whether it is a small program or a large program, is going to be average in quality relative to the quality of the breeding stock that is being used. A small percentage of offspring will be below average in quality and a small percentage will be above average in quality. A few will be extraordinarily poor and a very few will be extraordinarily high quality.

This is the law of the Bell Shaped Curve.

During earlier centuries, horses were bred by mass production, and genetic entropy was not so much of a problem. Hybrid vigor was at work to combat entropy. There was a place and a use for almost any horse, regardless of quality and regardless of genetic composition. Any horse could pull a cart or a trap. Agriculture was based on the use of horses to haul wagons and pull ploughs. Today the horse is a collector's item, and the Quarter horse and Thoroughbred industries have the same problem that Arabian horse breeders have: the problem of the "average" horse that is difficult or impossible to sell.

Unlike Egyptian Arabians, the Thoroughbred horse, Quarter horse, and Saddlebred businesses have flourished in recent decades. This is because these three breeds are heavily involved in racing and the associated gambling activities that accompany racing. Horse racing does not exist in these three breeds because curious and random onlookers like to watch horses run very fast. Nor, for that matter, are Arabian horse shows attended by random groups of curious people who came to see pretty horses.

Arabian horse racing has been attempted, and is still attempted, but its success has been very limited.

While there are a few independently wealthy individuals around the world who can breed Egyptian Arabian horses with no need to ever sell them, the vast majority of breeders do have to deal with the harsh economic realities of daily life. If breeders cannot make a profit by breeding and selling their produce, they would at least like to avoid losing money in the process. Horses must be housed, fed, and they must receive proper medical care. All of this is expensive.

Every serious horse breeder, whether oriental or occidental, is intensely goal-oriented. The process of horse breeding may be driven by commercial interest or by personal interests. In either case, the intention of the breeder is predictability.

In planning a mating, the breeder must decide which stallion is most suitable to breed to his mare. The breeder may wish to produce a foal that is similar to the mare, or he may want to produce a foal that is similar to the stallion. He may want to produce a foal that is similar to neither mare nor stallion. The type of foals that the breeder wishes to produce varies from person to person, but the goal in making breeding decisions is the same: to achieve predictable outcomes.

This is not a new problem.

> In establishing a breeding program, one must first of all determine the unalterable principle of what type of horse one wishes to breed, and

> according to which guidelines the horses are to be bred, maintained and fed, for any deviation from the established principle is opposed to a well ordered breeding program.
>
> (Cuirassier Captain Joseph Czekonics, on founding Babolna Stud in Hungary in 1789)

The motivation that energizes the research of Arabian horse breed specialists and researchers into the *purebred* Egyptian Arab horse's origins is and always has been a desire to obtain genealogic information that could be helpful in formulating a strategy that will make breeding more predictable. The breeder's goal is to produce livestock in a consistent manner over many generations, without loss of quality or type. Experience has shown that *purebred* foundation stock must be used in order to obtain reproductive consistency and predictive accuracy.

The pedigree is a record of the previous generations of forebears. To identify *purebred* stock, the pedigree must be analyzed, and parameters of genetic purity must be understood and established. Experienced breeders know that the visual evaluation of an animal in the flesh is not especially helpful in assessing purity or in predicting the offspring that can be expected. They also know that the genetic elements far back in the pedigree are more predictive of future success. Each breeder must arrive at a working definition of the meaning of the word "quality" and the meaning of the word "*purebred*". These two concepts often do not coincide. And they are both unfortunately subjective in nature.

To make matters worse, neither term has a universally accepted definition.

The fact is that nature does not provide Arabian horse breeders with the results that they want. Breeders want every mare to be pregnant every year. Breeders want each pregnancy to result in a specific pre-determined type of individual. This does not usually happen.

Animal reproduction, whether it is the reproduction of chickens or people, results in offspring that possess a wide range of qualities. Nature produces offspring with qualities that statistically can be described as a bell-shaped curve, ranging from very poor quality to superior quality. All features of an animal are defined in this way. Characteristics like height and weight and overall quality follow this law. The trait of tractability is also inherited in this way.

The offspring of the repeated breeding of a particular mare and stallion are not of uniform quality. They will exhibit wide variations and deviations from the parents. The parents are never reproduced. Sometimes the grandparents or great-grandparents are reproduced. This is true in dog breeding, and it is true in human breeding.

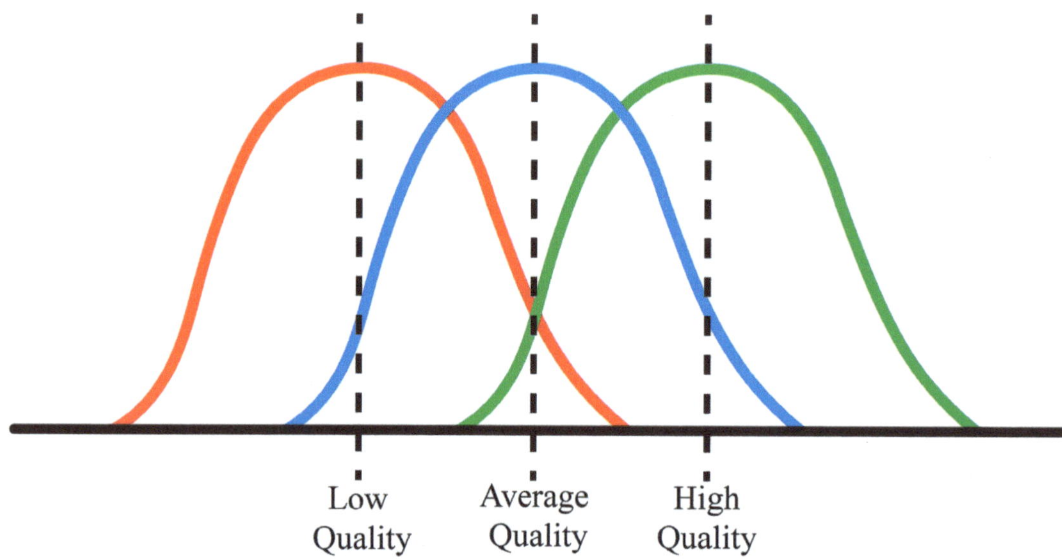

Figure 2 - *The range of quality in the progeny that can be expected to result from a long term large scale Egyptian Arabian horse breeding operation: the law of the bell shaped curve.*

Those Arabian horse breeders whose goal is the production of high quality stock find the practice very difficult and frustrating. Every breeder, whether breeding large numbers of mares or small numbers of mares, has the same intention; to produce high quality foals. But the principles of breeding remain the same, whether the breeder breeds 3 horses a year or 300 horses a year. The principle is valid if applied to a closed group of horses bred for a few years or for fifty years.

These three hypothetical curves shown above represent, for the sake of illustration, the ultimate outcomes of three types of breeding programs carried out over ten generations, or the sum production of a single Egyptian Arabian breeder who has bred a consistent core group of horses continuously for 50 or 60 years.

The blue curve represents the progeny resulting from a long term breeding program that used average quality sires and dams from the beginning, using average quality stallions from outside the herd over the years. Most of the progeny will be average quality, while some will be very poor quality. Only a very few horses will be an improvement on the quality of the founders.

The red curve represents the progeny of a breeding program that was based on poor quality foundation stock, using poor quality stallions over the ten generations, either from within the herd or from outside the herd. Most of the progeny will be poor quality, and some will be extremely poor. In this program, few progeny will be better than average.

The green curve represents the results of a long term breeding program that was based from its inception on the best quality foundation stock, using only high quality stallions from outside the program over the ensuing ten generations. Most of the offspring will be of high quality and a few will be of extremely high quality. The worst quality offspring in this herd will still, for the most part, be no less than average.

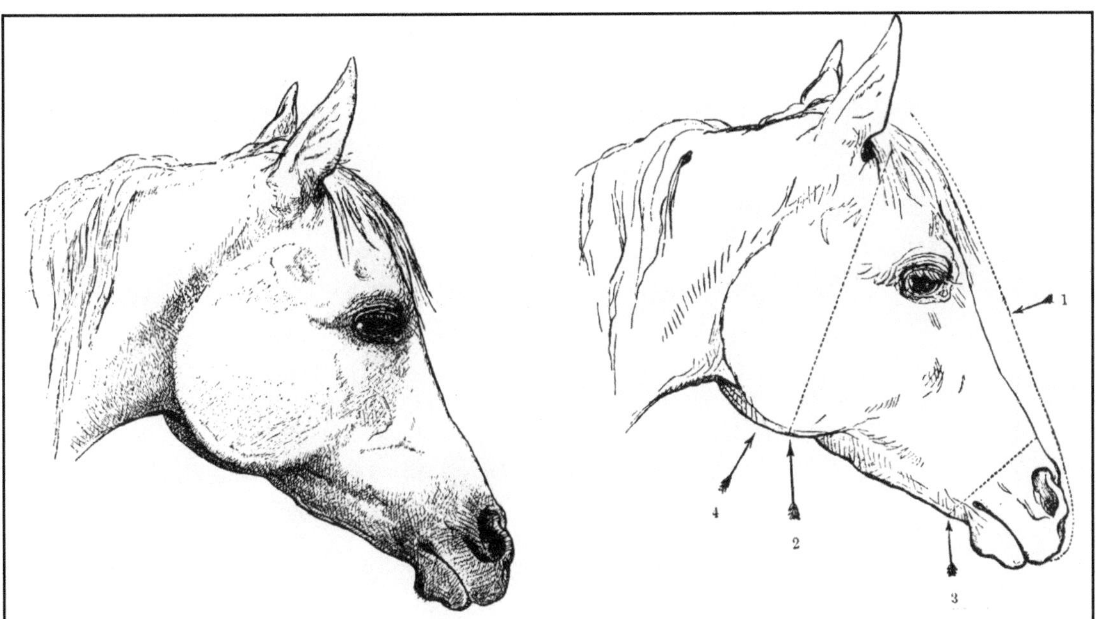

Figure 3 - *Sherifa's head.*

But what is meant by the term quality? Several issues are involved, but in Egyptian breeding, none is more important than "type". Type refers to the beauty of the horse; not just the beauty of the head but the beautiful image resulting from the harmonious symmetry of the animal as a whole. Many breeders have made the attempt to quantify beauty or perfection. Blunts, Wilfred and Lady Anne considered the head of the mare Sherifa to be the quintessential ideal. They measured the mare's various head dimensions, but could not arrive at a reliable scientific formula for excellence.

But the interpretation or inferences that have been made from an assessment of "type" can be misleading. Is a horse with superior type most likely a genetically stainless *purebred Arabian horse*? Certainly not. Is a horse of superior type likely to be a good breeding animal? Perhaps, perhaps not. Does the fact that a horse is seriously off-type mean that he is not a *purebred Arabian horse*, free from genetic contamination but non-Arabian blood? Certainly not. Does the fact that a horse's physical appearance deviates widely from what is considered true Bedouin type mean that the horse is not likely to be a superior breeding animal? Certainly not.

Figure 4 - *Ibn Rabdan, an influential Egyptian sire. Photo from Pearson.*

It is a generally accepted fact that there have been "off-type" animals in the Egyptian Arabian horse past that have made important contributions to the breed. Consider Ibn Rabdan and Rose of Sharon. On the other hand, there is a long list of national champion stallions that never succeeded as breeding animals. For the breeder, interpreting type in a breeding context is a frustrating and perplexing dilemma.

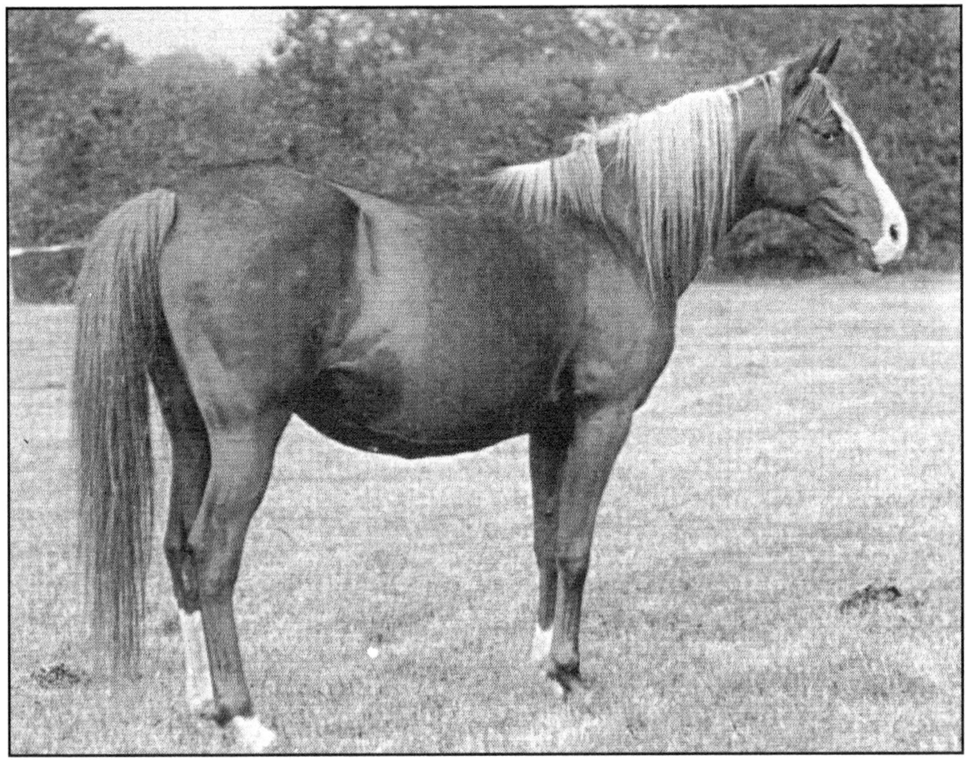

Figure 5 - *Rose of Sharon, an influential Egyptian mare. Photo from Pearson.*

The three daughters of Moniet El Nefous sired by Sid Abouhom illustrate the concept of the bell-shaped curve as seen in Egyptian Arabian horse breeding. Of the importance of the two sisters Mabrouka and Mona, no more need be said than that their descendants form the backbone of Egyptian Arabian horse breeding as it exists today. But there was a third full sister, less well known, named Lubna. She was a plain and unremarkable mare herself, and she produced off-type foals (three stallions and one mare in her lifetime), the most well known of which was a stallion named Sultan, sired by Sameh.

Figure 6 - *"Moniet El Nefous"("Shahloul"x"Wanisa"). Photo from The Raswan Archives.*

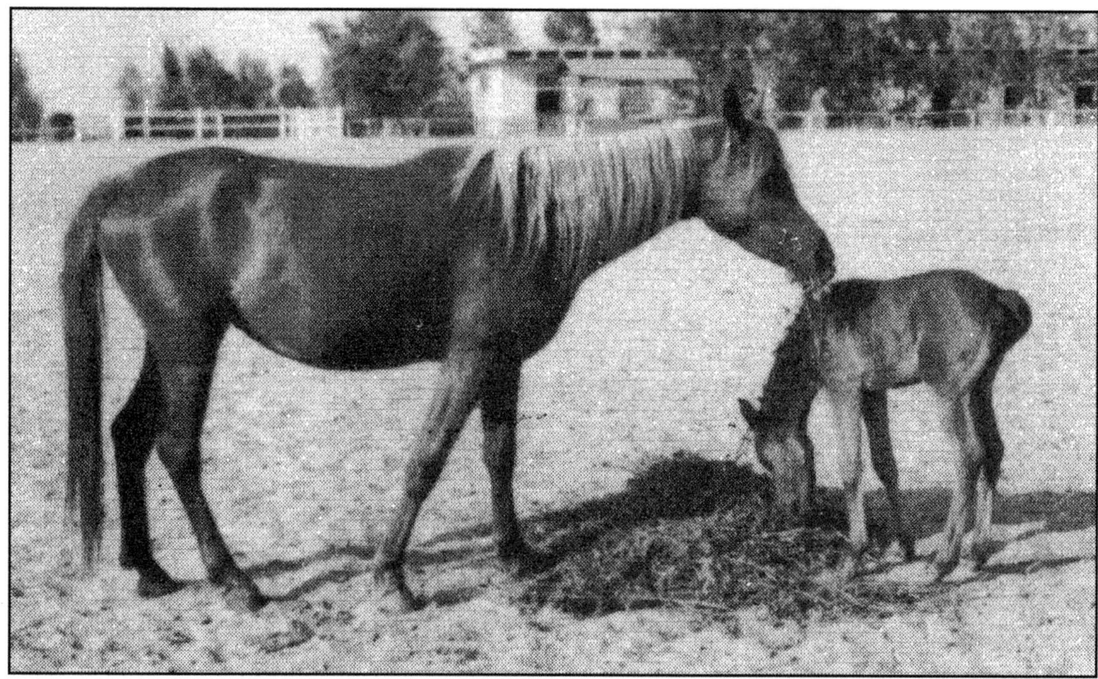

Figure 7 - *Moniet and Mona, photo from The Raswan Archives.*

Figure 8 - *Sultan, re-named Sultann in the US. Photo used with the permission of Johnny Johnston.*

Sultan was a powerful and dynamic horse, possessing all of the cranial features that the classic Arabian breeder wishes to avoid; straight profile, a long snake-like face, small eyes (sometimes called human eyes) with noticeable visible white sclerae, a shallow and narrow jaw and poorly conformed jibbah. His ears were long and he had excessive amounts of pink skin of both the upper and lower lips. He produced the yearling filly (see Figure 10), in which all of these undesirable characteristics of the Egyptian Arabian horse are seen. This filly's mother was a beautiful and classically bred Egyptian mare.

Born in 1961 at the EAO, Sultan was used at stud for nine years before being sold to an Egyptian horse breeder in the U.S. Sultan's appeal as a breeding stallion was based on his ideal racing conformation and his coat color. He was chestnut. Then, as now, breeders were concerned about the high percentage of grey horses in the

Egyptians as a whole, and any bay or chestnut stallion attracted breeders who wanted to breed color into their predominantly grey herd. The choice to use a bay or chestnut stallion often required a compromise in quality.

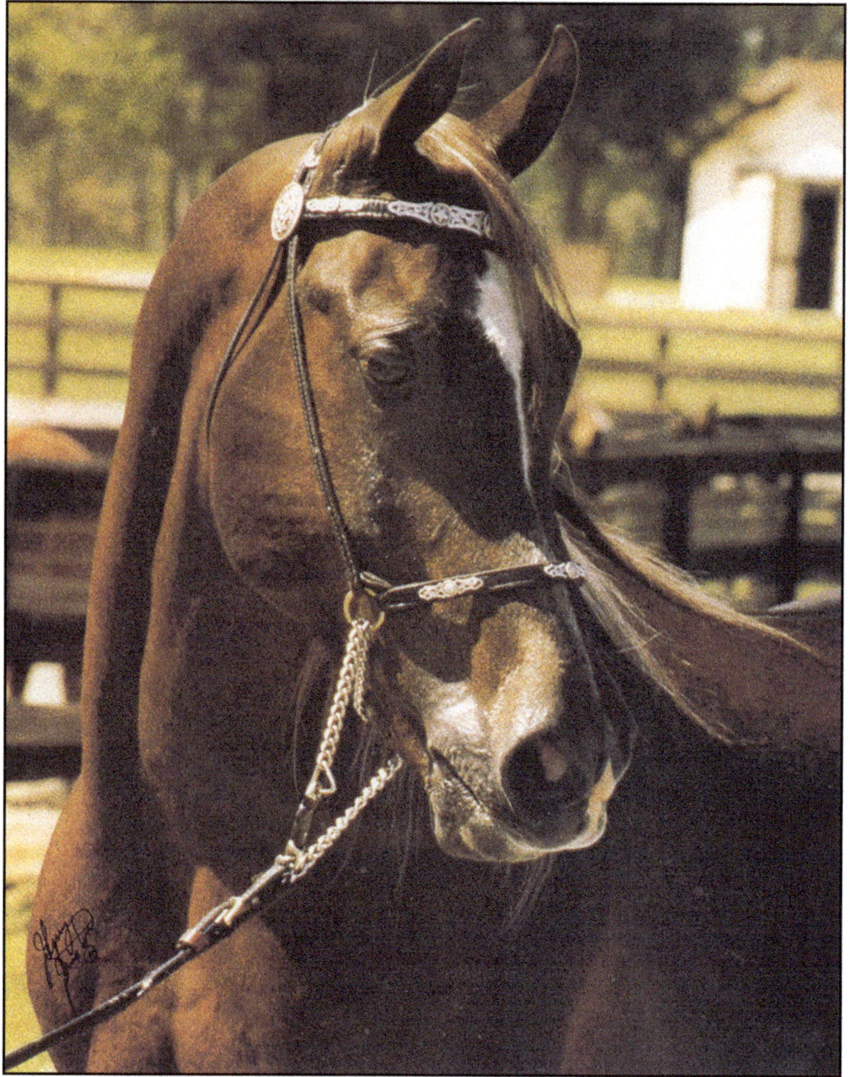

Figure 9 - *Sultan (Sultann) - Photo used with the permission of Johnny Johnston.*

Sultan was not without quality. He was an exceptional athlete. He had impressive and powerful gaits; he was conformed to run. When observed in an open field at liberty, Sultan's fluid freedom of movement at the gallop was exceptional. The forearms stretched forward, far forward, easily, and the enthusiastic propulsive thrust from the hind quarters was powerful. He floated. In life, at liberty, he seemed to be flying. It all appeared effortless. It was a breathtaking sight to see.

He was not, however, pretty. This presented a problem.

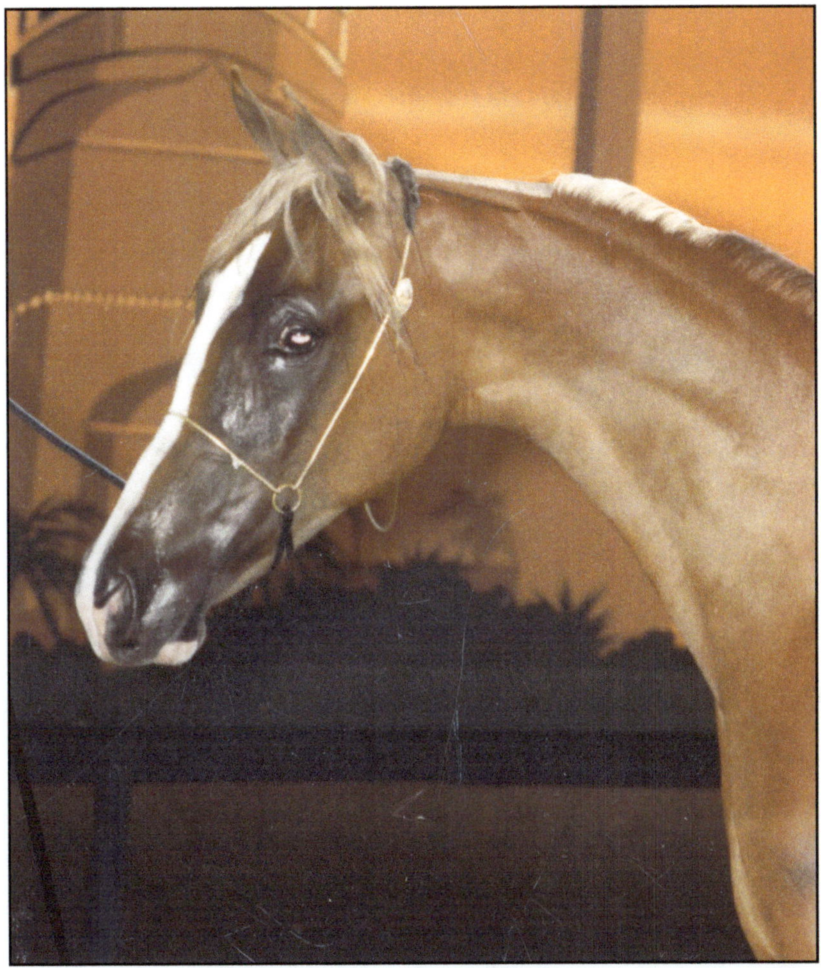

Figure 10 - *A Sultan daughter winning a ribbon at the 1987 Egyptian Event, Lexington, Kentucky. Photo from the author's collection.*

As a sire, his abilities were limited. His offspring did not distinguish themselves, with the exception of the stallion Sakr, a 1968 grey stallion and Asad, a 1971 chestnut stallion. Both horses became popular show horses in the U.S. and both failed, although they produced hundreds of foals in the U.S., to be distinguished sires themselves.

Even with the same identical breeding behind them, the three sisters and their progeny illustrate the phenomenon of the bell shaped curve.

Thus we have one extremely good mare (Mabrouka), one very good mare (Mona), and one average to poor quality mare (Lubna).

Of course, Moniet el Nefous had other foals. She had a colt named Badrawi by Sid Abouhom that did not live to maturity. She had three Nazeer foals; Ibn el Sheikh, Bint Moniet el Nefous, and Fakher el Din. She had three foals by Alaa el Din; Manar, Manaya and Soufian. She had one Morafic son, Ibn Moniet el Nefous. She had a son by El Sareei named Tuhotmos. Her last foal was a colt by Galal (Nazeer x Farasha) named Ameer. He was considered to be stallion quality and was retained by the EAO and used heavily at stud until 1987.

The offspring of Moniet el Nefous varied greatly in type. It has been observed that the type and quality of the produce of the "Queen of the Nile" depended to a large extent on the stallion used. She never reproduced herself. To the Egyptian Arabian horse community, she remains a cult figure. But even she conformed to the Law of the Bell Shaped Curve.

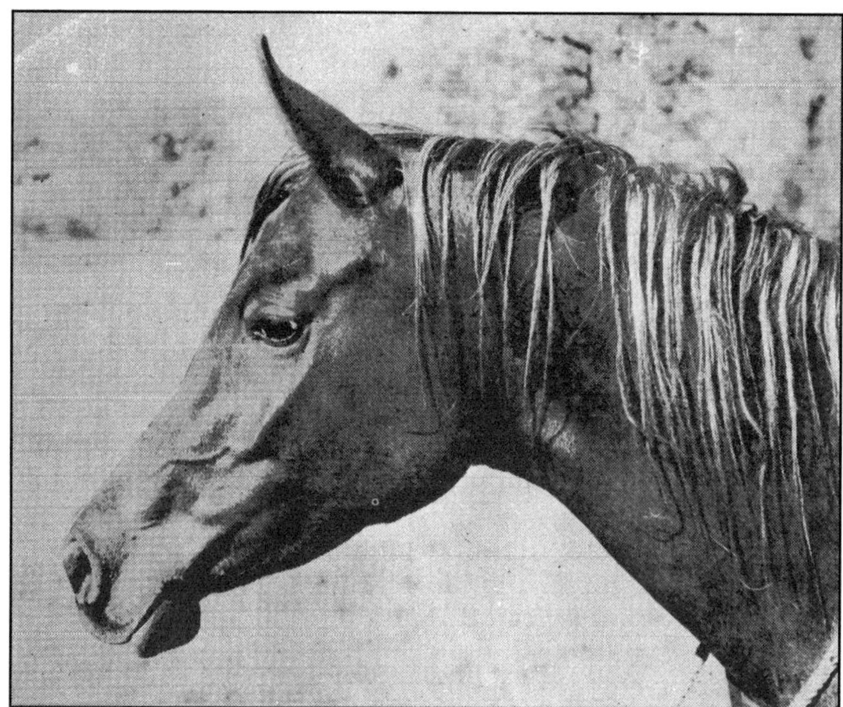

Figure 11 - *Moniet el Nefous.*

So, given these conditions, what is a new breeder to do to increase the probability that he will produce a higher percentage of superior horses and a lower percentage of poor quality horses? Here the breeder has to grapple with the thorny relationship between quality and genetic purity. He must deal with the law of the bell shaped curve.

The first question faced by a young breeder is whether or not there is in actuality any connection between Arabian horse quality and blood purity. Quality is evident to anyone who has spent even a small amount of time around Egyptian horses. Genetic purity is not clearly evident to anybody.

Phenotype does not necessarily relate to genotype. It is a common mistake to presume that an off-type horse must have an undesirable family member hidden in his pedigree. This is not generally true. It may be true, but many horribly off-type horses have been produced by a horse with very reliable pedigrees. And some of the Egyptian foundation horses consist of some individuals that would not be presentable in the show ring today. Even to this day, Egyptian Arabians are very rarely shown in national and international competitions. That venue is now dominated by Polish, Russian, and other mixed breed Arabian horses.

It is also an error to presume that an elegant and beautiful horse is, by virtue of being beautiful, purely bred and genetically pure. This is not a valid presumption. Some of the most lovely Egyptian related mares and stallions have been the result of pure EAO stallions being bred to high quality generic mare stock.

Veteran Egyptian Arabian breeders know the importance of using only the best quality horses for breeding purposes. Margaret Dickinson Fleming, daughter of General J.M. Dickinson, owner of the well known Traveler's Rest Arabian Stud in Tennessee wrote:

> To me, many of the Arab horses as we know them today are mockeries of the ancient breed. There are two forces at work to produce these mockeries. One is the "back yard" breeder who knows nothing of prepotence or of the qualities of various bloodlines. These people harm only through ignorance. By far the more dangerous and influential breeder is the man who does know what he wants. Unfortunately

> this wanting is not necessarily in the best interest of the Arab. This man is taken by graceful high stepping horses of other breeds, or by heavy, low to the ground horses of other working breeds. He tries to emulate this in the Arab instead of accepting the Arab as he was meant to be.

There are four methods for a young breeder to use in starting out on the road to developing a long term breeding program:

1. Buy mares and stallions that are consistent winners at upper level horse shows, and breed them to each other.

2. Buy mares and stallions from a well-recognized long-established farm that is respected by others in the industry.

3. Buy mares and stallions that the young breeder likes simply based in his own personal tastes and assessments and then breed them to each other.

4. Buy mares and stallions based solely on pedigree and breed them to each other. This method requires that the young breeder do considerable pedigree research, and develop his own opinions about the relative value of certain male and female lines.

This method requires considerable pedigree research, and the beginner must develop his own opinions about the relative value of particular male and female lines, and the relative usefulness of the pedigree.

None of these methods have been shown to be superior to the others and none have been shown to produce reliable results. None of these methods has been shown to be consistently successful.

Of special importance to the novice breeder, is an appreciation of the necessity of culling. It is one of the most difficult and painful elements of breeding horses. The bell-shaped quality curves are only valid when aggressive culling is carried out. The veteran breeder knows that most of what he produces will be unsuitable for his continuing breeding purposes. Many cannot be sold. He must cull, that is, re-home those horses that cannot be sold.

Conscientious re-homing of unwanted stock requires the breeder's full attention and careful assessment of the people to whom the horses are to be given. Care must be taken to avoid sending unwanted horses to the "slaughter pipeline". Mrs. Ott of Blue Book fame required that home inspections be carried out on potential customers before she would even sell them a horse. She also required a letter from the customer's veterinary to determine if he possessed the skills necessary to manage horses. While this idea seems quaint and impractical today, it is still important that breeders provide for the humane care and placement of unwanted stock.

Livestock breeders are well aware of the fact that the low quality individuals that they produce should be removed from the breeding population. Most breeders agree that of the remainder, only the best mares and the very exceptional stallions should be retained and used for continued breeding. This means that the breeder must cull. Failure to cull properly leads to general degeneration of the herd. One colt out of 100 is stallion quality. 50% of fillies are not suitable for continued breeding.

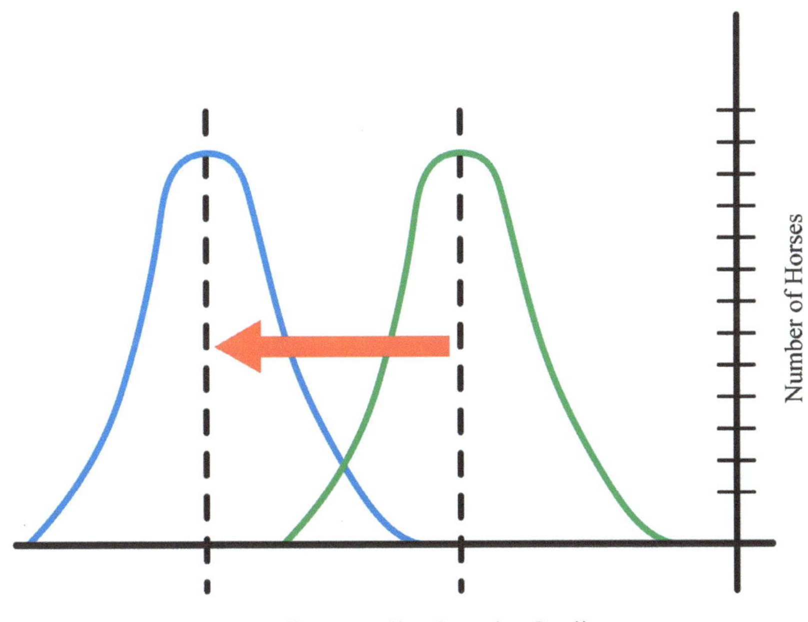

Figure 12 - *This diagram illustrates the effect of failure to cull in a breeding program over time. In this theoretical demonstration, the effect of failure to cull a herd of originally superior quality horses is shown. After breeding this herd as a closed population for 5 to 10 generations, the overall quality of the horses has declined. Instead of a herd of generally above average mares and stallions, the breeder is left with a herd that contains very few good quality animals. The herd now contains a substantial number of below average quality horses.*

Failure to cull disables Nature's most essential and most powerful tool for the maintenance of breed stability. Nature is ruthless in culling out weak or deformed individuals. Of interest is the fact that Nature does not cull an individual because it is not pretty or handsome, but only because of its degree of fitness. Failure to cull nullifies the law of nature: it leads to survival of the frailest. In the hands of humans, animals that are domesticated and bred in captivity are not subjected to the forces of Nature and therefore, if no culling is done, all specimens survive, both the fit and the unfit. This practice does Nature no favor. It does the Egyptian Arabian horse breed no favor. Only the strong, that is to say, those horses of superior quality, ought to be used as breeding stock. The rest will make excellent pets.

GENETIC PURITY

The matter of genetic purity requires much more reflection and study than the matter of phenotype. What you see is often not what you get. In fact, what you see is usually not what you get. An Arabian breeder who is convinced of the validity of the Bedouin theory of breeding will do his best, through research of the pedigree literature, to select Arabian horses that are "*purebred*" or, failing that, horses that are, as far as can be determined, as *purebred* as possible. He may end up concluding that EAO horses, Al Khamsa horses, Asil Club horses, or Pyramid horses are as close as he can get to "purity". He may conclude that the best he can do is to obtain the "purest of the impure". This is done through pedigree research.

This dilemma is due to the fact that to date, there is no scientific definition of the term *purebred Arabian horse*; there is no genetic test of purity or contamination. Some skeptics go so far as to claim that there never was a wild type desert horse, living for eons as an isolated landrace of horses in the central Arabian Peninsula.

Breeders who use the "pedigree method" make their breeding decisions based on their knowledge of the ancestors of the mare and stallion under consideration, and then make educated guesses about the likelihood that the result of such a breeding will be a foal that will meet their requirements. And then they breed.

The fly in the ointment of course is the reliability of the pedigree. Research has shown many cases in which the accepted pedigrees of some Arabian horses contain errors. This fact discourages young breeders.

But what, really, is the evidence that thorough knowledge of the pedigree of an Egyptian Arabian horse really matters in making breeding decisions? The only evidence prior to the introduction of DNA analysis was the "blood will tell" evidence. But the goal or intention of the breeder may be such that the pedigree is unimportant. There are many types of breeders with differing intentions.

If the pedigree of an Egyptian Arabian horse contains inaccurate information which conceals the presence of an ancestor of alien non-Arabian origins, then the expected bell shaped model will break down entirely.

The traditionalists, also known as the pedigree purists or the "strict constructionists", adhere to the Bedouin system of breeding. They maintain that if an individual in a horse breeding program contains even small elements of non-Arabian blood, then the entire bell shaped curve becomes sharply displaced to the left for that individual's offspring. This may not be apparent immediately, and may take several generations to become evident. Overall quality will suffer; blood will tell. Some extremely poor quality animals will be born, previously masked hereditary defects will be revealed, the average quality of the herd will fall, and no exceptional horses will be produced.

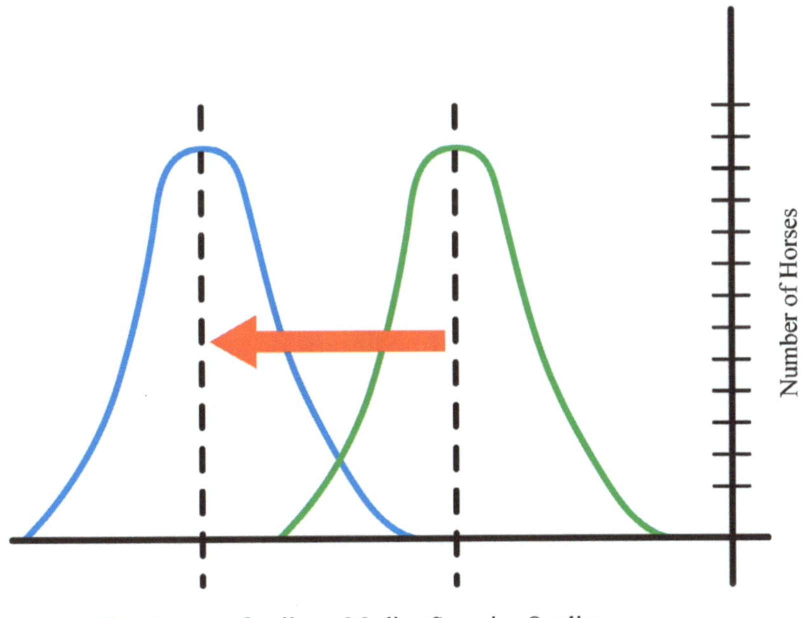

Figure 13 - *This graph illustrates the effect of a horse with an erroneous or falsified equine pedigree being added to an established herd of above average Egyptian Arabian horses. The herd will, if bred over time as a generally closed population, deteriorate in quality. This effect can occur in as little as one generation if the new individual is a stallion. This illustration is simply another way to express the adage "blood will tell".*

As an aside, it is ironic to consider the fact that an incorrect pedigree may just as easily conceal the presence of an exceptional individual of high genetic value. An erroneously identified ancestor may be one of great value, just as well as one on ruinous potential. Errors of record keeping caused by clerical errors, lapses of memory, or lacunes in the historical records do not necessarily conceal an undesirable alien or non-Arabian ancestor. They may just as easily conceal a superior quality ancestor. The Bowling research concerning the case of the two Yamamas is one such example. The horses thought to belong to the Jellabi strain were in fact descendants from the Saklawi matriline of Ghazieh I of Abbas Pasha. Either way, the bloodlines both represent exceptional pure desert blood of reliable provenance once owned by Abbas Pasha himself.

On the other hand, pedigree errors that are made intentionally are always intended to conceal the presence of an undesirable ancestor in the pedigree.

The goal of registering a foal with an incorrect pedigree is financial gain. A superior sire may be given as the father to conceal the fact that the real father was an inferior or non-Arabian horse. A superior mare may be given as the mother of a foal to conceal the fact that the actual mother was an inferior or non-Arabian horse. Before the days of DNA requirements for foal registration, this could be done easily, as long as the nefarious breeder had ownership and AHRA registration of the horses in question.

The reason that pedigree specialists pursue their work is to discover which pedigrees are correct and which are misstated pedigrees. They do not do their work out of academic zeal or intellectual curiosity; they do this type of work because they believe in the truth of the Bedouin horse breeding principle, based on long experience, that even a small amount of genetic contamination in an Arabian horse will result in deterioration of the subsequent generations produced from that horse. They work to clarify the validity of accepted pedigrees or to expose errors. They may be fighting a losing battle.

Not everyone agrees.

Carl Raswan agreed with the breed purists.

> For this reason we must save the few (truly) PURE Arabians which are still available in America, Europe, and Egypt so that we can re-

> tain part-bred Arabians (but which cannot be retained indefinitely without periodical infusions of PURE Arabian blood). We cannot afford any longer to mate our PURE Arabians with part-bred Arabians as otherwise the PURE Arabians will become an extinct breed of horse.

But how does one know which horses are *purebred* and which are not?

Arabian horse breeding relativists, the pedigree cynics and skeptics, typified by WAHO, do not believe that absolute genetic purity is important. The fly in the ointment does not bother them at all. They expect to find flies in the ointment. Purity, they contend, is a relative matter. They state that no Arabian horse can be proven to be absolutely pure genetically.

The pure RAS/EAO horses, they contend, are, like all other Arabian horses, only relatively pure. The Egyptian Arabian horses that are living today are only relatively pure-blooded, say the skeptics; they are undoubtedly contaminated in some small way by non-Arabian blood, albeit very far back in their ancestry. Having a pedigree that is 100% valid say the skeptics, would be useful, but the evidence supporting claims of absolute purity simply does not exist.

For the Bedouin, blood purity was the issue, and it was the "unalterable principle" on which horse breeding was based. Purity, it was known, was essential to breeding success. Purity in bloodstock was well known to be the key to predictable and reproducible outcomes in a breeding program. Genetic contamination was known to be fatal to a breeding program; most of the resulting progeny of a contaminated herd would, sooner rather than later, exhibit features of poor quality, and would have little or no economic or functional value. To the Bedouin, the term "*purebred* horse" meant one thing and one thing only. They were adamant. Once a non-Arabian was introduced into a breeding herd of *purebred Arabian horses*, the genetic contamination was ineradicable.

While this classical definition of the *purebred Arabian horse* is clear and absolute, the matter becomes very tenuous when applied to the matter of present day living Arabian horses. In theory, the *purebred Arabian horse* of today should trace in all lines to Deserta Arabia, but this assumption has been found in practice to be generally insupportable. The pedigrees of Arabian horses living today are not uniformly accepted as verifiable statements of fact, and considerable mainstream skepticism exists regarding the genetic purity of the modern Arab horse. Many breeders are critical of the wholesale acceptance of the written pedigree. The question of authenticity of the Arabian horse is widely debated, and even the meaning of the term "*purebred*" was re-evaluated and re-defined. This issue remains controversial and unresolved.

Figure 14 - *A Bedouin elder of the Sinai. Photo used with the permission of the Library of Congress.*

At one time in America, the unquestioned authority regarding pedigrees was the Arabian Horse Registry of America. The AHRA was clear on this point. They asserted that all *purebred Arabian horses* registered with them were traceable in all lines to the Arabian Desert. When the AHRA was dissolved and replaced by the Arabian Horse Association, this assertion was dropped. The AHA definition of the *purebred Arabian horse* became "any horse that is found in our *purebred* registry". This circular reasoning was adopted from the WAHO credo.

The objections to the belief in the purity of Egyptian Arabian horses are well founded. The problem is an epistemologic one: "How do we know what we know?" We know what we know about the ancestry of the Arabian horse by the process of hearsay, regrettable as that fact may be to the purist. For example, no RAS director was present at the birth of Ghazieh I: her pedigree is based on what others wrote about her.

All historical ancestral records are based on the testimony of one person that was given to another person. Testimony can be flawed. The onus of proof of authentication rests on the shoulders of the purists. Legally, hearsay is inadmissible evidence in court. Evidence that is spoliated, or intentionally fabricated or altered, is also inadmissible. Circumstantial and corroborating evidence however is admitted, and it is on these bases that the modern EAO pedigrees are based. Even in the beginning of documented Arabian horse breeding, there was no irrefragable proof that the Bedouin orthodoxy was correct. Curiously, there was no evidence that it was not correct.

Arab horse enthusiasts are divided on this question, with pedigree skeptics on the one hand and pedigree purists on the other.

From the skeptical relativistic viewpoint, the pedigrees of Arabian horses are patently full of flaws, errors and misrepresentation. As evidence, skeptics point to inaccuracies in published pedigrees resulting from the mendacity of horse breeders and traders, surreptitious and intentional alterations of the historic record, incomplete and missing documents, entire pedigree fabrications, and errors of judgment made by historians of the breed. The pedigree skeptic does not claim to possess the truth concerning Arab horse pedigrees; he simply asserts that the truth is unknown and unknowable. He considers the revised view of the term *purebred Arabian horse* to be the only rational way to deal with the matter. The relativist sees himself as making the best of a complicated and exasperatingly uncertain situation. The defense of the relativists' position requires some manner of argument to nullify or invalidate the Bedouin principles regarding the essential value of breeding *purebred* horse to *purebred* horse. This is very difficult to do.

Figure 15 - *The Giza Pyramids. Photo from the Library of Congress collection.*

On the other hand, pedigree purists assert a commitment to the primary importance of genetic authenticity. They accept as true the Bedouin theory that the effects of foreign genetic contamination in an Arabian horse, no matter how small that amount may be, will sooner or later produce defects and deterioration in the quality in the progeny of that horse. They accept as true the Bedouin theory that any group of genetically stainless Arabian horses that is bred together in isolation will, generation after generation, maintain its type and quality. If bred this way, the breed will neither deteriorate nor degrade in quality. They believe the Bedouin theory that the introduction of even one contaminated individual into a closed breeding population will eventually cause that herd to decline in quality and type. They have an uphill battle to fight against the relativists.

Figure 16 - *Bedouin Warrior, early 20th century. Photo from the Library of Congress Collection.*

The purists have behind them a very long history of reputable authorities who have supported this position. Lady Anne Blunt was among the first Western Arab horse breed purist. She wrote:

> It is essential that we should preserve a foundation stock of perfectly pure Arabian blood of the highest type and strains, rejecting vigorously all doubtful blood." Near the end of her life, she formalized this mandate into a diary entry "to leave these horses absolutely to Judith (her daughter, the future Lady Wentworth) urging she should maintain

> as a permanent stud them or their produce to be called the Heirloom Arabian Stud." "I wish to have a try at making it a permanent institution."

The cause of the breed purist was then, and has remained, only a belief, beyond absolute proof. Precisely which horses are of pure Arabian blood? What exactly is doubtful blood?

The controversy which has existed for the past 50 years between the pedigree skeptic and the pedigree purist is succinctly expressed by Dr. Hans Nagel, a widely respected author, Arabian horse breeder, and noted breed authority, who wrote the following concerning the validity of the pedigrees of Egyptian Arabian horses in his book Hanan:

> Thus not only the pedigree requires a large amount of trust, even the beginning itself is a test of faith and uncritical acceptance.

Regarding the purist breeders, he wrote:

> Thus many sensible people may rightly advise the extremist Arabian enthusiasts to stop regarding perfectly natural processes as something special and to reconsider their vocabulary instead of constantly repeating terms like 'purity' ad infinitum.

There is much truth in this.

Hanan is essential reading for the student of the Egyptian Arabian horse pedigree. Dr. Nagel goes into great detail about the history of the mare Hanan. He describes the problems that can be caused by a pedigree when Arab horses and their breeders interact with each other in a very public forum. The problems can be serious, and are compounded when considerable commercial value is attached to the success or failure of the contestants. The story of Hanan shows ways in which a horse's pedigree can be used as a weapon in the highly competitive and high stakes world of international Egyptian Arabian horse breeding.

Hanan was born at El Zahraa Stud, Cairo, Egypt, in 1967, and was bought by Dr. Nagel as a youngster on a "sudden, unpremeditated impulse". Dr. Nagel saw the 7 month old weanling in the paddocks at El Zahraa and, not even knowing her name, felt "A spark was ignited, a vague idea formed in the middle of nowhere, a vision of a horse, a breeding farm of one's own."

Figure 17 - *Alaa el Din By Nazeer out of Kateefa, photo by EAO.*

As it turned out, the weanling was Hanan. She was sired by Alaa El Din, a stallion of unimpeachable quality and purity who regularly produced exceptional daughters.

But Hanan's dam was a mare called Mona. This was going to be a problem. She was sired by Badr, a 1947 Inshass stallion whose sire was Beshir el Ashkar. Badr's dam was Badria. Both of these horses were given to King Farouk just after World War II by a Palestinian named Bisharat Bey, who had obtained them from a group of Jordanian Bedouins. Details of the gift and the identity of the donor were scant. Mona's mother was Mahdia, by Hamdan and out of Mahrousa, a 1937 mare sired by El Zafir.

Mahrousa was the daughter of El Shahbaa, founder of the El Obeya Om Grees matriline at the RAS, making her a significant figure in the overall picture of Egyptian Arabian horse breeding today. According to the information found in Hanan, El Shahbaa was gifted to King Fouad of Egypt in 1926 by a man named Al Haq Ibrahim Mohamed. The identity of the donor otherwise unknown and no director of the RAS, Inshass or the EAO had any recollection of the man or the name. Pedigree research has turned up no leads.

Pearson states in Arabian Horse Families of Egypt that a mare named El Obeya Om Grees was given to the King of Egypt by King Aziz Ibn Saoud, around 1931. Her date of birth is not known. She left only one recorded offspring, and this one horse was not bred at the EAO.

Her daughter was El Shahbaa, and she is said to have been a grey mare born in 1925, sired by El Hamdani el Nasiri, a non-RAS stallion of whom absolutely nothing is known. The name of the breeder is given in EAO records as M. Ibrahim. It is not known whether the mare was given to the King as a gift or was purchased from M. Ibrahim for use at Inshass. El Shahbaa was eventually sold to an unnamed party in 1942.

Thus the provenances of three of the horses in the pedigree were very limited, and this fact left them open to charges of "unproven purity". None of the horses in this group had previously produced anything that even approached the quality of Hanan. She matured into what was unquestionably one of the finest specimens of the Egyptian Arabian horse. The competitors of the Nagel breeding program were jealous. The lack of clarity and detail in the documentary evidence left the critics of the horse much ammunition. The lack of clarity and detail, of course, was in no sense reliable or valid information to be used as ammunition, but it was used all the same.

Lack of clarity does not establish impurity of blood. It simply establishes the fact that there is lack of clarity. The truth remains unknowable.

As Hanan matured and began producing excellent quality foals, commercial considerations began to play a role. Nagel's competitors and critics pointed suspiciously, and in public, to the pedigree of Mona, suggesting that her pedigree could not be trusted. They mainly questioned the likelihood that Bisharat Bey's gift horses were asil. Mona had made no significant impact as a brood mare at El Zahraa. She was completely unknown to Arabian horse breeders outside of the EAO.

El Shahbaa, like all Inshass horses, was criticized by the *purebred* horse intelligentia as a mere "gift horse" to the King of Egypt, and thus open to charges of unproven bloodlines. This charge of questionable quality and asalah at the Inshass stud was not new.

On further and more mature reflection, the fact that a horse is given to the King of Egypt by the King of Saudi Arabia does not in and of itself lead inevitably to the conclusion that the horse is de facto of poor quality and of doubtful purity. This charge has been made so often, for so long, by so many well known experts, that it has become difficult to step back a few paces and consider the true merits of the case.

Certainly, many authorities who have made this charge have seen and are intimately familiar with both the EAO and the Inshass horses. Their opinions are informed, but like all opinions are subject to personal bias and hyperbole. An opinion on the overall quality of any collection of horses leaves out the fact that there are likely a few good horses lumped in to the blanket assessment. A generalization, by definition, leaves out more than it reveals.

For the charge to be credible, one must assume that no King would ever give away a high quality asil horse from his own stud. This is a broad assumption. Mistakes in identifying horses have occurred. The possibility also exists that a mare of average quality might turn out to be a superior brood mare.

Besides, there is the case of the Egyptian mare Maaroufa being sold by Prince Mohammad Ali Tewfik to Henry Babson in the U.S. as a Kuhalayh Jellabi descendant when she was in fact descended from the Saklawi mare Ghazieh I of Abbas Pasha. And they were both highly regarded lines. Mistakes in pedigrees do occur. A mistaken assessment of the quality and future breeding potential of any given horse does at times occur.

It is not inconceivable that a King would give away a horse as a diplomatic gift that he thought was of inferior quality but was in fact not of inferior caste. The King could be mistaken; his advisors could be mistaken.

It is also not inconceivable that a King would actually give a prized and valuable Arabian horse to a fellow potentate out of sincere good will. The gift could be a true expression of friendship and generosity. Such things have been known to occur.

A gift horse is not of necessity an inferior animal. It may be a horse of inferior quality, but its status as a gift does not make it so. An American oil engineer who spent time in Arabia and at the Saud stud at Al Kharg in the 1950s wrote that the King simply instructed his herd manager to send a horse to King x,y, and z. The gifts were in appreciation for support in military or diplomatic matters carried out by the receiver. The Saud king was normally not involved in the actual selection of the horses that were given as tokens of regard and appreciation.

The rise to international stardom of Mona's grandson Jamil led to increased attacks from Nagel's competitors. Jealousy, writes Nagel, led his opponents to target Jamil with particular ferocity. The owner received threats from anonymous sources that accelerated as the stallion's star rose higher.

Concern over the safety of the horse was genuine, and not without foundation. In Europe there was the precedent of Galero. The Spanish stallion Galero, bred by and owned by the Spanish Yeguada Militar, was a rising star in the international scene, siring champions with increasing regularity. He came to dominate the European show ring through his offspring, who won the Spanish National Championship and the World Cup Championship.

On the night of March 18, 1982, the stallion was reported missing from his stall. A country-wide police search throughout Spain was mounted, but the stallion was never seen again. The abduction and death of the stallion remained unsolved. Officials considered foul play to be the motive. Many assumed that the stallion was stolen and killed by jealous competitors.

One critic and competitor of Jamil went to the extreme measure of filing a law suit, attempting to ban Jamil from registration in Germany. This failed. The trail of doubt and uncertainty that resulted from the very public law suit, however, persisted. It was then that Dr. Nagel began to ask himself the question "Is it necessary to have a beautiful pedigree to be a beautiful horse?"

A full discussion of this perspective can be found in Dr. Hans Nagel's book Hanan, published by Alexander Heriot & Co., 1998. Dr. Nagel makes the following points about Arabian horse pedigrees and about the Bedouin themselves. The following points are quoted from his book.

Figure 18 - *Jamil. Photo used by permission of Jerry Sparagowski.*

1. Any pedigree which purports to prove the pure desert origin of a given horse is to be regarded with doubt, since the conditions required for such a document to be created and transmitted would be difficult to fulfill given the illiteracy of the nomadic Bedouin society and the incompleteness of records outside Arabia.

2. The testimony of those who produced or provided pedigrees of desert horses cannot be uncritically accepted. Competent, deliberate deception would hardly have been noticed and is sure to have been practiced wherever it led to advantages and the seller was not too hampered by his conscience.

3. The taller, faster Arabian horses found in northern "Greater Arabia" (Syria and Lebanon) likely resulted from the unacknowledged use of Anglo-Arab stallions or other non-Arabian breeds found throughout the Ottoman Empire. One would have to be hopelessly naïve to deny such a possibility. Lady Anne Blunt wrote "No Nedjean believes in the purity of blood amongst them except in such families as . . . who kept aloof both from wars and from the Turks. For centuries, northern Arabia was one of the most attractive and richest cultural centers of the world, nothing like a nomadic hinterland. There was active commerce . . . for selling wool, meat, and horses." Horses with heavy heads and bones may be commented on in a polite and tolerant manner, but this will not alter the deplorable fact that one or more of their ancestors did not belong to the Arabian elite or might even trace to an outside origin.

4. Not a single Arabian horse exists whose purity of blood can be historically proved.

5. Honest they were, without question, in dealing among themselves because it was in their best interests to preserve their breed and to live up to their own traditions as far as possible. But with regard to strangers, the horses that were sold abroad disappeared from the picture and could do no more harm. So why should certain dealers or smart traders not have taken the opportunity as it presented itself? Thus every attempt was made to produce what was required, even in writing ...the Bedouins had a rather dubious reputation among the northern Arab settlers. They were regarded as sly, exceedingly stubborn, and unreliable, beholden only to their families. Nobody would have trusted a Bedouin with their money or property. This became all the more true as the old traditions were eroded and the resulting needs often led to dubious methods of self-preservation...Burckhardt, who traveled through Arabia and Syria in 1812 and 1817 remarks disparagingly in his report: "They all

stole and robbed, from friends as well as enemies. As a result, suspicion and secrecy were the most obvious characteristics of this part of the population, and these traits were indeed necessary if they wanted to protect their own interests. To arouse desire in someone else could have dire consequences."

In their dealings with livestock, the Bedouins were described as disorderly and cruel even by their countrymen.

6. Simple believers . . . try hard to keep alive the idea of purity in Arabian breeding. They forget that the social order of Arabia is self-contained and closed toward the outside. It is a great mistake to underestimate this fact and shows unforgivable ignorance.

7. The theory of purity is supported neither by facts nor by history and thus cannot be sustained. It is the wrong premise and it doesn't get any better by being repeated over and over again with fanatical conviction. The faster and more deeply it is buried, the sooner the way opens for a rational view. A single instance of contamination in a pedigree is not harmful, especially if this individual occurred many generations back in the pedigree. It is insignificant if in the past a stallion or mare was introduced into this group that did not correspond properly. As long as neither phenotype nor genotype deviates from the desired standard, such elements will have no disturbing influence.

8. All studbooks today are closed. The entire Arabian population of the world is a closed society, controlled and determined to a degree never known before. Arabian horses are registered in almost 50 countries according to identical rules established by the World Arabian Horse Association (WAHO) which was founded in 1972. All horses entered in these studbooks may claim to be *purebred Arabians*. Whether it is acceptable to groups of people who claim to be concerned about the breed's "purity", is irrelevant for the breed as a whole and of no account worldwide.

The scientific test of this theory is now being run. The Bedouin theory is just that, a theory, and until it is tested in a clear and rigorously defined scientific manner, it will remain a theory.

The purists would say that the theory has already been tested, carried out by the desert dwellers for thousands of years. This opinion, however, does not constitute scientific rigor. To quote the British aesthete Walter Pater: "we should be forever testing new opinions, never acquiescing to a facile orthodoxy."

The WAHO conglomeration of horses that it considers to be Arabians is now being bred, one to the other, and will continue to be bred, with no regard for degrees of genetic authenticity. This is a valid and objective test of the Bedouin theory, since this group of horses contains such large percentages of non-Arabian blood.

Figure 19 - *Al Zahraa Stud, The Egyptian Agricultural Organization, Cairo Egypt 2011. Photo from the author's collection.*

There are now horses in the WAHO fold who have no pedigree at all, registered as the offspring of unidentified desert bred parents. Based on WAHO principles and definitions, WAHO adherents will see no reason to consider the degree of genetic authenticity when selecting mares and stallions for breeding. They will have the criteria of phenotype as a guide, but they will have nothing else on which to base their decisions. They can only hope that what they see is what they will get.

If the Bedouin, theory is incorrect, then the results of breeding these horses together will have no adverse effect on the outcome. WAHO sanctioned horses of two hundred years from now will look and act just the way that they do today.

The experiment is now running. If the Bedouin theory is correct, then the horses resulting from WAHO-guided breeding will, in a few hundred years, begin to deviate from type, and breeders will notice that genetically determined characteristics and diseases that are inapparent in the horses today will begin to express themselves.

Time will tell.

**Trusting men is like trusting water in a sieve.*

CHAPTER TWO

CHAPTER TWO

THE ROYAL AGRICULTURAL SOCIETY CONCEPTION OF THE GENETIC AUTHENTICITY OF THE PUREBRED EGYPTIAN ARABIAN HORSE

أنت تريد وهو يريد والله يفعل ما يريد

The earliest documented instance of *purebred* desert sourced Arabian horses in Egypt was the house of Qalaoun which ruled Egypt during the 14th century. El Nacer Mohammed Ibn Qalaoun collected and bred large numbers of these horses, building large stud farms for their care and breeding. He also built a magnificent and costly hippodrome to display his horses' beauty and athletic abilities. His source of desert horse stock was the desert Bedouin tribes of Arabia; particularly the Beni Mouhanna, the Beni Fadl, and the Al Murrah. It is said that at the time of his death, he left a collection of over 7800 *purebred Arabian horses*. The herd was lost over time. There are no surviving records, only memories passed down through the centuries.

Egypt has had a splintered history. When the pharaonic millennia of Egypt came to an end, a vacuum was created that was filled by the growth and expansion of the Greeks. An era of Hellenization culminated in the control of Egypt by Alexander the Great. Following the death of Alexander in 323 BCE, Egypt continued to be dominated by degenerating Greek rule, leaving it as a weak nation, both economically and politically, for several centuries.

The Ottoman Empire gained control of the nation in 1517, and Egypt was part of that Empire for several centuries. The French invasion of Egypt in 1798 resulted in a brief period of freedom from the Turks, but this came to an end when the British expelled the French from the country in 1801.

After the expulsion of the French, a four year period of internal chaos political and economic turmoil ensued. From 1801 to 1805, Egypt existed in a state of leaderless barbarism and political flux. Factional wars for control of the country were fought between the Mamelucks, the Ottoman forces, and a newly emerged force in the region, the Albanian military mercenaries, a military auxiliary attached to the Ottoman army. The Albanians were ambitious and became a force unto themselves.

The Egyptian Arabian horses that exist today are largely the result of the acquisitions of desert horses by a man who began life as an obscure foreigner to Egypt, a Macedonian mercenary soldier. He rose from a life of poverty in his homeland to a position of absolute power in Egypt. He was, in a sense, a popular leader, highly regarded by the people of Egypt. He brought change to Egypt, modernizing the country and turning it into a significant political and economic force in the region. He was also ruthless and cruel with regard to obtaining political power.

This ambitious foreigner created an Egyptian dynasty built on might of arms that would govern the country of Egypt for over 100 years. For five generations, his dynasty ruled the country and, in the process, became extremely wealthy, building lavish palaces and acquiring extensive farms and land holdings, all at the expense of the working Egyptian populace. During a series of Ottoman backed wars against his neighboring desert dwellers of Arabia, he also extracted hundreds of authentic desert Arabian horses from the desert Bedouin. He and his descendants, entranced by the beauty of the animals, collected and bred these Bedouin horses, building elaborate and costly stables for them in and around Cairo. These acts of rapacious appropriation of the Arabian Desert horse from the Bedouin tribes would prove to be its salvation.

He was an Albanian named Mehmet, better known to history as Mohammad Ali the Great.

Arabian horses had existed in Egypt for millennia, but they have left no traceable living descendants. The crucial difference in the case of Mohammad Ali the Great and his dynasty was that his horses did survive into the modern day, being well documented and well identified. And this bloodstock is bred to this day.

The unstable conditions of 1801 to 1805 ended with the success of the rapacious Albanian; he achieved absolute power in Egypt. He declared himself supreme ruler of Egypt, and thereby established a dynasty that would rule Egypt until 1952. Although he was recognized officially by Constantinople as an Ottoman "viceroy" he was in truth a law unto himself. The Mohammed Ali Dynasty, as it came to be called, ruled Egypt according to the sentiment expressed in the old Arab saying, "the Sultan's sword is long." So too was the sword of the Kavali Mehmet.

Figure 1 - *Massacre of the Mamelucks at the Citadel in Cairo. Mohammad Ali watches the carnage. The scene depicts the last hurdle facing Mehmet in his desire for dominance, the destruction of the Mamelucks. The ambush of his rivals cleared the way for his ultimate ambition. Image by Carle Vernet, from the National Gallery of Art, Washington D.C.*

Figure 2 - *The Mameluck's Leap as he escapes from the ambush at the Citadel, from the London Illustrated News.*

Mehmet was born to Albanian parents, not in Albania but in Kavala, a small coastal town in Macedonia, which was at that time part of the Ottoman Empire. His birthplace was not far from Pella, the birthplace of Alexander himself. But Mehmet was not Greek. He was the son of a local police chief, and he spoke Albanian his entire life, learning Turkish as a second language. His first attempt at a career was as a tobacco merchant. Failing in this, he obtained a commission as an officer in the Turkish division of the Ottoman Army. He was posted to Egypt where he rose rapidly through the ranks, coming to the attention of the Sultan in Constantinople as a capable military commander.

Figure 3 - *The city of Kavala as it appeared in 1900. Kavala is located in northern Greece, and has been the principal seaport and trading center for Macedonia for centuries. It is on the Bay of Kavala which opens on to the Aegean Sea. The name of the city is derived from the Italian word cavallo which translated into English means "horse".*

The importance of Mehmet to Egyptian Arabian horse breeding is due to his fondness for the authentic desert horse of the Arabian Bedouin. This interest, coupled with his son's willingness and eagerness to plunder the finest quality asil horses of the desert Bedu tribes of Arabia during the 19th century formed the basis of the finest collection of asil desert Arabian horses ever assembled.

Kavalali's grandson exhibited an even more fanatic and maniacal hunger to collect and breed the finest and the highest caste desert Bedouin-bred horses; some of them he bought from their Bedu owners, and some he took. His grasp on sanity was at times tenuous. He was ultimately murdered by disgruntled servants.

His name was Abbas.

Figure 4 - *Arabian Horses in Egypt, by an unidentified artist. Photo used with the permission of Dr. Georg Olms.*

> The original breeders of the true Arab horse were the Bedouin. It was bred under unhampered conditions, free to wander unrestrained over vast tracts of desert, and occupying a place as an important member of the Bedouin family. The special characteristics of the *purebred Arab* horse – strength, speed, intelligence, and form – are definitely traceable to the conditions under which it was bred and lived.
>
> From very early times the Arabs have been reputed for the preservation of the pedigrees of horses as a precaution against any possible intermingling of their strains.
>
> (H. E Fouad Pasha Abaza, Director General of the Royal Agricultural Society, Egypt.)

The foundation stock that was gathered by the Royal Agricultural Society to create the Egyptian Arabian horse came from sources that were believed by the program's directors to be 100% genetically *purebred Arabian* Desert-bred horses from the Bedouin tribes that lived there. From its very beginning, genetic purity was the focus for the RAS, and every effort was made to assess the blood purity, and by implication the genetic purity, of every horse chosen for inclusion in the breeding program. The organizers of the RAS wanted to obtain pure-blooded Arabian horses for preservational purposes, not for general horse breed improvement or manipulation.

Whether the individual horses that they selected were pure blooded Arabians is still an open question, but the fact remains that they did fervently believe that they were collecting the last remnants of a previously populous landrace of noble desert horses that, they believed, had been present on the Arabian Peninsula for ages.

They wanted to form a centralized herd of *purebred Arabian horses*, and they believed that they were selecting 100% genetically pure Arabian horses. The directors used the following criteria in their selection of stock:

1. Horses were selected whose pedigrees traced, either directly or indirectly, in all lines to the stock bred by the Bedouin tribes of the Arabian Desert, and the central highlands of Nejd in particular. It was assumed that horses that

met this criterion were "*purebred*". This assessment was inherently problematic in that it depended on the evaluation of the fragmentary written and oral provenience of each horse, which extended back in time only as far as the era of Abbas Pasha, who died in 1854. Most of the horses had even less documentable provenances. There were no surviving continuously recorded pedigrees of Egyptian Arabian horses prior to those collected by Abbas Pasha.

In some cases, horses were included in the early iterations of the RAS based on the reliability and high standing of the breeder from whom the horse was obtained. Members of the extended royal family of Egypt and certain of the desert sheikhs, the Khalifas for example, were regarded as reliable sources simply because of their long standing reputations as breeders of high quality asil horses, and their reputation for honesty and accuracy in providing valid pedigrees for horses which they sold or donated.

2. Physical examination of the individual horse. Points of behavior and anatomy were assessed. Horses exhibiting strong Arabian traits were chosen. Those horses exhibiting traits that were thought to indicate the presence in the pedigree of a contaminating non-Arabian influence were rejected. These points included overall height, head morphology, neck length and set, back conformation, tail carriage and leg conformation.

3. The quality of the offspring of those candidates for inclusion was evaluated as well. Emphasis was placed on detecting the emergence of flaws and faults in the subsequent generations that could indicate the presence of genetic impurity in the parent.

The first criterion proved difficult to determine with any degree of certainty due to the fragmentary nature of the historical record. The second criterion proved to be limited by the personal biases and opinions of the succession of directors and the influence of the shifting cultural contexts in which they lived. The third criterion was likewise limited by its subjective nature.

For the early horse breeders of Egypt and the RAS, the concept of the importance of blood purity was based on their implicit acceptance of the ancient practices and traditions of the Bedouin of the central deserts of Arabia. Egyptian horsemen saw for themselves the proof of this tradition; the horses of Abbas Pasha, an unimpeachably reliable source of pure Nedjean horse stock, were a testament to the Bedouin's success in breeding. These horses, preserved for the Arabian horse breeders of the late 19th century by Ali Pasha Sherif, were considered to be of the highest quality.

It was an article of faith among the horsemen of this era that the only true *purebred Arabian horse* was that which was bred by the Bedawi in the central Arabian highlands of Nejd. It was accepted dogma that these horses were indigenous to Nejd, domesticated and bred there continuously since before recorded history. Before the time of the Prophet, few Arabs could read or write. This changed with the spread of Islam. Nedj was considered to be the original homeland, the fountainhead, the source of the true blood. However, objective proof or documentation to support this conviction was not in evidence.

The Horse Breeding Commission was established by the Egyptian government in 1892 with the goal of improving the quality of the general horse stock in Egypt. This was superseded by the Royal Agricultural Society which was founded in 1908. Its original mission was a practical one; to breed and produce Anglo-Arab horses for agricultural use in rural areas of Egypt.

In 1912, a group of concerned individuals met in Cairo for discussions about the possibility of creating a government-sponsored *purebred Arabian horse* breeding program in Egypt. Under the patronage of H.H. Abbas Hilmi Pasha, the Khedive of Egypt, this movement was called the International Society for the Preservation of the Arab Horse. The members included H.H. Prince Mohammed Aly, (the Khedive's brother), Prince Youssef Kemal, Prince Alexandre Sherbatoff, and Prince Kemal el Dine Hussein (the Khedive's cousin).

The group did not arise out of a vacuum nor did they work in a vacuum. They represented the last few men with the knowledge and appreciation of the *purebred Arabian horse* who at the same time had access to the last few living specimens of this noble creature. They were a unique group of men in a unique historical setting. Their decisions and dedication were to have lasting effects on the survival of the last vestiges of the once great assemblage of *purebred Arabian horses*.

They were all well educated men of a serious demeanor who were familiar with and relied heavily on the expansive number of historical and scholarly manuscripts concerning the *purebred* horse. The nature of the *purebred Arabian horse* and its origins were the subject of well known treatises for many centuries. For the Western observer, these primary source materials were unavailable, since the majority of them were written in the Arabic language of the Middle Ages. The Raswan Index provides a list of over three hundred Arabic scholars who left behind academic treatises on the features, attributes and origins of the *purebred Arabian horse*. Most of these classics remain untranslated to this day.

As background for the assertion that the Nedjean horses were the only truly *purebred Arabian horses*, Raswan mentions a number of non-Arabic language authors well known to the Western student of the asil horse. These included the British breeder Wilfred Blunt, the Italian explorer Guarmani, Gayot, Prisse d'Avennes, who was a French artist and accomplished paleographer, Burckhardt, who was a fully Arabized scholar and traveler, Damoiseau, Hammer-Purgstall, Niebuhr, Palgrave, Upton, Tweedie and Davenport. The book Das Pferd Bei Den Arabern, published in 1856 by Hammer-Purgstall, stands apart as being both authoritative and well informed on the subject of the *purebred Arabian horse* as a result of the author's close contact with desert sources.

Joseph von Hammer-Purgstall was singularly well qualified to write on the subject. Born in Syria, he was educated in Vienna and on maturity entered the Austrian diplomatic service. He served in various capacities in the Middle East for about 8 years. He was a prolific translator of Arabic texts which covered a wide range of topics. He was knighted in 1824.

The Raswan Index is replete with references to Arabic language writers and works such as the well known manuscript of Abbas Pasha, written by Al-Lallah. A contemporary of Abbas, Bujayrimi Ash-Shaykh Futuh Al-Bujayrimi, wrote a book on the evaluation of the *purebred Arabian horse*, in which he discussed the traits, strengths and weaknesses of Abbbas' horses. Ibrahim Pasha wrote a brief work on horses. Sharaf Ad-Din Abd Al-Mumin Ibn Al-Khalaf wrote a textbook on the qualities of the desert horse in the 14th century entitled Kitab Fadl Al-Khayl. There is the treatise by Imam Ali, written at the time of the 4th Caliph, now in a Vienna library and a book by Abdullah Ibn Amru written in the 9th century CE. The 10th century author Muhammad Ibn Abbas El-Jesidi wrote about desert Bedouin horses.

There is a volume written by Abu Abdallah Muhammad Ibn Ziyad concerning the equine genealogy of horses on the Arabian Peninsula. Abu Al-Ash'at wrote a treatise on *purebred* horses that were found in the region of Mecca in the 10th century. Abu Zakariyah Ibn Muhammad Al-Andalus wrote a book on Arabian horse husbandry in the 12th century. The treatise Kitab Fil-Furusiyah Wa Istikhraj Al-Khayl Al-Arabiyah is a treatment of the topic of the *pure blooded Arabian horse*.

Abu Muhammad Abdullauh Ibn Muslim Al Rufi, writing around 883 CE, wrote four books on the subject of the *purebred Arabian horse*. Abu Sabit Ali wrote a treatise on the physiology of the Arabian horse around 838 CE.

Apart from the body of literary works about the *purebred Arabian horse*, the committee members who drove the enterprise had a vivid image of the physical appearance and phenotypic attributes of a *purebred* desert Bedouin-sourced Arabian horse as depicted by artists of the 18th and 19th centuries.

Figure 5 - *One of two pages from a Turkish manuscript on the training and care of warhorses from the collection of the Walters Museum of Art. The book was written by Ahmed Ata Tayyarzade in 1854.*

Examples of a few of the more well known art works featuring Arabian horses of the 18th and 19th centuries will be presented later in the book.

The group of Egyptian horse preservationists began the task of assembling the seed stock with which to begin a breeding program consisting of unimpeachably *purebred* horses. Apart from the Russian Prince, this group represented the last remaining vestiges of the formerly great house of Ali. The distilled wisdom of five generations of avid and passionate Arab horsemen was concentrated in these men. They possessed the written and remembered history of what the Arabian horse was, and what it was not. More importantly, they personally owned the few remaining authentic asil horses that were left following the bloody civil wars of 19th century Arabia.

A wealth of graphic images of the authentic Arabian horse was created by artists of the 18th and 19th centuries. The series of books by Dr. Georg Olms, Asil Araber, contain reproductions of the many of the prints, paintings and etchings from the 18th and 19th centuries, depicting the asil Arabian horse. The early directors of the RAS were familiar with many of these works and used them in their evolving conception of purity in the selection and breeding of desert sourced horses.

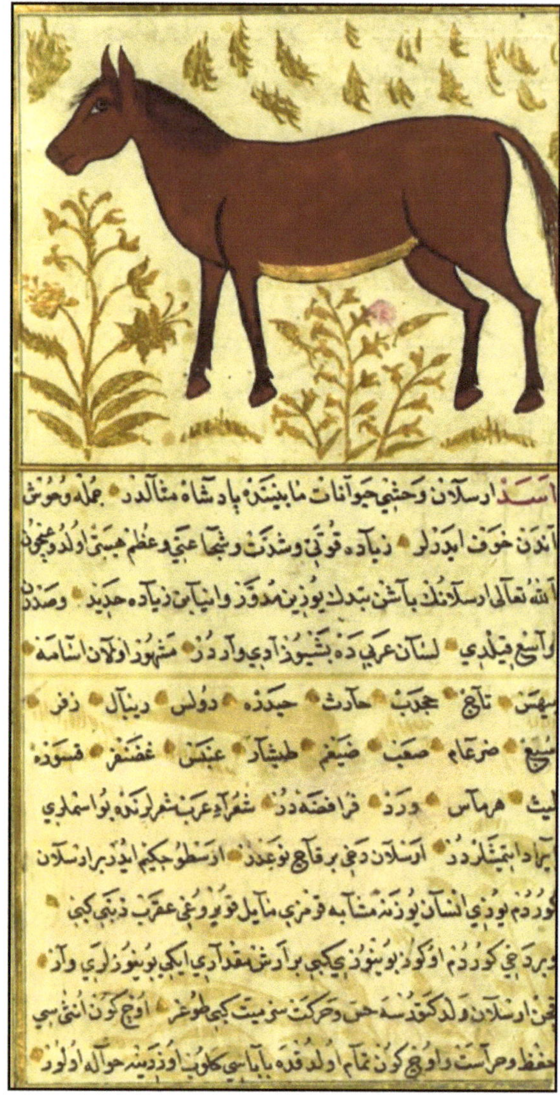

Figure 6 - *One of two pages from a Turkish manuscript on the training and care of warhorses from the collection of the Walters Museum of Art. The book was written by Ahmed Ata Tayyarzade in 1854.*

Figure 7 - *Khedive Abbas Hilmi II.*

Figure 8 - *Prince Yusuf Kemal.*

Figure 9 - *Prince Kemal El Din Hussein.*

Figure 10 - *Prince Mohammad Ali Tewfik (transliterated also as Taufiq).*

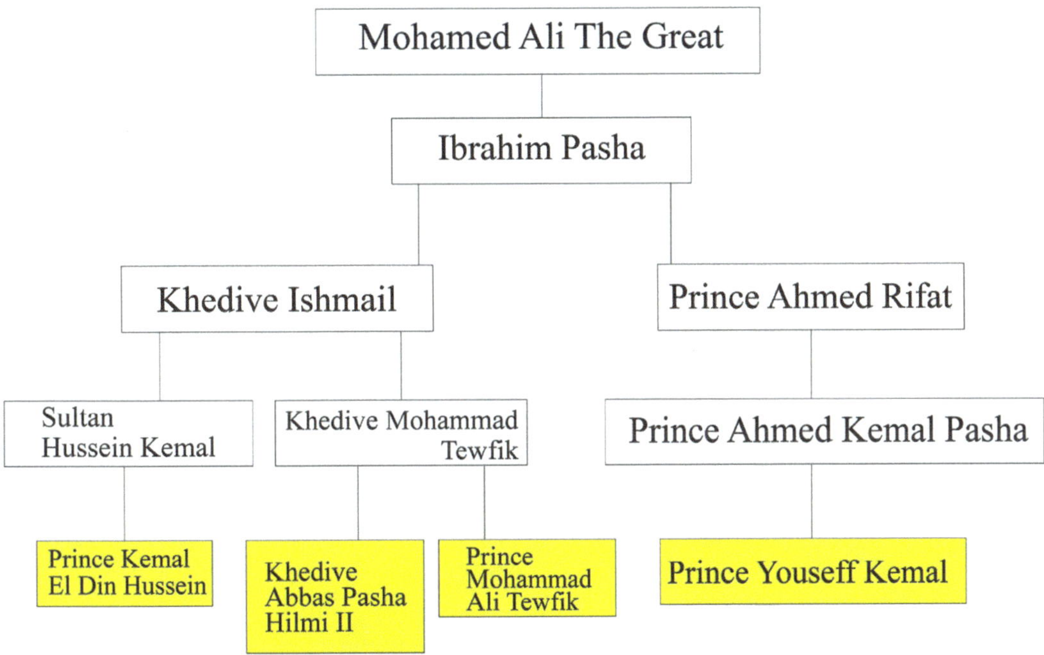

Figure 11 - *The relationships of the founders of the Royal Agricultural Society.*

Based on these images, the modern Arabian horse breeder can see exactly what the genuine Arab horse of that time looked like. When comparing the Arabian horses of today with the images of the 19th century asil authentic horse, a striking contrast emerges. This difference between the features of the horse in the past and the present requires some reflection.

The committee had to address the question of which horses should be included in the *purebred* breeding program of the Egyptian government. Since biological and scientific criteria regarding genetic purity did not exist at the time, subjective assessments made by the most highly qualified specialists were the sole criteria. These decisions, however, were not made in a vacuum.

The decision to include any horse in the formation of the core herd was made with deliberate and careful considerations of all available information. The decisions were made by consensus. No one man was responsible for any of the early decisions.

It was also the policy of the RAS to evaluate the offspring generation after generation in order to detect defects that may indicate the presence of detrimental genetic characteristics in the lines. The emergence of defects could lead to the termination of a line of horses at the RAS. This was an especially important element in stallion line evaluation.

The opinions of this group of men shaped the eventual formation of the RAS, and their influence is still being felt today. Their opinions were informed by a number of factors; Bedouin lore, the family tradition of the members of the royal family, and preserved images of desert horses that illustrated the physical appearance of the authentic desert horse in the days before photography.

Many of the members of the new organization in Cairo had seen the horses of Abbas Pasha breeding themselves, or had seen their descendants in the collection of the preservationist Ali Pasha Sherif. They all owned and bred horses which they believed to be *purebred Arabian horses*, although Abbas Hilmi Pasha also experimented with the use of English stallions on his Arabian mares. They were united by one common characteristic; a consuming love of beauty, whether it was found in works of art, palaces, women, or horses. They shared a second commonality; great wealth.

HH Abbas II Hilmi Pasha, great-great-grandson of Mohammad Ali, was Khedive of Egypt from 1892 to 1914. Born in 1874, his youth was spent in Europe, receiving an excellent education in Geneva and Vienna. He was fluent in English, French and German. Hilmi became Khedive at age 18, upon the death of his father. His administration was characterized by constant quarreling with the British, and he eventually began to secretly oppose them. His real passion, however, was raising horses, and he kept breeding stable at Qubbah, near Cairo, and at Muntazah, near Alexandria. At one time or other he owned or bred Bint el Bahreyn, Bint Hadba el Saghira, Bint Gamila, and Obeya. Among the many high caste Arabian horses that he owned, the mare Yamama stands out as she figures prominently in the confusion of the "Yamama bay, Yamama grey" matter which is discussed at length later in this book.

Prince Aleksandr Grigorievich Sherbatov was a cultured and well educated Russian nobleman with a highly developed artistic sense. He had become entranced with the idea of the *purebred Arabian horse* from the writings of Lady Anne Blunt, and in 1888 had traveled to Syria to obtain horses for a breeding project of his own. Following the course described in the Blunt journal, he began in Damascus, aided by Shaikh Nasr ibn Abdullah of the Resallin tribe, a division of the Al-Sebaa. He evaluated horses in Karietein and visited the Hesenneh and the Beni Wali tribes. On the whole the horses that he saw were of poor quality and the prices asked for them were "scandalously high". In Tadmor the results were similar. He eventually reached Deyr on the Euphrates, and then returned to Damascus. Along the way he was able to find and purchase enough mares and stallions of sufficient quality to return to Russia and begin breeding horses himself.

Using this stock, he established a *purebred* desert Arabian breeding farm on his estate at the foot of Mount Zmeika in the northern Caucasus of Russia. By the time of the 1912 meeting in Cairo he had 24 years of experience breeding the desert Arabian horse.

The Count died in 1915. The horses that he had bred for 27 years were destroyed during the violence of the Russian Revolution of 1917.

Prince Yusuf Kemal, born in 1888, was the grandson of Ibrahim Pasha. His father, Prince Ahmed Pasha Kemal, was a long time avid collector and breeder of Arabian horses, obtaining most of his stock from Ali Pasha Sherif. On Prince Ahmed's death in 1907, his son inherited a herd containing such outstanding horses as the mares Roda, Dalal and Dahma as well as the stallions Rabdan and Sabbah. In 1908, Youssef Kemal sold the horses at auction.

He was an aesthete first and foremost. He was a nationalist at a time when this sentiment was unpopular with the British who controlled every aspect of Egyptian political life. Kemal placed himself and his fortune into public service, focusing on education for the public and promoting art for the masses. He was extremely intelligent, well educated, and a proponent of advanced education for the Egyptian public. He was influential as one of the founders of Cairo University and the Fine Arts School, heading the University from 1916 to 1917. He was adventurous man, having traveled to Tibet and Northern India. His aesthetic appreciation of nobly bred Arabian horses was widely respected.

Prince Kemal el Din Hussein, son of Hussein Kemal, was born in 1874, was also of the House of Mohammad Ali. He was virulently anti-British and declined his rightful succession to the throne as long as Egypt remained a British protectorate. He was an explorer, a sportsman and a collector of Oriental antiquities and works of art. As a breeder of *purebred Arabian horses*, he is best remembered as the breeder of Bint Serra, who was exported to the U.S. by Henry Babson. The Prince often used the Blunt stallion Rustem on his mares, and he produced Bint Dalal and Bint Bint Dalal.

A thorough discussion of Prince Mohammad Ali Tewfik and his horses is found later in this book.

Lady Anne Blunt, 15th Baroness Wentworth, was an essential contributor and consultant to the members of the committee, providing an aesthetic sensibility to the discussions among this group of experts. In 1912 she was 75 years old, living alone at Sheykh Obeyd, which was a 32 acre apricot orchard near Cairo that had been converted into a horse breeding facility. She had turned it into a breeding farm for her private hermitage. She had received an excellent liberal arts education as a youth in England and was fluent in French, German, and Arabic. She was a gifted artist and violinist. Her violin, known as the "Lady Blunt Stradivarius", was sold at auction in 2011 for USD 15.9 million, an unprecedented record.

Figure 12 - *Manial Palace on the Isle of Roda in the Nile River at Cairo, The Raswan Archives.*

The study of Prince Mohammad Ali Tewfik at Manial on the isle of Roda near Cairo served as a meeting place for visitors interested in the Arabian horse. The Prince lined his study walls with reproductions of many of the classic works of art of Arabian horses, and his visitors were invited to compare the Prince's horses with those horses seen in the works of the European artists. The Prince's library and works of art were described by Carl Raswan as a "treasure of incomparable value". This study was probably the site of many of the discussions that shaped the ultimate selection of the foundation stock of the RAS.

"I am an artist myself," wrote the Prince, "my perfect horse must have all the symbols of an antique pedigree, a horse that might have stepped from some rare or ancient picture."

Figure 13 - *Photo of the study of Prince Mohammad Ali, from The Raswan Index.*

A visitor to Manial Palace wrote:

> The Prince had a strong sentimental interest in continuing the Arab blood which had been identified with his family so many years. The Prince made a collection of old prints and drawings showing the

> traditional Arab horse, the horse of poetry and romance. These serve as his guide, and it is his aim to breed to this standard. Every sire and dam, as well as their progeny, is studied with this ideal before him and the tests are applied with almost mathematical precision. Any that fall short in the minutest way are ruthlessly weeded out and sold.

The following sets of reproductions were displayed on the study walls, and represented the Prince's ideal vision of the Arabian horse. They were taken from his book Breeding of Pure Bred Arab Horses, published by Paul Barbey's Printing in 1935. He often entertained visitors, both foreign and domestic, with displays of his own horses, attempting to demonstrate the close similarity that his own horses bore to these models of Arabian horse excellence suggested by the 19th century prints.

Figure 14 - *Horse of His Highness Abbas Pasha, by Alfred De Dreux.*

Figure 15 - *Horse of His Highness Abbas Pasha.*

Figure 16 - *Horse of His Highness Abbas Pasha.*

Figure 17 - *An Arab Horse of Nejd.*

Figure 18 - *A Typical Arab, by Victor Adam.*

Figure 19 - *Horse of Napoleon, by Victor Adam.*

Figure 20 - *A Horse of Napoleon, by Baron Antoine Gros.*

In 1914, progress toward realizing this goal began in earnest. The RAS was able to put together a small group of *purebred Arabian horses*, which were intended to be bred as a genetically closed population. The nucleus of the asil Arabian horse collection consisted primarily of horse that had belonged to Abbas Hilmi, who had been forced by the British into exile at the outbreak of World War I. The Khedive intended to support the Ottoman Empire in the war, and the British would not allow this to happen.

In early 1914 Khedive Abbas Hilmi was seriously injured in an attempted assassination while he was in Constantinople. He received lengthy and complex medical treatment. The British took the opportunity of his absence to declare Egypt a "Sultate under British protectorate". Abbas Hilmi was deposed in absentia and his uncle Hussein Kemal was made ruler of Egypt. Abbas was stripped of all personal property and banned from Egypt for life. Thus his excellent collection of asil horses that had been passed down to him by his family became available to the royal family and to the RAS.

In 1917, Lady Anne Blunt presented the struggling Society with her stallion Jamil, and two mares, Jamila and Radia. Lady Anne was by then very old and frail. Her husband was in residence at Newbuildings with his new young lady companion. Lady Anne's daughter Judith had occupied the family estate in England and now ran the Crabbet Stud, ruling over the remains of the once-great stables with unabashed arrogance, ruthlessness, and tyranny. It was a sad occasion when Lady Anne left England after her final brief visit to Crabbet in 1915. She returned to Sheykh Obeyd, living in seclusion. A few months later she donated her horses to the RAS.

Figure 21 - *The Blunt residence in England, viewed from the lake.*

The next year she was stricken with dysentery and was admitted to the Anglo-American Hospital in Cairo. In her frail state and with her advanced age, the disease proved fatal. She died, on December 15, 1917 watched over by a friend Phillip Napier. She felt, even to the end, irreconcilably divorced from her daughter and her husband, who learned of her death only later.

She was buried in the nun's cemetery on Jabal Ahmar. Her headstone reads "here lies in the Egyptian desert which she loved LADY ANNE BLUNT." It was composed by Wilfred.

Figure 22 - *The final resting place of Lady Anne Blunt in Egypt.*

Prince Kamal el Din Hussein served as the first president of the Royal Agricultural Society from 1914 to 1932. Omar Tousson followed him in the position of president beginning in 1932.

The first manager of the RAS stud farm was a Scotsman, Dr. A.E. Branch, a highly regarded horseman and veterinarian. His first task was to address the lack of sufficient numbers of high-quality stallions, and in 1919 he went to England to purchase horses from the Blunts, seeking those horses that were of Ali Pasha Sherif bloodlines. He bought 19 stallions, most of whom, but not all of whom were discarded after arriving in Egypt. The reason for this is not known. It appears that they Egyptians found them to be unsuitable for breeding purposes. However, he also bought two mares descended from Rodania that would have a significant impact on the program for the next 100 years. Dr. Branch served in this capacity until his retirement in the mid-thirties. Branch was followed by Dr. Ahmed Mabrouk as director, and he, in turn, was followed by Dr. Abdel Alim Ashoub. In 1949 the directorship passed to the European horseman General Tibor Von Pettko Szandtner. This was to be the golden age of Egyptian Arabian horse breeding.

Figure 23 - *The Pink House, by Sheykh Obeyd, the last home of Lady Anne Blunt.*

The collection began to grow in size during the 1920s, with the addition of horses from the Crabbet Stud in England in 1919 and gifts from the Manial stud of Prince Mohammad Ali Tewfik. The leadership of H.H. Prince Kemal el Din Hussein proved to be invaluable and the enterprise began to flourish.

Originally, the horses were stabled at the RAS experimental station at Bahteem, a small village at Heliopolis, near Cairo. In 1926 the Board of Directors made the decision to move the herd to a newly constructed facility in a desert setting, built on fifty acres at Alzahra, outside of Cairo. The new facility was named Kafr Farouk in honor of the son of the Egyptian King Fouad. Breeding plans and nutritional regimens were modernized. Current scientific methods pertinent to horse care and breeding were sought out and utilized. Experimental stabling using double layered adobe roofing was constructed to combat the extremes of heat in the summer. The main clients of the RAS in the early years were private American breeders and the Italian government.

Figure 24 - *Dr A. E. Branch. Photo used with the permission of the Forbis collection.*

Figure 25 - *Stallions from the Kafr Farouk Stud, Egypt, from the Journal of the Arabian Horse, 1936*

Figure 26 - *Stallions from the Royal Agricultural Society, Cairo, Egypt, from the Journal of the Arabian Horse, 1936.*

THE BEDOUIN AND HIS HORSE

Figure 27 - *Egyptian desert Bedouin at Giza. Photo from the Library of Congress.*

The directors of the Royal Agricultural Society adopted their conceptions of equine quality and assala (nobility) from a number of sources, one of which was the desert Bedouin themselves. Most of this knowledge was handed down from generation to generation through oral transmission. The textual records of Bedouin practices were fragmentary and incomplete, and those sources that do exist have less to say about the horse itself and more to say about the attitudes, opinions, and practices of the Bedawi and their highly cultivated feeling for the horse.

Bedouin traditions and folk history reflect a keen interest in both the origins and purity of the desert horse. Desert lore preserves tales of the long and often romantic association between the "children of the desert" and the authentic desert horse. The 19th century Bedouin himself saw nothing romantic or exotic about life in the desert; he called it a "death in life". However, it was his birth right, and he preferred his freedom and defiant independence in the wastelands of the peninsula to a life of relative comfort and safety in the cities under Ottoman rule.

Figure 28 - *Bedouin leading an Arab mare and foal. Photo used with the permission of the Raswan collection.*

Bedouin legend maintains that the horse has been bred under their protection for ages. The Arabic phrase describing a *purebred Arabian horse* of the desert is "al-hail al-arabia al-asila" or "the noble horse of Arabia". The Bedouin also referred to the desert horse as "*atiq*", a term applied to the horses of the first order of quality, faultless and born of noble parents. The noun "*asalah*" translates as "nobility". Among the Bedouin there was no ambiguity or vacillation when defining the term "*purebred* horse".

Figure 29 - *A Bedouin warrior. Image from the Library of Congress.*

The Arabic term "asil horse" had a very precise and specific meaning. To the desert Arab, an asil horse was one which traced in all lines in an uninterrupted manner to Bedouin *purebred* horses that originated in Nejd in Deserta Arabia, containing no genetic material from horses originating from outside the Arabian Peninsula. This commitment resulted from the desert dweller's innate awareness that the preservation of this horse with its valuable characteristics could only be achieved by maintaining genetic purity in breeding. Steadfastly avoiding the introduction of alien equine blood into his breeding population, the Bedouin was an insightful breeder. His success with the Arab horse was the result of acute sensitivity and intuitive appreciation of the rules of horse breeding necessary to maintain the desert horse in a state of genetic equilibrium. To deviate from the path of pure breeding was for the Bedouin the path to ruin.

The Bedouin concept of blood purity consisted of three elements: first, the *purebred* horse must be asil, noble and authentic. A horse must be 100% asil, not "mostly asil" or "probably asil". The entire pedigree must contain only asil horses whose ancestors were known with certainty by the elders of the tribe – known to them as far back in time as collective memory permits. The presence of even one, albeit remote, antecedent of foreign origin in a pedigree rendered the horse kadish (forever contaminated).

Secondly, the horse must belong to a recognized rasan, which is a "rope" or strain. This consists of a historical connection to a recognized female line of Arabian horses that is acknowledged by tribal authorities to be of stainless quality. This means that the horse must descend in the tail female line from a respected strain, recognized by the desert authorities as pure.

Figure 30 - *A Bedouin portrait. Image from the Library of Congress.*

Lastly, the horse must have a legitimate marbat, literally, the "place where the rope (rasan) is tied". This term is usually a personal name and refers to the specific individual or tribe who bred the horse.

The breeding of a mare was a serious matter to the Bedouin, and it was given considerable thought and care. The Bedouin himself, even from the earliest of times, was dependent on the reliability of orally transmitted pedigrees in making his horse breeding decisions. However, even the unlettered Bedouin realized that his assessment of the accuracy of a pedigree was of necessity very limited, relying as it did on the honesty and thoroughness of all those unknown breeders who came before him. This sense of uncertainty and distrust has been present in the Arab horse pedigree discussion from the very beginning, and probably accounts for the exaltations and compensatory protestations of honesty and truthfulness found in the orally transmitted history of each horse. While often over blown and highly fanciful these claims were made by the breeder of the horse and passed on to their subsequent owners as proof of provenance.

Figure 31 - *A desert halt. Image from The Raswan Archives.*

Among the Arabs, part-bred horses were common. There was a unique term used by the pre-Islamic Arabs to describe the cross-bred offspring resulting from breeding an asil mare to a foreign horse. This type of horse was called *hajin.* The foreign stallion could be a Turkish horse, a Kurdish horse, or a Turkoman horse, among others. In more modern times, the foreign sire might be an English Thoroughbred. These half-breeds were produced either for sale as stock which would be bigger and stronger than the asil mare herself, or may be produced to sell to unsuspecting outsiders as if they were "*purebred*". This allowed the breeder to avoid paying a large stud fee. Classical Arab horse breeders considered the practice despicable. The man who practiced *Hajjana al-khayl* was considered to be in violation of community standards.

The concept of the hajin had a different meaning when applied to the Arabs themselves. A hajin was a person who was the son or daughter of a slave and whose father was a high borne *purebred Arab* of noble lineage. Since the slave mother was usually a black African or Ethiopian, the half-breed child would be dark skinned, with the hair that was characteristic of an African native. He would be conspicuous among his people as a "mongrel". There were, of course, many lighter skinned and even white skinned slaves among the Arabs, including some of those kept by the Prophet himself. Hajin children produced from these women would be less detectable in their social spheres. Regardless of skin color, the social status of the hajin was considered to be low, along with slaves. Rarely, the father would publicly recognize the child as his own. However, this did happen on occasion. After the rise of Islam, the father was generally required to acknowledge paternity.

There were three very famous hajin Arab poets who lived during the pre-Islamic period. They were Antara, Kufaf (also called Ibn Nabda) and Sulayk ibn al-Sulaka. They were referred to as Aghribat al-Arab, or "ravens of the Arabs" because of their dark skin color. Their poetic works are still admired, read, and recited today.

Any discussion of the beliefs and attitudes of the Bedouin needs to include the caveat that the term itself refers to a very heterogeneous population of desert-dwelling pastoralist Arabs. Each Bedouin tribe was an isolate, and each differed in many ways from his fellow desert inhabitants. Some were very poor, and some were very wealthy.

Some were honest, others were not. Johann Burckhardt, traveling in the desert in the early part of the 19th century, often remarked on this fact. There are very few broad generalizations that can be made about the Bedouin that are quite accurate.

There was a distinct class structure among the Bedouin. The lower classes were those who herded sheep and goats, and the upper noble classes were those who raised camels. Powerful and influential leaders who became sheikhs were often so wealthy and influential that they were given the title of "Emir" or "prince." But the defining features that unified all of these individuals were commonality of blood, pastoralism, and a shared history.

Figure 32 - *An Arab Bedouin. Image from the Library of Congress.*

Figure 33 - *A Bedouin chieftain and tribal elder. Image from the Library of Congress.*

The Bedouin were a passionate, whimsical people, deeply rooted in the spiritual habits of pre-Islamic paganism. Their beliefs, mores, behaviors and practices were not uniform. They saw the influence of the supernatural in all of the phenomena around them. The djinn and the animistic spirit world were very real to them. They were superstitious about objects and places and consequently they gave names to everything that they encountered, since the ability to name potentially malignant natural forces gave the Bedouin a perceived degree of magical control over them.

Figure 34 - *An Arabian Horse, by Carle Vernet. Image used with the permission of the Olms Archives.*

> The Bedouin of the present day continue to bestow on the smallest hill, projecting rock, or little plain, a distinct and particular name; which circumstance renders the history of Arabia often obscure, as the names have, in the course of ages, sometimes changed.
>
> Johann Burckhardt

Even as late as the 20th century, the Bedouin were uniquely innocent. Carl Raswan wrote that in his early travels in Arabia the desert dwellers still considered themselves to be "Children of Ishmael" and believed that angels in disguise often came to visit them, blessing the horses and the family living in the tent. For this reason unexpected guests were never questioned about their personal lives, since such questions would be considered unsuitable and intrusive to a potential angel.

They were free, frank and bold people, warm to their friends and ruthless with their enemies. While they were an intelligent people, they tended toward craftiness rather than the exercise of reason and rational thought patterns. They were a very proud people, high-spirited, and generally good-natured and hospitable to each other and to strangers.

Figure 35 - *Arab Sheikh, by Leon Bonnat. Image used with the permission of theWalters Art Gallery.*

Figure 36 - *A desert warrior, "Drinker of the Wind," from The Raswan Archives.*

At the same time, the Bedouin were a long suffering people, hard bitten and pessimistic. Life for a pastoralist was not romantic and it was decidedly savage, and at times even vicious. The Bedouin was realistic about his lot.

The Bedouin was a dignified individual whose sense of self-worth and self-importance was clear and unshakeable. The Bedouin believed in his own resilience and capacity to endure suffering, and despite the brutality and poverty of life in the desert, he maintained his belief in himself and preserved the dignity of his family. He measured himself by his own yardstick.

The Children of Ishmael were proud people. The Bedouin held himself in the highest regard, superior even to the Sultan. When fights and disputes among the Bedouin occurred, but these were usually easily interrupted by a more pacific companion with the oft-repeated phrase "God has made us great sinners, but he has bestowed upon us the virtue of easy repentance." The main factors that united the Bedouin were their distrust of city dwellers and their hatred for the government in Constantinople.

But the horse was the central point of reverence for the Arab.

The rigor and dedication to the blood purity of the Arabian horse was conditioned by a religious element for the desert people in the post-Islamic period. In the Muslim world, the value of the *purebred Arabian horse* was echoed in the attitudes and the instructions from the Prophet Mohammed, who urged his followers to keep and maintain the asil desert horse in reverence and respect.

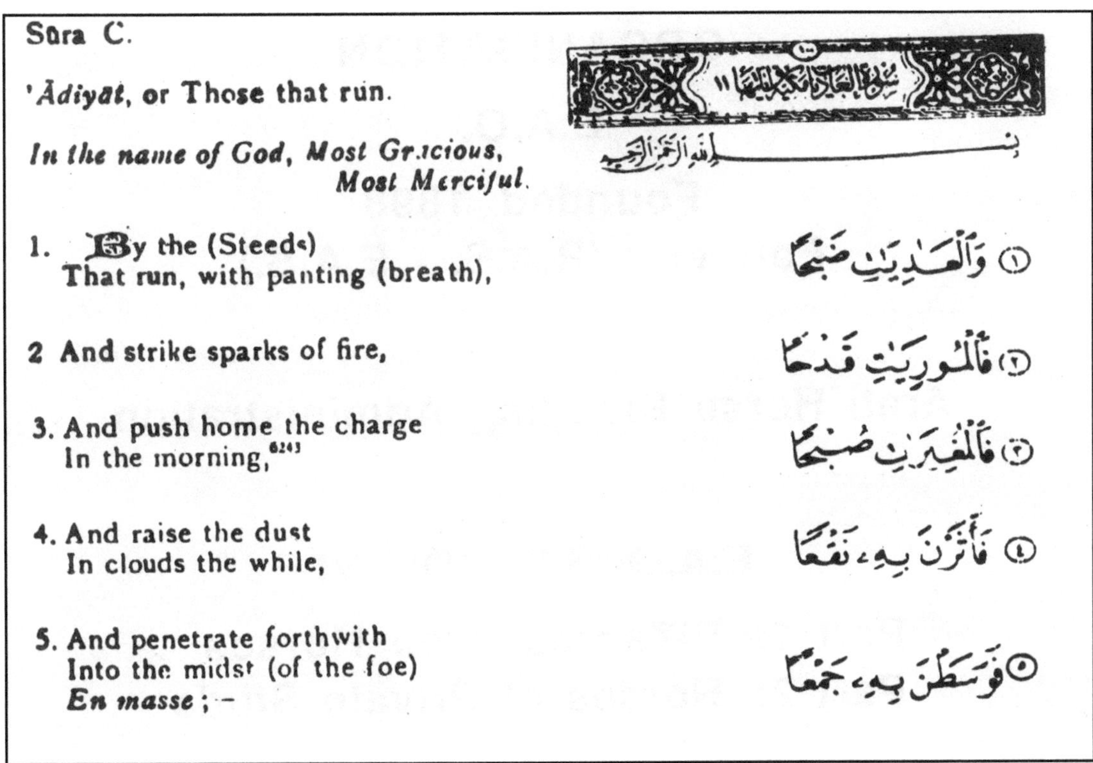

Sûra C.

'Âdiyât, **or Those that run.**

In the name of God, Most Gracious, Most Merciful.

بِسْمِ اللَّهِ الرَّحْمَٰنِ الرَّحِيمِ

1. By the (Steeds) That run, with panting (breath), — ① وَالْعَادِيَاتِ ضَبْحًا
2. And strike sparks of fire, — ② فَالْمُورِيَاتِ قَدْحًا
3. And push home the charge In the morning,[6243] — ③ فَالْمُغِيرَاتِ صُبْحًا
4. And raise the dust In clouds the while, — ④ فَأَثَرْنَ بِهِ نَقْعًا
5. And penetrate forthwith Into the midst (of the foe) *En masse*; – — ⑤ فَوَسَطْنَ بِهِ جَمْعًا

Figure 37 - *Sura 100, from the Qur'an. Image from the EAO* Stud Book, *Volume 6.*

From the time of the prophet Mohammed (570 CE – 632 CE), the written record establishing the central importance of the horse to the Arabs has been expansive. While the Quran itself contains no specific mention of a *purebred* or noble horse, the hadith, sayings and legends that are attributed to the prophet Muhammad, contain frequent references that illustrate the value that he attached to his horses.

He decreed that the breeding of the Arabian horse was a religious obligation. Mohammed referred to the asil horse as "the steed of the island of the Arabs".

In times of war, the Prophet allowed his warriors to pray on horseback.

The Prophet said that a believer who kept a *purebred* horse for God's sake will save his mother from the fires of Hell on the Day of Judgment.

A great blessing, he said, are a faithful wife and a pregnant mare.

Nothing leads a man more quickly to perdition, said the Prophet, that the selfish possession of a beautiful horse and a beautiful woman.

Mohammed forbade cross-breeding the noble horse of highest caste to common horses; he placed a high value on purity: "As many grains of barley as thou givest thy horse, so many sins will be forgiven thee."

The Prophet himself kept horses. Classical scholars wrote that he owned seven: Sakb, Sahab, Murtajiz, Lizaz, Durab, Lahif, and Ward.

In the Hadith, Mohammad is said to have encouraged very Muslim to raise and maintain as many horses as he could afford.

According to the hadith, gambling was forbidden among the pious. To the Prophet, gambling was cheating, which he understood to be betting on a sure thing. Horse racing was permitted. If a horse owner was challenged and knew that his horse was superior to the challenger's horse, then one may not bet money on the outcome, since this is gambling. If the challenger's horse is completely unknown to the other party, then one may bet money on the outcome as this is not gambling.

The hadith contain the following:

> The purely bred horse is a reward for one man, and protection for another, and a burden for a third man. The third man, the man who buys and keeps horses for the sake of pride, only in order to show them off, will find that they become a burden. "The horse shall be the reward of the man who lets it run free in meadows and gardens," and "The Evil One dare not enter into a tent in which a *purebred* horse is kept."
>
> The Prophet's depiction of Paradise states that "When thou shalt enter into Paradise thou shalt meet a ruby colored horse, and it will take thee up and fly with thee whithersoever thou wilt."
>
> According to legend, the Prophet spoke of his horses as he drew his final breath. As he lay dying, cradled in Aisha's arms, he reminded his followers of the value of the asil horse; "For a noble and courageous breed of horses are true riches."

Arab poets of the time wrote effusive passages describing the might and grandeur of the bands of Islamic warriors. They wrote of:

> The dark shadows thrown from the shining crowd of mounted men gliding across the face of the desert, while the hum of their prayers rise above their spears and lances like a cloud that has found a voice to speak and chant to the Unseen.

The teachings of the Prophet and the spread of Islam brought a new interpretation of the horse and its origins. In pre-Islamic Bedouin lore, legend held that God decreed the creation of the first desert horse, and instructed the angel Jibrail to present this horse to Ishmael, the man of the wilderness. Jibrail descended to earth and caused a whirlwind of red dust to appear to the sleeping Ishmael, a formless form, burning the sand with its feet, scattering the sand with blasts from its nostrils, and screaming with ferocity. Out of this confusion there was condensed a creature, the most handsome creature that man had ever seen, prancing and running, noble and free: the world's first Arabian horse.

Figure 38 - *Al Buraq, the mythic steed of the Prophet, a heavenly creatures said, only by later traditions, to have the face of a human. He was tall, and white. Traditions also state that he had two wings on his thighs. The Prophet rode Al Buraq on the Night Journey through the seven heavens, receiving instructions from Allah. The Qu'ran states the Al Buraq "did take His servant for a journey by night from the Sacred Mosque to the farthest Mosque, whose precincts We did bless…" Sura 17.*

Jibrail then spoke to Ishmael:

> "This noble creature of the dark skin and painted black eyes is a gift of the living God to serve you as a companion in the wilderness, and reward you, because you have not defiled yourself with pagan gods, but remained in the faith of your Father Abraham."

Figure 39 - *The Prophet Mohammed receiving the first revelation from the Angel Gabriel as he meditated in seclusion on Mount Hira, during the month of Ramadan.*

Creation legends abound. Another says:

> "When God was about to create the horse He said to the south wind: become solid flesh, for I will make a new creature of thee, to the honor of My Holy One, and the abasement of my enemies, and for a servant to them that are subject to me."

The prophet Mohammed gave his followers this explanation of the creation of the Arabian horse:

> When God wished to create the horse, he called the south wind and said to it, I want to form a creature from thee; condense thyself and abandon thy liquidity, and the wind obeyed. Then God took a handful of the now tangible material, breathed upon it, and the horse emerged.

The Qur'an relates the creation of both man and horse:

> He created man from a clot . . . and horses too . . . for you to ride upon and for an ornament. (Chapter XVI)

The Bedouin was above all a practical horse breeder and attached great importance to matrilineality, the tail female line of descent of his horses. The stallion lines were relegated to a secondary position in the considerations of purity. Because of the conditions under which horses were kept in the desert, it was not unlikely that a stallion could breed a mare without the owner's permission. This could even take place without the owner's knowledge. The possibility of sire confusion was evident. When a foal was born, it was very unlikely that tribal members would become confused about the identity of the mother. Therefore, the Bedouin tacitly acknowledged this point of potential confusion by emphasizing the continuity of the female line as the most reliable and accurate element of the pedigree.

The Bedouin's relationship with the horse was paradoxical. The Bedouin did not breed the horse for beauty, although it was a thing of beauty. He did not breed the horse because it was profitable business, although it could be that too. He did not breed the horse because it was a noble creature coincident with his own values of honor, integrity, and self-reliance, although it was that too. He bred the horse as a matter of personal survival, and he used the animal skillfully, pressing him to the limits of endurance in the never-ending struggle to survive the raids and retaliatory wars that characterized life among the tribes of the Arabian Desert. The Arab horse was beautiful, valuable and noble, but to the Bedouin he was first and foremost a tool of evolutionary adaptation. The Bedouin used the horse as a tool for his own survival. Life in the desert was, among other things, Darwinian. And it was harsh. The tribes rarely existed in a state of peaceful tranquility. More often than not, inter-tribal enmity led to constant raids and counter-raids, some to recover booty and some to regain the honor of a disgraced tribe.

Figure 40 - *Bedouin on a ghazu. Photo used with permission of the Raswan collection.*

The swift durable horse was not only a defensive and offensive tool of inter-tribal warfare; it was also a deterrent to aggression. Those tribes that possessed the largest number of the hardiest horses would not likely be attacked by their neighbors since the capacity for swift and severe retaliation was so evident. To the Bedouin, there was only one primary measure of quality, and that was the capacity of a horse to endure hardship and perform under duress. A warrior simply could not entrust his life or the lives of his family to a war mare of inferior endurance. He must have an asil mare as a matter of survival.

The Arabian Peninsula natives continue to feel this strong affinity for the Arabian horse today. Lt. Col. Khaled Hamad el Merri managed the police and Royal stables in Ras al Khaimah, and expresses the feeling this way:

> "An Arabian horse is just perfect, in its figure, in its proportions, in its endurance, its loyalty . . . they are among the most intelligent, proud animals that I have come across."

Based on this information, one can infer that:

1. The horse was an integral part of Bedouin life at least as early as the 6th Century CE.

2. The Bedouin firmly believed that they were practicing pure breeding of the desert horse. That is to say, they recognized that they existed in isolation from the outside world and valued this isolation. They went to great lengths to keep their horses from being bred to non-Bedouin horses.

Figure 41 - *An example of the Arabian war mare in the central Arabian Desert. This mare was photographed in 1927 after having survived many weeks of long distance raids against a neighboring tribe. Though badly underfed and thin, she possessed the vigor and durability of an asil Arabian horse.* The Raswan Index.

Figure 42 - *Attala, a member of the Schamar tribe, from The Raswan Archives.*

Figure 43 - *A Bedouin of the Nefud Desert."Only the night and my mare know me"from The Raswan Archives.*

Figure 44 - *The pedigree of the tribal founder Magran. The Bedouin regarded their ancestry with reverence and attention to detail. Their horses were given the same attention to ancestry.*

Figure 45 - *From Oppenheim, Der Bedouin. The principal warriors of the Schamar.*

Figure 46 - *Mnahi, chief of the slaves of Amir Fuaz, from The Raswan Archives.*

THE WESTERN DISCOVERY OF THE HORSE OF INNER ARABIA: THE WRITTEN HISTORICAL RECORD

Explorers, journalists, and travel writers came to Arabia from ancient Greece, embarking on daring and dangerous journeys to map the width and breadth of the unknown world. They came during the Renaissance as adventurers, living in a European world that was re-inventing itself by discovering the wisdom of the ancient world. They were seeking the manuscripts of the ancients, the Arabic scholars, and the fabled libraries and universities of old Arabia. They came during the Victorian era seeking excitement, glamour and glory, hoping to make ethnologic and archaeological discoveries that would assure their fame as authors of exciting travelogues. Many of these adventurers were supported and financed by the Geographical Societies of Germany, England and France. Some of them were diplomats from Europe, assigned to the Middle Eastern embassies. Some were soldiers of fortune. Their works form the body of literature on the discovery of inner Arabia.

The early directors of the RAS had some limited knowledge of the Arabian interior. Beyond the legends and tribal lore regarding the *purebred Arabian horse*, the directors of the RAS had very little in the way of objective guidance in their analysis of the question of the purity of the horses in question. They had little textual basis for defining the concept of "genetic purity", and little objective evidence to support claims of blood purity. They did have the writings of European travelers and journalists who had traveled the length and breadth of the peninsula.

Some authorities, however, suspected that the Arab horse originally came to Arabia from central Asian or African sources, and fragmentary historical texts tended to support this idea. If the Arab horse was introduced into Arabia by human efforts, then the case for purity became more muddled. If the horse originated in central Asia and was later introduced into Arabia and Egypt, then it would be difficult to assert that the horse was in any sense a "*purebred*", given the number of differing horse types that were then present in central Asia. The surviving texts from the ancient world shed little light on the problem.

The real problem in researching the written record left by Arabs from this period was its nearly complete absence. The interior of Arabia had existed in complete isolation from the rest of the ancient world, and until recent centuries virtually nothing was known about the people and places of the interior. No enduring records or histories were written; the natives of the interior were pre-literate. Figure 47 illustrates the magnitude of this isolation. At a time in history when the geography and cultural dimensions of Egypt and the Levant were well known to Herodotus, the geography of the interior of Arabia was a complete blank. There simply was no recorded history of the land, the people, the customs, or the horse. The island of the Arabs was impenetrable.

Figure 47 - *Geography of the Near and Middle East at the time of Herodotus.*

The earliest surviving textual fragments written by Arabs concerning the Arab horse are only a few centuries old. A few examples of fragmentary pre-Islamic Arabic poetry have survived from the first millennium. The 5th century CE poet Amralkais wrote: "At the first light of dawn I get up and mount a short coated noble steed, which overtakes the animals of the forest at the gallop." The 6th century CE poet Amru, whose documentary fragments survive, wrote: "We inherited the noble horses from our ancestors, and after our death, our sons will own them." These two references warrant special attention as they represent the earliest written mentions of the horse in Arabia, written by Arabs.

Histories written by scholars of the ancient world contain occasional references to the Arabian horse but provide no solid clues as to its origin. It is difficult to find evidence that the Bedouin of prehistory had domesticated the horse at all.

The land of Arabia itself was isolated and largely unknown to ancient Asians and Europeans. Herodotus, writing in the 5th century before the Common Era, was intimately familiar with the geographic details of Egypt and North Africa, but was nearly silent on the subject of Arabia. He describes the seas that border the country but says nothing of the interior of the peninsula. He noted that during the 6th century BCE, the Arabs had granted safe passage through their land to Cambyses as the Persians sought to subjugate Egypt. The Persians showed no interest in the conquest of Arabia and remained at peace with the Arabs. He records that the Arabians annually paid 1000 talents of frankincense to the Persian Empire as tribute. The normally prosaic Herodotus was entranced by the perfumes and scents of the peninsula, writing that "the whole country exhales a more than earthly fragrance." Herodotus never mentions the presence of horses among the Bedu.

Writers in the ancient world rarely mentioned the Arabian horse although they occasionally wrote about the Arabian people. The Greek travelers and writers Diodorus and Strabo notably do not even mention the horse in their discussions of the Arabian people, but always portrayed the Bedu as a camel breeding and camel riding people. However, authors writing after the first century CE do begin to note the presence of the horse and its importance to the Bedu. The Greek historian Oppian, writing in the second century CE, speaks of the value of the horse in Bedouin life, commenting on its usefulness in hunting. This is the first surviving written account of the Arab horse.

The Roman historian Ammianus Marcellinus wrote during the 4th century, and describes the Bedouin of northern Arabia as "appearing out of nowhere on their fast slender horses". Another 4th century writer, a veterinary researcher and author named Vegez, describes the horse of the Sappha-rene (Dhofar, southern Arabia) as being of very high quality. The latter is the first evidence in the documentary record in which the desert-type horse is described as being present in southern Arabia.

A reference to the desert horse has come down through Buddhist literature. A surprising and peculiar description of the desert horse occurs in the Fo-Sho-Hing-Tsan-King, a Chinese translation of a life of the Buddha, written in India by Asvaghosha Bodhisattva about 80 CE, in which he describes the miracles that surrounded the nativity of Siddhartha, son of Suddhodana, about 500 BCE. From all corners of the world great and glorious honors and tributes appeared at the feet of the infant in a mystical profusion, spontaneously presented at the feet of the Buddha: "Without noise, of themselves, (they) came; not curbed by any, self-subdued, every kind of colored horse, in shape and quality surpassingly excellent. With sparkling jeweled manes and flowing tails came prancing round, as if with wings; these too, born in the desert, came at the right time, of themselves." This reference from the 1st millennium BCE is remarkable for the characterizations of the desert horse that would become so common in the literature and texts of later centuries in the Middle East.

Figure 48 - *Syria and Mesopotamia.*

MODERN ARCHAEOLOGY AND THE SEARCH FOR ORIGINS OF THE ARABIAN HORSE

Researchers and archaeologists have continued to sift through the record in search of further clues concerning the origins of the Arab horse. The answer as to whether or not the light desert horse was endemic to the Arabia Peninsula remains unclear. Researchers have pointed to the development of the light chariot (from the old French chariot, derived from the Latin *carrum*, car) and suggested that it was probably powered by a "light horse", perhaps an "Arabian-like" horse. This is, however, a matter of speculation. The archaeologic record has revealed that the military light chariot with spoked wheels and a single axle first appeared in Khazakstan in central Asia about 2000 years BCE, and the concept spread rapidly, reaching China by 1600 BCE and reaching India, Anatolia and Europe by 1500 BCE. The type of animal that powered it is not known.

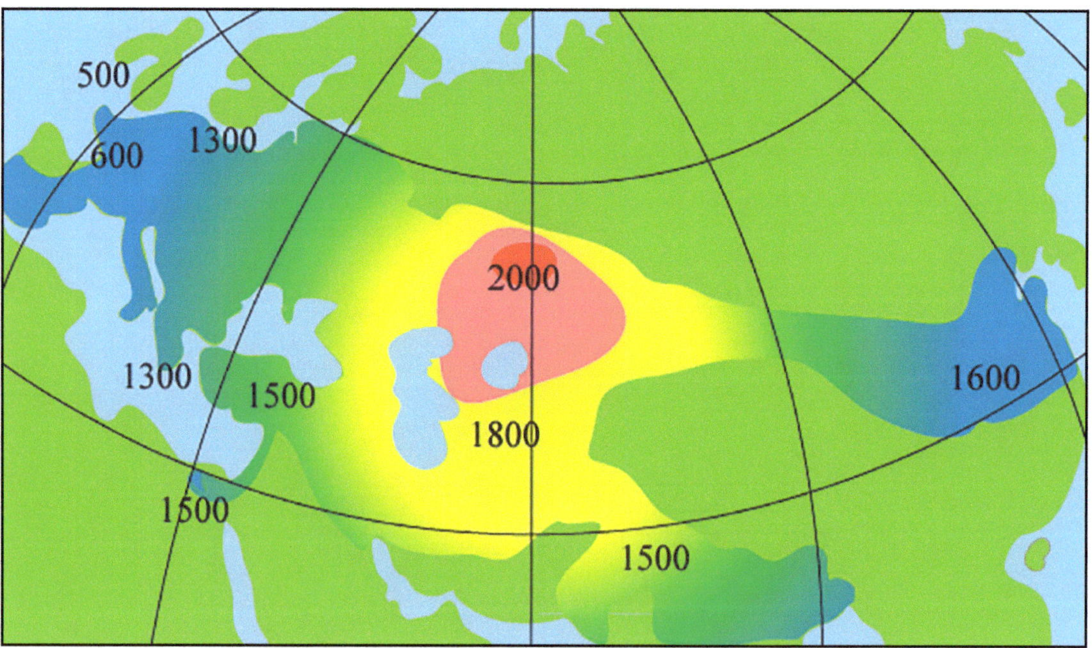

Figure 49 - *The spread of chariot technology in the ancient world (time expressed in years BCE).*

Figure 50 - *The standard of Ur associated with the Sumerian King Ur Pabilsag, who died about 2500 BCE. The chariots are cumbersome carts, with two axles, the wheels are made of solid wood, and onagers are used to pull it.*

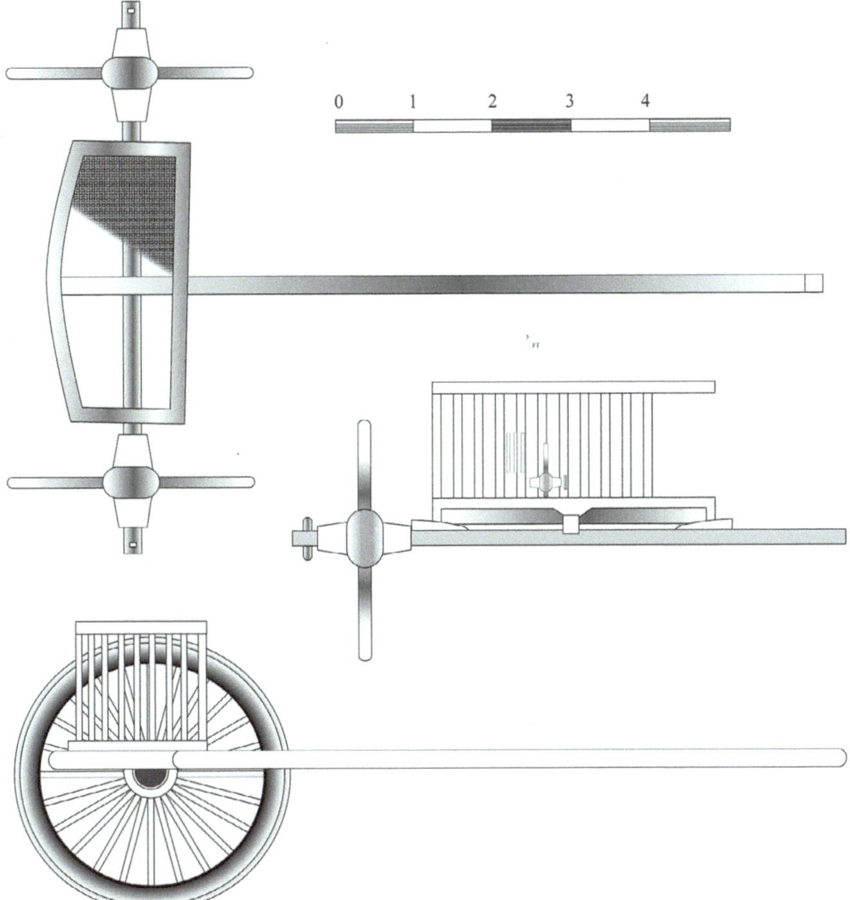

Figure 51 - *Chariot construction from the earliest examples uncovered in Khazakstan, produced approximately 2200 BCE by the Sintashta-Petrovka Proto-Indo-Iranian civilization.*

The earliest clear evidence of the presence of the Arabian horse in the Near East is found in Egypt. There have been long historical associations between the light desert horse and Egypt; the light chariot existed there by 1500 BCE. The desert horse has been associated with Egypt in the lore of the Near East for several thousand years. Homer wrote of the significance of the horse in Egypt in 1200 BCE. In The Iliad, Achilles refers to the "Thebes of the hundred gates, whence sally forth two hundred warriors through each, with horses and chariots". However, theories concerning how the light desert horse was introduced into Egypt remain speculative.

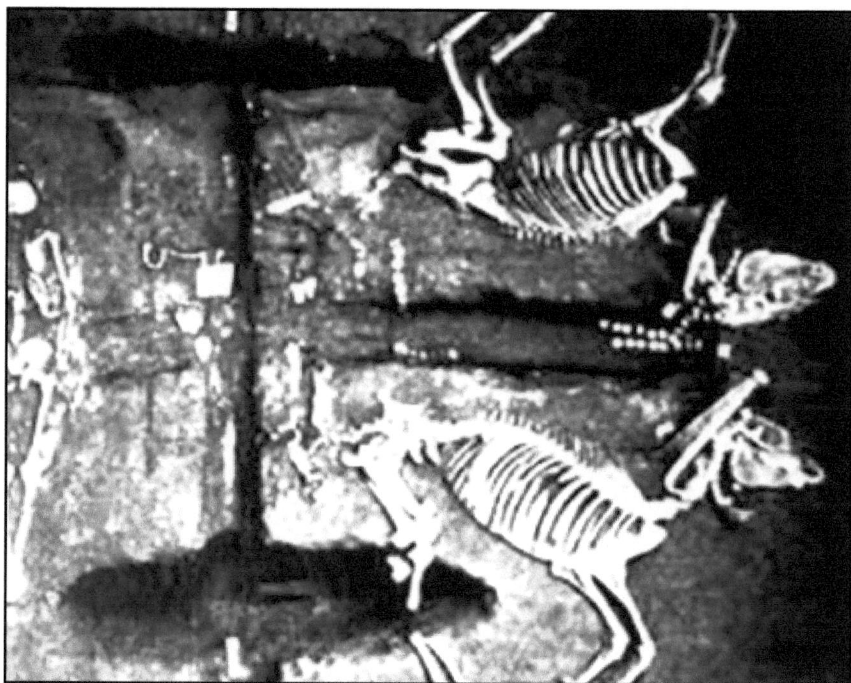

Figure 52 - *Funerary tumulus contents from this same time period, showing the remains of the driver, the chariot, and two horses.*

It is known that the Hyksos possessed the light desert horse and single axle war chariot prior to their occupation of Egypt. They were a Semitic language-speaking people. The Jewish historian Titus Flavius Josephus, writing in the first century CE, speaks of them as "Arabians", while other writers of the time associated them with the Amalekites (desert Arabs); it is plausible to assert that they were. Raswan referred to the Hyksos as the shepherd kinds of Arabia who brought their horses to Egypt. Whatever the case, they were not a "conquering horde of Asiatics", but they did rule Egypt from their capital in the Nile Delta, known as Avaris, from 1783 BCE to 1550 BCE. Their rule was based on a gradual long-term infiltration of the Egyptian civilization which was weakened at the time by dynastic struggle. The evidence suggests that the Hyksos overwhelmed the Egyptians not by armed struggle but by the mere threat of arms. The Hyksos brought with them new and intimidating military technology: horse-drawn chariots, metal helmets, chain mail, composite bows, and improved lethal metal arrow tips. They possessed a significant technological edge.

Figure 53 - *A mural graphic from the tomb of an Egyptian official during the 12th dynasty, showing unidentified Asiatics thought to be Hyksos.*

Recent archaeologic work in Egypt has shed light on the period of Hyksos rule. In 2009, excavations at Avaris, the Hyksos capitol, were carried out by the Austrian Institute in Cairo and have revealed necropoli containing the inhumed skeletons of a number of female horses, along with their human royalty, making these the oldest known horse specimens found in Egypt. There is, however, no graphic depiction in the Hyksos records or monuments to illustrate the type of horse brought by the Hyksos into Egypt. It can be inferred but not proven that the type of horse that they brought was a "desert type" or Arabian horse.

The subjugation of the Egyptians by the Hyksos was brought to an end by King Ahmose, who led the Egyptian forces out of Thebes and in a series of military campaigns succeeded in expelling the hated foreigners from their lands. The sole remaining documentary evidence of this victory is found on the walls of the tomb of one of Ahmose's warriors: "How I was decorated with gold. During the siege of Avaris the King noticed me and promoted me, and we took Avaris. I carried off four people, a man and three women. His majesty let me keep them as slaves."

While there is no documentary evidence that the true Arabian type horse was present in the central Arabia deserts before the turn of the millennium, there is ample proof that this type of light horse was an integral part of the life of the royalty and nobility in Egypt from a very early date.

Figure 54 - *Ramses II portrayed in his funerary temple in Luxor. He is depicted as the warrior king at the Battle of Kadesh. Depictions of chariot horses are consistent with the phenotypic features of the Egyptian Arabian horse of today.*

The Egyptians produced no durable pictographic evidence of the horse and chariot prior to the Hyksos occupation, but the Egyptian pictographic evidence began to clearly show that the Egyptian had adopted single axle chariot technology after the Hyksos overlords were expelled from the country. The horse portrayed in Egyptian temple art after the expulsion of the Hyksos is morphologically and phenotypically reminiscent of the desert horse, the horse known today by the name "Arabian".

The earliest pictorial representation of the horse of distinctly Arabian features appears in Egyptian temple friezes during the reign of Ramses II, during the 18th dynasty (1580-1350 BCE).

Figure 55 - *Mural art depicting a caparisoned horse, from the mortuary temple of Ramesses III, at Medinet Habu on the west bank at Luxor, from the 20th dynasty, about 1200 BCE. Photo by Polly Knoll.*

While there is evidence of earlier horse domestication in the Middle East, the portrayals of these early horses suggest that the stock was of stocky and short legged tarpan-type body habitus, not the dry, gracile, desert type of the Arabian horse.

The British Museum collection contains a tempera painting on papyrus of a desert horse, present in Luxor around 1400 BCE. The same collection contains a depiction of a desert horse of noble bearing in a fragment of a painting on plaster from the Luxor tomb of Sebekhotep, a treasury official, about 1400 BCE. A similar frieze from the tomb of the nobleman Nebamun in Luxor shows a desert horse of distinct Arabian type.

The temple parietal carvings of Medinet Habu in Luxor depict a similar horse, present during the reign of Ramses III, about 1186 BCE.

The seven images that follow were taken from The Monuments of Egypt; The Napoleonic Edition, published by the Princeton Architectural Press and used here with their permission.

The Napoleonic fleet sailed from Toulon on May 19, 1798, bound for Egypt. Along with the 13 warships and 17,000 troops, Napoleon ordered a veritable army of engineers and artists to accompany the expedition. This research arm of the expedition consisted of 151 civilians who were to document every detail of the ruins and artifacts of Egypt. There were 8 artists, 14 surveyors, 14 civil engineers, and 27 printers among the group. There were also 4 architects and 2 archaeologists. The following plates were among the results of that undertaking.

The horses of the King are depicted in uniform fashion; dished face, large dark eye, delicate muzzle, arched swan-like neck, and a flagging tail.

Thebes was inhabited as early as 3200 BCE, and is the site of the modern city of Luxor. The Karnak Temple Complex in on the east bank of the Nile, and Medinet Habu, the Mortuary Temple of Ramesses III, is on the west bank.

Figure 56 - *Bas relief from the palace gallery at Thebes, Medynet Abou.*

Figure 57 - *Detail of the harness of the horse from the Karnak Palace at Thebes.*

Figure 58 - *The Hunt, from Medynet Abou, Thebes.*

Figure 59 - *Battle scene depicting the King fending off an attack from the sea, from Medynet Abou at Thebes.*

Figure 60 - *The King at War, from the Temple of Karnak at Thebes.*

Figure 61 - *The King at War, from the Temple at Karnak, Thebes.*

Figure 62 - *Image depicting the King engaged in a military action, from the Palace at Karnak at Thebes.*

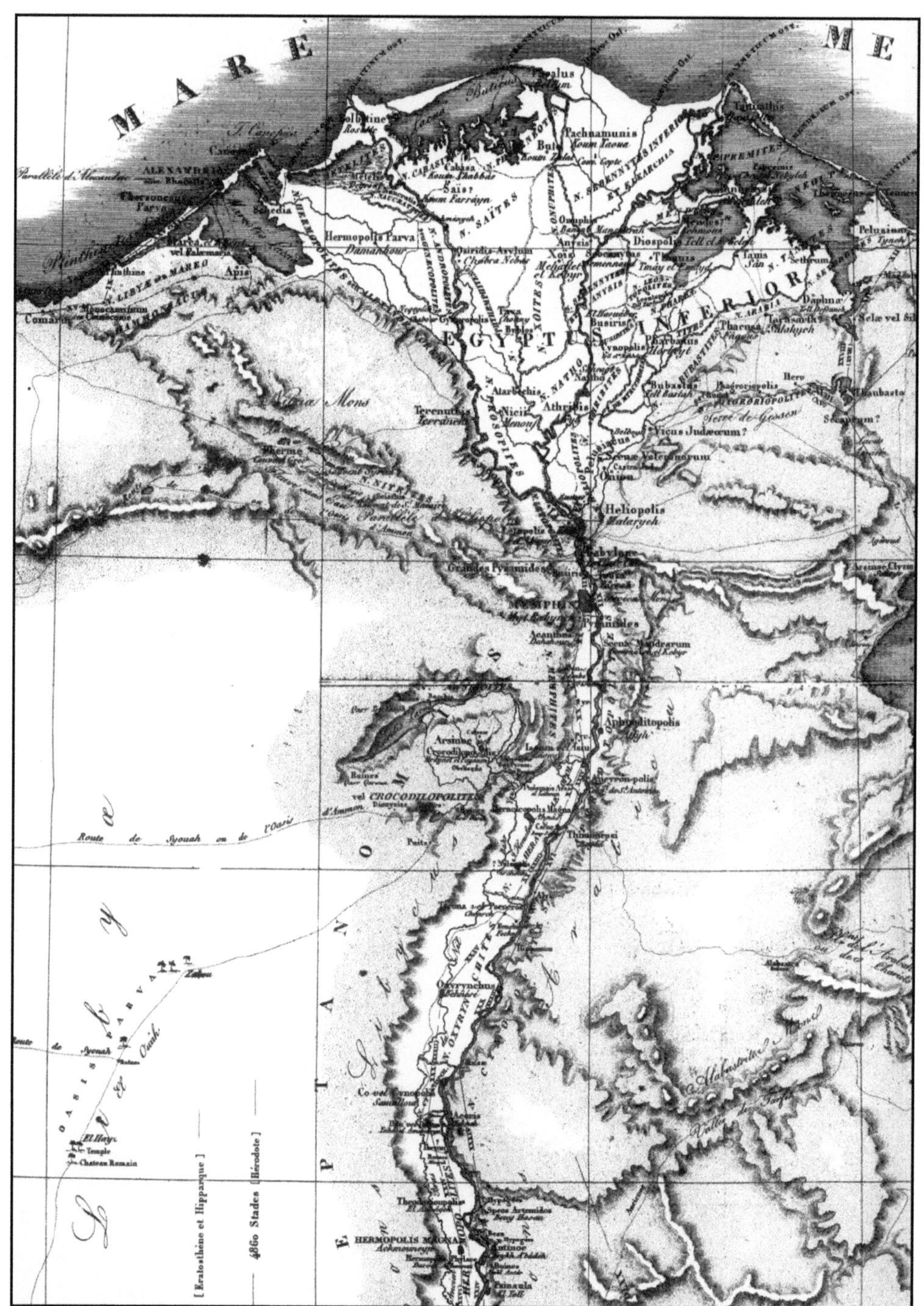

Figure 63 - *Map of Egypt, from the Napoleonic survey.*

Figure 64 - *Asiatics presenting tribute to an Egyptian King, from the tomb of Sobekhotep at Thebes, from around 1400 BCE. From the collection of the British Museum*

Modern research also turns to the examination of rock art in Arabia. The region of Shuwaymis is located about 370 km southwest of Hail in southern Arabia. Extensive petroglyphs have been found there and many of these depict horses. However, it is very difficult to date the ages of rock art, and these finds do not yield any definitive proof of the presence of the horse in Arabia in prehistory.

In 2011, archaeologists with the Saudi Antiquities commission reported excavations in Al Maqar which revealed evidence of Neolithic settlements and evidence of horse domestication 9,000 years before the present. Research Professor Ali Al-Ghabban stated that DNA analysis and C14 dating established both the age of the find as well as genetic information about the human inhabitants. The artifacts were discovered near Al-Maqar in the Asir region of southern Arabia but the results of analysis have not been formally published. If these finds are substantiated, this site will predate any known sites of early *equid* domestication in Asia by several thousand years.

The al-Magar finds do little to expand on the inquiry into the origin of the Arabian horse. The stone figures of al-Magar represent *Equus asinus*, the wild ass, not *Equus caballus*.

Figure 65 - *Al-Magar stone figure, Saudi Commission for Tourism and Antiquities.*

SCHOLARSHIP AND THE QUESTIONS OF THE ORIGINS OF THE ARABIAN HORSE

More substantial information concerning the origin of the Arabian horse began to come to light with the flowering of Islam. Arab scholarship was maturing and Arab explorers had mapped most of the known world. The earliest surviving documentary reference to the Arab horse comes from the Arabic-speaking explorer Ibn Battuta, (1304-1369) who, while traveling in southern Arabia, noted that the town of Dhofar in present day Yemen was a center from which thoroughbred (*purebred Arabian horses*) were regularly shipped to India.

The Muslim geographer and historian Leo Africanus was an Arab (born Al Hassan Ibn Muhammad Al Wazzan, in Granada) and reported his personal observations in Cosmographia Del' Africa, published in 1520 CE, where he states:

> "Horses of the greatest swiftness and agilities are in the Arab tongue called throughout all Egypt, Syria, Asia, Arabia Felix and Deserta by the name of Arabian horses. Historiographers affirm that this kind of wild horse ranged up and down the Arabian Desert, ever since the time of Ishmael have so exceedingly multiplied and increased that they have replenished the most part of Africa whose opinion savoureth of truth, for even at this present (1512 A.D.) there are great store of wild horses found both in African and Arabian Deserts."

As late as the 1500s, it appears that wild Arabian horses freely roamed the vastness of the Arabian Peninsula. Regarding his experience with the Barb horse in North Africa, he wrote:

> The horses of Barbary are no different from any other common horses;

> the choice, fast race horses, on the other hand, are Arabians, that is, they are bred by the Arabs who are living in Africa, and their origin is traced back to Arabia.

The German orientalist K. W. Ammon wrote Historical Reports on Arab Horse Breeding in 1834, in which he summarized a number of very rare reports then available from Europeans who had witnessed the customs and practices of the Arabs first-hand during the 17th and 18th centuries. In his book, Ammon states emphatically that the peninsular Arabs did not have horses in the pre-BCE period, citing a review of the literature on the subject. He surmises that the desert horse came into the possession of the desert Arabs from elsewhere, in the centuries early in the Common Era. He bases his opinion on the absence of references to the horse in the preserved writings of the Sumerians and the Hebrews. Donkeys, camels, and asses are mentioned but not the horse.

An exception to this generalization would seem to be found in the Hebrew Bible in the *Book of Job*. The author of the *Book of Job* is not known, but it is known that the book was written several centuries after the events described in the book. The book contains a passage concerning the war horse, clearly, an *Equus caballus*, describing the animal with vivid clarity. But the author uses this description only as a metaphor to describe the power of Jehovah and, by contrast, the feebleness of mortal man. Perhaps the author could have seen such a sight in another time, and in a different part of the Middle East. The passage does not suggest that either Job or his companions had ever seen a horse, much less raised or used horses. Or perhaps it was simply a later addition to the book.

He paws in the valley, and rejoices in his strength;
He gallops into the clash of arms.
He mocks at fear, and is not frightened; Nor does he turn back from the sword.
The quiver rattles against him, the glittering spear and javelin.
He devours the distance with fierceness and rage;
Nor does he come to a halt because the trumpet has sounded.
At the sound of the trumpet he says 'aha!'
He smells the battle from afar, the thunder of captains and shouting.

(Job 39:21 New King James Version)

Figure 66 - *Job's Evil Dream, by William Blake.*

Job was in all likelihood an Arabic Emir, says the author, who had donkeys, onagers, and asses but no horses. Job lived in the land of Uz, a desert region in northern Arabia coincident with the Syrian Desert. He seemed to have quite a large number of oxen. And Job was a man of immense wealth.

The Madianites (Midianites) in the age of the Judaic Judges were peninsular Arab raiders who frequently invaded Palestine, but they had no horses, only camels.

Ammon further supports his conclusion by referring to the barren terrain and hostile wastes of Arabia.

> "Certainly our Creator in his wisdom provided each species initially with a home offering in abundance everything it required for its nourishment and subsistence, and most likely He did not place the horse originally in the barren deserts of Arabia, where the grass it must feed on is very scarce, and the water often cannot be found."

Figure 67 - *A Syrian Bedouin horseman from Damascus, early 20th century, image from the Library of Congress.*

Modern documentation of the customs and values of the aboriginal Arabians began with the progressively more daring and detailed incursions into the interior of the Arabian Peninsula by Swiss and English adventurers and travelers. Their reports frequently mention the Nejd as the archaic home of the Arab horse. During the 19th century Europeans began to explore the Arabian interior, some out of a desire for adventure and some to pursue amateur anthropologic research. These travelers were not horsemen, but all were affected by the nobility and beauty of the horses of the Bedouin and wrote on the subject at length. They were especially struck by the importance that the desert natives attached to the subject of blood purity.

Figure 68 - *William Palgrave, National Portrait Gallery, London.*

William Gifford Palgrave's book Narrative of a Years Journey Through Central and Eastern Arabia (1862-1863) was one of the most significant and controversial works of the time. Palgrave's work helped shape the European horsemen's ideas about the Eastern horse. He was regarded as an authority by his contemporaries since he was the first European to enter the walls of both Riyadh and Hail. This was all the more remarkable when viewed in the context of the ongoing radicalism of the Wahhabis and the bloody reprisals of the Ottoman government. His enthusiasm on the subject of horses, however, influenced the subsequent journeys to Arabia by Col. Guarmani, and Lady Anne and Wilfred Blunt.

Figure 69 - *Bedouin warrior, early 20th century, photo from the Library of Congress.*

> "Nejed is the true birthplace of the Arabian steed. The primal type, the authentic model...The genuine Nejdean breed is to be met with only in Nejd itself, nor are these animals common even there; none but chiefs or individuals of considerable wealth and rank possess them...Nejdee horses are especially esteemed for great speed and endurance of fatigue; indeed in this last quality none come up to them...The rider on their back really feels himself the man-half of a centaur, not a distinct being."
>
> William Gifford Palgrave

The advances of science in the 19th century provided researchers with new tools for developing hypotheses on the origins of the Arabian horse. Modern speculation about both the origins and the purity of the Arabian horse began with the English Victorians, as developments in archaeology and evolutionary biology began to open the field to scientific examination. The fundamental laws of genetics were known and the fields of comparative anatomy and paleontology were maturing. Cambridge Professor of Archaeology William Ridgeway (1853-1926) postulated that the Arab horse was derived from North African native stock originating in Libya. He assigned the name *Equus libycus* to the horse as a result of his analysis of coat color differences that were observed among living Arab horses. Based on faulty genetic assumptions, his theory was not broadly accepted.

In his day, however, Ridgeway was considered an authority on the subject. The British Museum curators followed his opinions. In Museum collections, the Arabian horse was described as the Southern horse, *Equus caballus asiaticus, Equus caballus libycus*, and *Equus africanus*. The Museum exhibited the skulls of the Blunt horse Jerboa and the Prince of Wales horse Dwarka. Skull features said to be characteristic of the Arabian included the short skull length, the sinuous profile of the face, great width between the eye sockets, the narrow muzzle, and the great width and depth of the lower jaw. The articulated skeletons of Arabians and the Thoroughbred stallion Eclipse were on permanent display. The anatomic plane of the pelvis of the Arabian was set more horizontally than in other breeds. This feature was regarded as being related to the speed of the horse and, when transmitted to the breed known as "Thoroughbred" accounted for its remarkable speed in short distance track racing.

Lady Wentworth (Judith Blunt-Lytton, 16th Baroness Wentworth, 1873-1957), noted English Arab breeder and tennis player, claimed that the Arab horse was a unique species, endemic to the Arabian Peninsula and different from all other *equids*. She assigned the name *Equus arabicus*. Her claims were based on insupportable osteologic comparisons and found little support in the scientific community. Lady Wentworth had no scientific education and a fanciful imagination.

Figure 70 - *Wilfred Scawen Blunt, National Portrait Gallery, London.*

Wentworth's father, the British Arabian horse breeder and authority Wilfred Scawen Blunt, subscribed to the more broadly accepted theory. He wrote in his diary on July 13th, 1909:

> Spent the day at Caxton's (Newbuildings) with Professor Osborne, head of the Natural History Museum in New York, and his daughter. He is a highly intelligent man, a pupil of Huxley's and is writing a book on horse history. He has read my controversy with Professor Ridgeway on the origin of the Arab horse in Arabia and takes my side in it, holding the Arab to have been the descendant of a distinct wild breed in the peninsula.

Figure 71 - *Anne Isabella Noel Blunt, 15th Baroness of Wentworth - Lady Anne Blunt.*

Wentworth's mother, Lady Anne Blunt, expressed the same opinion held by her husband Wilfred; the *purebred Arabian horse* existed in a wild and feral state in the interior of Arabia before being captured and domesticated by humans.

Elements of ancient Bedouin history have been preserved by the Arabian horse historian Carl Reinhard Schmidt (later Raswan), a German author born in 1893 who spent 30 years living and traveling among the desert Arab tribes. His collection of traditional testimony gathered from the desert tribes relates that the ancient Arabian horse originated before the time of recorded history in southwestern Arabia in the region of Imran (now called Yemen) near Wadi Dawasir, Wadi Nejran, and Wadi Jauf. According to this legend, the wild horse then migrated from Yemen and Asir, propagating throughout Arabia and northeast Africa. Tradition calls these horses asil or noble creatures. Tradition maintains that these horses are *purebred*, free from alien or non-Arabian blood.

Figure 72 - *TheYemeni countryside.*

Figure 73 - *Bedouin falconer, early 20th century. Photo used with the permission of the Library of Congress collection.*

Explorer and academic Alois Musil, Professor of Oriental Studies, Charles University, Prague, and a contemporary of Raswan's, studied the desert Bedouin in considerable detail and published The Manners and Customs of the Rwala Bedouin in 1928. He states that:

> "The Bedouin assert that no horses were created by Allah in Arabia. According to their tradition, they brought their first horses from the land of the settlers whom they raided. In the inner desert the horses have no place and would perish, did not the Bedouin look after them better than they look after their own children. A runaway horse cannot live long in the desert."

These opinions and observations shared the common feature they all lack substantial facts upon which to build theories. The questions of the purity and the origins of the Arab horse remained unanswered. These bits of information provide no clear evidence that these horses were biologically or genetically related to the ancient "Arabian" or the modern day "Arabian" horse. Furthermore, there is a complete absence of factual information concerning the presence or absence of the desert horse in the central deserts of Arabia before or during this period.

The possibility remains open that the horse *Equus caballus* was indeed endemic to the peninsula, and was actually the source of the fast, light-boned equine blood stock that was later known to exist in central Asia and Africa.

A few facts that are known; the authentic desert horse was well established among the nomadic Bedouin of the Nejd plateau from great antiquity and was an integral part of their pastoral existence. Nothing more can be said with certainty about the time and place of origin of the Arabian horse. No conclusions can be reached concerning the meaning of the term *purebred Arabian horse*.

Factual information about the horse during this period is lacking as a result of the physical and ethnologic barriers that existed between the Arabian interior and the outside world. The creation myths do not provide factual information and the opinions of authorities do not lend themselves to verification. The historical information paradoxically is even less helpful. The origins of the Arab horse remain unclear, and the search for answers continues.

The quest for the source of the *purebred Arabian horse*, indeed the quest for the meaning of the term "*purebred*" itself, remains elusive.

**Man proposes and God disposes.*

CHAPTER THREE

CHAPTER THREE

THE ARABIAN HORSE IN THE FINE ARTS

As discussed in earlier chapters, the desert Bedouin of Arabia had a very clear definition of the asil horse. The concept of asalah was deeply imbedded in their cultural collective psyche. Likewise, the men who selected the horses that were used to create the Royal Agricultural Society had a very clear idea of the classic Arabian horse, and they selected horses that were, in their estimation, purely bred classic Arabian horses. The conformation and the general appearance of the horses was their guide. The Bedouin knew, and the founders of the RAS knew, what a *purebred Arabian horse* looked like. This point is critical to a full understanding of the meaning of the term *purebred Arabian horse*.

Yet the definition of this term continues to be elusive.

This difficulty is based on the reliable and well documented findings of pedigree researchers that many Arabian horses living today contain non-Arabian progenitors in their pedigree. The difficulty is compounded by the physical appearance of the Arabian horses, officially recognized by the Arabian Horse Association as registered Arabian horses, that dominate the Arabian horse market today. The registration certificates issued by the Arabian Horse Association in the U.S. claims that its registered horses are "*purebred*" but does not define this term in its bylaws. The Arabian horses that are winning national and international competitions are registered AHA horses, but they exhibit physical features, or type, that suggests the presence of non-Arabian blood in their genetic past. They do not look "Arabian". Most Arabian enthusiasts consider this to be unimportant.

Breed purists however, do not consider the matter unimportant. In an attempt to arrive at a definitive understanding of the physical characteristics or "type", of the *purebred Arabian horse*, researchers often study the images of the Arabian horse as it was portrayed by artists of the 18th and 19th century. This type of analysis is based on the theory that Arabian horses from this past era, as compared to the present era, were more likely to be actual authentic, that is to say *purebred*, desert horses, uncontaminated by cross-breeding with non-Arabian horses, principally Barb, Turkish, and English Thoroughbred. This line of reasoning is founded on the assumption that before the year 1800 CE, horses on the Arabian Peninsula were for the most part isolated from non-peninsular horses.

The technique of comparative analysis using depictions of the Arabian horse in the fine arts began with Prince Mohammad Ali Tewfik. In his study at Manial Palace in Cairo, he displayed fine art prints depicting 19th century Arabians. He encouraged his many foreign visitors to compare his horses to the classic models displayed in the art works. The Prince thought his horses to be excellent examples of classic Arabian type.

The reasoning is straightforward. If one can arrive at a set of morphologic data that define the *purebred Arabian horse* as it existed before the potential genetic contamination of the authentic desert horse began on a large scale, then one could use this information to identify those Arabian horses living today that may be truly pure bred.

Perhaps by identifying a uniform list of morphological features that provides definitive evidence of what a *purebred Arabian horse* looked like before 1900, then it would be possible to identify those horses living today that fit that description. This line of reasoning, however, consists of a series of unsupportable assumptions, and is subject to the limitations of personal perception and opinion.

There are, however, some historical reasons to suggest that this line of reasoning is valid. The main source of purported genetic contamination of the *purebred Arabian horse* is from the English Thoroughbred, and the English had little presence in Egypt or Arabia prior to the opening of the Suez Canal in 1869. The French showed little interest in the region prior to the invasion of Egypt by Napoleon Bonaparte in 1798, and even then they were expelled from the Middle East in 1801 by the English. The region possessed no strategic value for either country. But western powers brought with them large numbers of English Thoroughbred horses and Thoroughbred crosses, and many were left behind in the Middle East.

Most significantly, the English mania for horse racing did not arrive in the Middle East until the annexation of Egypt by the British in 1886. Horse racing in Cairo and Beirut became very popular, and appears to have provided the motivation for Arabian horse breeders and traders to surreptitiously introduce the blood of the English Thoroughbred into native Arabian stock. In this way horses could be produced that could be registered at the race tracks as an Arabian but possess the superior track performance of an English Thoroughbred horse.

It is a mistake to analyze these images as if they were scientific documents of fact. They are not. It is also a mistake to study these images with the presumption that they are representations of the *purebred Arabian horse*. Some of the animals probably are, and some are probably not.

The quality of the art varies widely. The artists may be creating the image from life, or they may be using the works of others as a basis for their work. Some of the artists were prominent professional painters of their day, and some were skilled amateurs.

Nonetheless, it is useful to study these images, but with a critical eye. The general impression of the Arabian horses presented by artists in the days before photography became commonplace provides the researcher with a rather clear conception of what the authentic desert Arabian horse actually looked like during the18th and 19th centuries. Comparing the physical attributes of the Arabs from the past with the Arabian horse as it exists today provides significant and valuable insights.

However, analysis of these images does require some degree of discrimination. Some of the horses presented here were drawn or painted by the artist from life, as they saw them in various parts of the Middle East. Others were painted by artists using their fertile imaginations, working from pictures of the orient and horses that they had seen in European studs. There were large numbers of Arabian horses present in Europe at the time that had been directly imported from the Arabian and Syrian deserts. The horses were generally imported to meet military requirements. 19th century European artists had the opportunities to see, study, and paint images of horses that had been imported directly to Europe from the Arabian Desert.

It is important to bear in mind that the artists were Europeans, and in most cases not entirely conversant with the concept the *purebred Arabian horse*. Most of them had not traveled to the homeland of the desert horse. In many cases the artists took license with their portrayals. Some of the artists depicted Arabian horses with impossibly short and tipped ears and some depicted horses with improbably large eyes. Some of these artists never traveled outside of Europe, and some drew and painted horses that they saw in the stables of Europe, which they were told were *purebred Arabian horses*. On the other hand, some of these artists spent long periods of time living with the Bedouin and riding with them on raids. Some became fully Arabized.

Many of the examples shown here represent Mamelucks and their horses. There is no historical reason to suggest that Mamelucks rode purely bred desert horses. However, their close association with the Mohammad Ali dynasty suggests that this might at times be the case. The frequent appearance of the Mamelucks in these portrayals was due to the Victorian-era European fascination with and idealization of this Egyptian warrior class, whom they regarded as noble and valiant. Europeans of the time were avidly interested in orientalism, and the prints and lithographs sold well; hence, many of them were created.

The sculptures made during this period of Arabian horses are especially noteworthy. The conformation of the legs, portrayed with structural limb and joint deformities are subjects for reflection. It is difficult to find images of 19th century horses that did not possess structural deformities. They did not exhibit proper equine conformation, particularly leg faults, as it is understood today. This point is significant.

The size of the horses and riders portrayed also gives the modern day enthusiast useful information about the size of the *purebred Arabian horse* of past eras. The average Bedouin of the 19th century was 5 feet four inches tall. The height of his horse can be determined accordingly. These horses were small. This point is significant.

All things considered, the images presented here provide a clear idea of the general morphological characteristics of the Arabian horses that artists portrayed in the 18th and 19th centuries. For this reason, the images are useful to the researcher in the field. The reasons that the artists chose one horse over another in a particular painting will remain forever unknown. Whether the artists thought that they were portraying *purebred Arabian horses* in their works will also remain unknown. The most that can be said with certainty of the works in general is that the artists must have thought that their subjects would interest the general public, and that this would result in substantial sales of copies, prints and lithographs.

Victor Adam

Victor Adam (1801-1867) was born in Paris, son of the engraver Jean Adam. Victor received his early artistic training in the ateliers of Meynier and Regnault. At a young age he was commissioned to paint a series of pictures for the Louvre Museum, primarily consisting of a series of massive canvasses depicting historical French military victories. His reputation as a consummate technician was widely acknowledged. At the peak of his success he disappeared from public view for a period of eight years. His absence was not explained. In 1846 he was again seen on the Paris streets, but he was then devoting himself solely to small scale lithographic works, concentrating on depicting animals.

Figure 1 - *Arab Horse and Groom, by Victor Adam, from the Victoria and Albert Museum.*

Figure 2 - *Arab Horse and Groom, by Victor Adam, from the Victoria and Albert Museum.*

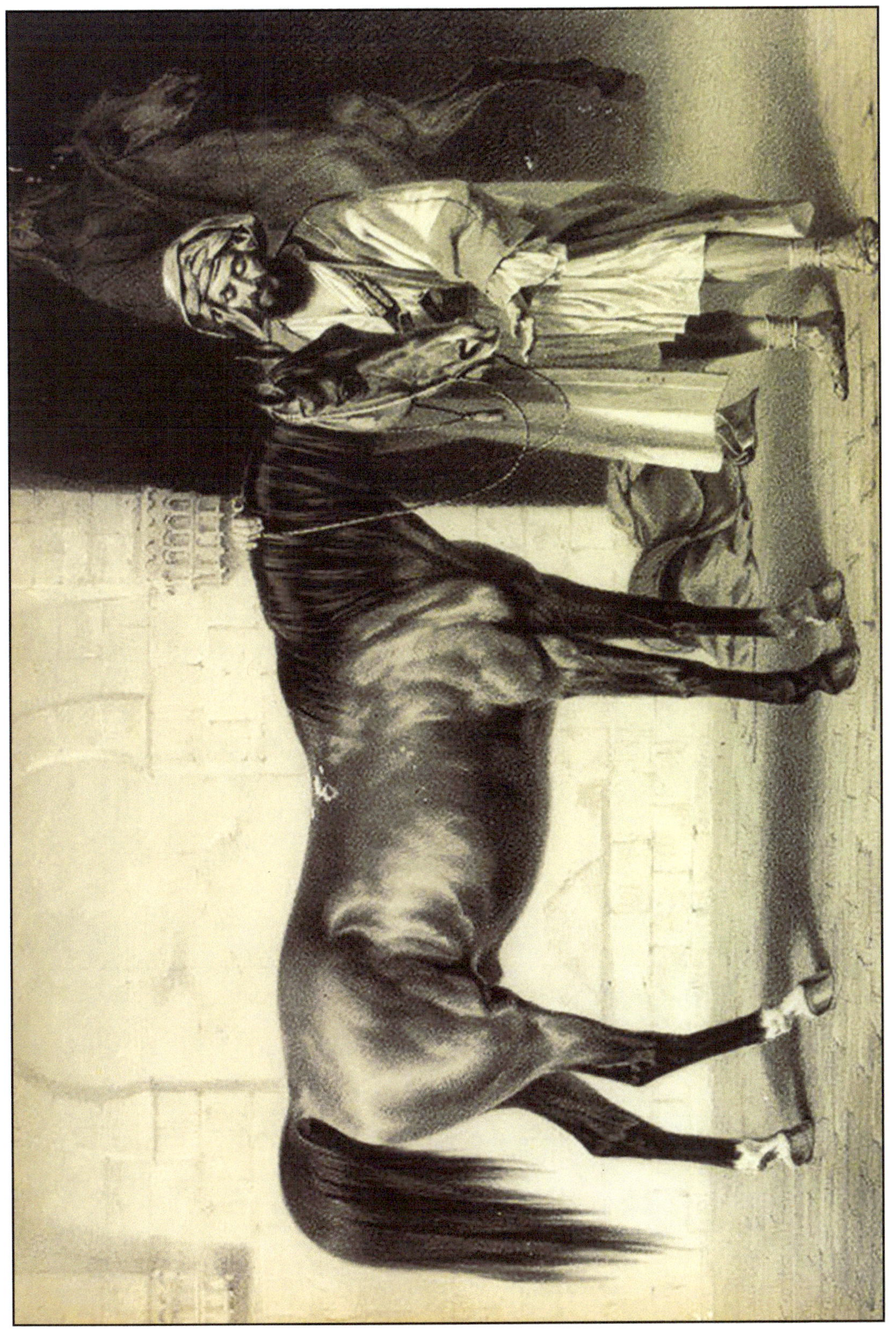

Figure 3 - *Arab Horse and Groom, by Victor Adam, from the Victoria and Albert Museum.*

Figure 4 - *Rearing Arab, by Victor Adam, from the Victoria and Albert Museum.*

Figure 5 - *Djodar Race, Etalon Arabe, by Victor Adam, from the Victoria and Albert Museum.*

Figure 6 - *Jument Egyptienne, by Victor Adam, from the Victoria and Albert Museum.*

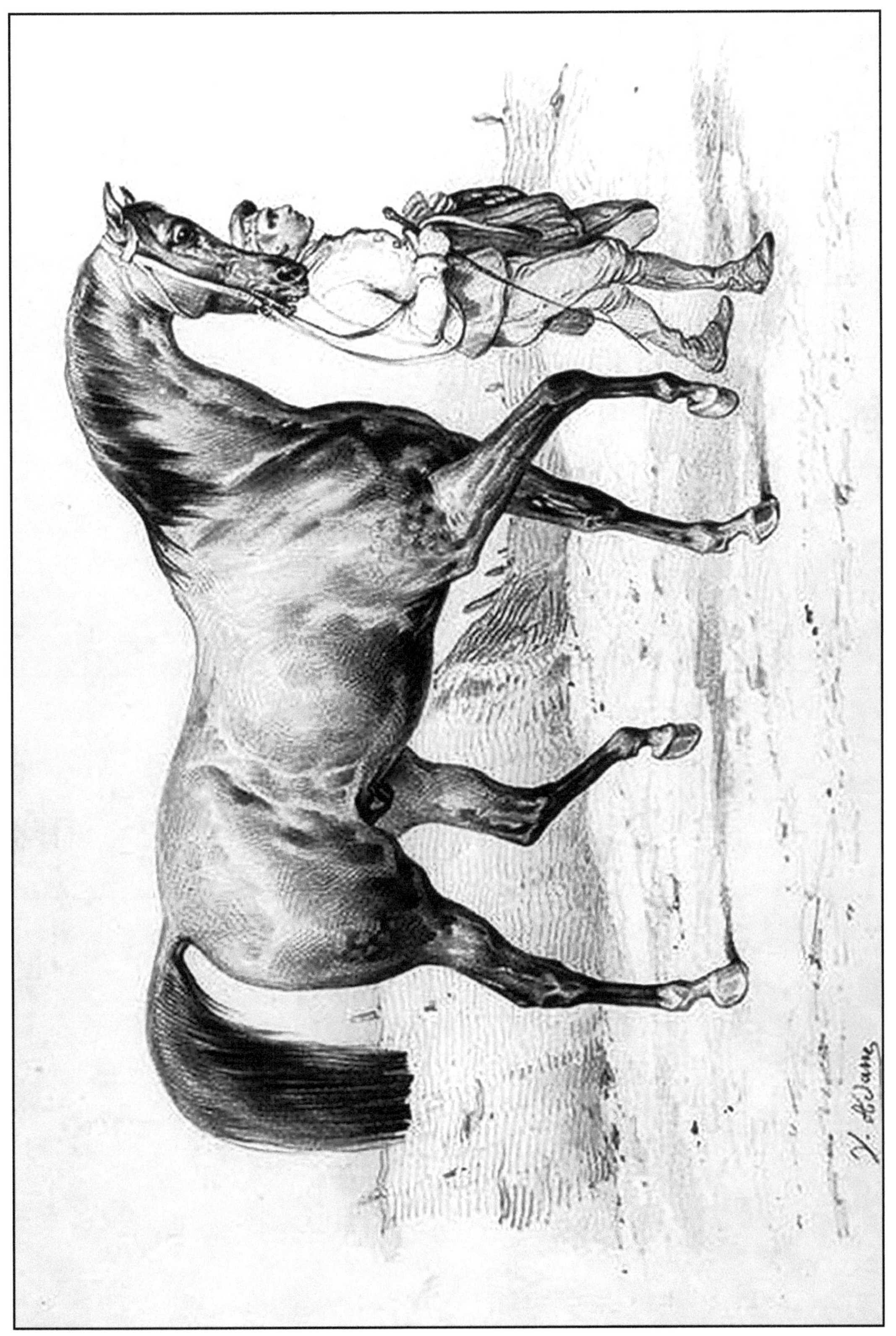

Figure 7 - *The Morning Walk, by Victor Adam.*

Figure 8 - *Arabian Horse, by Victor Adam, image in the public domain.*

James Ward

James Ward (1769-1859) was a British painter noted for his allegorical paintings of the Romantic period. He subsisted by obtaining commissions to paint the families and horses of wealthy patrons. His works can be seen in the Tate in London and in the Yale Centre for British Art. Regarded as one of the finest animal painters of his day, he was admitted to the Royal Academy in 1811. His later years were marred by a stroke, and he died in poverty.

Figure 9 - *Lions Attacking Men and Horses, by James Ward.*

Figure 9.1 - *Mr. Alfred Bonar's Arabian, Dare Devil, (oil on canvas), by James Ward / Private collection / copyright Arthur Ackermann Ltd. London / Bridgeman Images.*

Figure 10 - *A Grey Arabian Horse, the property of Sir Watkins Williams-Wynn, by James Ward, image in the public domain.*

Figure 11 - *Marengo, by JamesWard, image in the public domain.*

George Henry Laporte

George Henry Laporte (1799-1873) was the son of the well known London art instructor John Laporte, who was descended from French Huguenot stock. George followed in the family tradition, presenting his animal portraits in the various London exhibition halls. He was a frequent exhibitor at the Royal Academy. He specialized in scenes of rustic rural country life, portraying horses, harriers, and foxhounds. A regular contributor to the *New Sporting Magazine*, Laporte specialized in scenes of horse racing, steeple chases, and fox hunting. He was at one time appointed as court painter to the King of Hanover as well as to His Royal Highness the Duke of Cumberland.

Figure 12 - *Mameluck Purchasing an Arabian Stallion, by George Henry Laporte, Yale Centre for British Art, from Creative Commons.*

Figure 13 - *Arab Mare and Foal With Attendant, by a Ruined Temple, by George Henry Laporte, used with the permission of The Tate, London.*

Figure 14 - *Noblemen before Constantinople, by George Henry Laporte.*

Figure 15 - *Grey Arabian Mare and Foal With a Family, by George Henry Laporte, Yale Centre for British Art, from Creative Commons.*

Jacques-Laurent Agasse

Jacques-Laurent Agasse (1767-1849) was born in Geneva Switzerland, and discovered his artistic talent for depicting horses at a young age. He studied in Paris, first at the veterinary school to learn equine anatomy. A wealthy English patron sponsored the young man, who was said to have a "resistless force of natural genius". In England Agasse excelled in portraying English racehorses. He also painted several notable canvasses of Nubian giraffes that had been gifted to George IV King of England by Mohammad Ali, Viceroy of Egypt. His works were exhibited at the Academy from 1801 to 1845. He was considered to be among the most talented artists of his day. He died in England. In the end he was, like so many other gifted artists of the 19th century, destitute. He cared little for money, devoting himself entirely to his craft.

Figure 16 - *The Wellesley Grey Arabian Led Through the Desert, by Jacques-Laurent Agasse, Swiss, 1767-1849, from the Yale University Center for British Art, Yale University.*

Edwin Landseer

Edwin Landseer (1801-1873) was an English painter, well known for his pictures of animals. He is remembered primarily for his monumental sculptured bronze lions commissioned for the base of Nelson's column at Trafalgar Square in London. His successes came early in life and he was admitted to the Royal Academy at age 24. His *Monarch of the Glen*, the famous stag portrait, was widely sold in the form of engravings. He became a favorite of the Royal family, teaching both Victoria and Albert to sketch. One of his last works was a full size equestrian painting of the Queen. When he died, all England mourned. He was buried in St. Paul's Cathedral London.

Figure 17 - *Arabian Stallion, (oil on canvas), by Sir Edwin Landseer / Roy Miles Fine Paintings / Bridgeman Images.*

Figure 18 - *The Arab Tent, by Edwin Landseer, used with the permission of the Wallace Collection, London.*

Theodore Chasseriau

Theodore Chasseriau (1819-1856) was a French Romantic painter who often painted oriental and biblical subjects. He was born in El Limon, in the Dominican Republic; his father was French and his mother was a mulatto. The family moved to Paris where the young Theodore displayed an unusual talent for drawing. His early training was as a classicist, studying with the master Ingres. Later he became influenced by Delacroix, and his work exhibits the tension between the two styles. He traveled in Algeria to see the colors and forms of the oriental way of life first hand. His best known oriental work is the enormous painting *Ali-Ben-Hamet-Caliph* of Constantine. An entire room is dedicated to his works at the Louvre in Paris.

Figure 19 - *Cavaliers Arabes Enlevant Leurs Morts, Arab Horsemen Carrying AwayTheir Dead, by Theodore Chasseriau, 1850, used with the permission of the Harvard Art Museum.*

Figure 20 - *Ali Ben Hamet, Caliph of Constantinople, by Theodore Chasseriau, image in the public domain. Painting the same face on the three horsemen may have amused the eccentric artist.*

Figure 21 - *An Arab Mounting, by Theodore Chasseriau, from the National Gallery of Art, Washington.*

Figure 22 - *Combat Arab/Combat de Cavaliers Arabes, (Arab Combat)*, by Theodore Chasseriau, 1855, used with the permission of the Harvard Art Museum.

Eugene Delacroix

Eugene Delacroix (1798-1863) was a French painter of the Romantic period. His early training in the neoclassical style was in the studio of Pierre-Narcisse Guerin. Delacroix rejected this style of painting. He was instinctively drawn to exotic images and locations. As a young man he traveled in North Africa, where he found vivid examples of his strong identification with the forces of the violent and the sublime. He made hundreds of sketches of Arabs and their horses which resulted in finished works many years later. He was "passionately in love with passion" and his work was that of an individualist; no one had painted like him before. His expressive brushstroke and exploration of colors were unique. He was however deeply influenced by Gericault. Delacroix painted *Liberty Leading the People* in 1830, and the popularity of this work insured his immortality.

Figure 23 - *Arab Scouts, by Eugene Delacroix, used with the permission of The Hermitage Museum, St. Petersburg, Russia.*

Figure 24 - *Arab Horseman Giving a Signal, by Eugene Delacroix, used with the permission of the Chrysler Museum.*

Figure 25 - *Arab Saddling His Horse, by Eugene Delacroix, used with the permission of The Hermitage Museum, St. Petersburg, Russia.*

Figure 26 - *Arabs Skirmishing in the Mountains, by Eugene Delacroix, 1863, National Gallery of Art,Washington.*

Figure 27 - *Collision of Moorish Horsemen, by Eugene Delacroix,Walters Art Museum, Creative Commons.*

Figure 28 - *Recontre de Cavaliers Maures, by Eugene Delacroix, used with the permission of Musee Eugene Delacroix, Paris.*

Figure 29 - *Cheval Arabe au Picquet, 1857 after Delacroix, from the NewYork Public Library collection.*

Figure 30 - *Chasse au Tigre, by Eugene Delacroix, image in the public domain.*

Figure 31 - *Arab Horses Fighting in a Stable, by Eugene Delacroix, image in the public domain.*

Otto Stotz

Otto Stotz was born in Ludwigsburg, Germany in 1805, and died in 1873. He excelled as a painter and lithographer, and many of his best known works depicted Arabian horses. He made several fine drawings and paintings of the Arabian horses at the Babolna Stud in Hungary. His works are shown today in the Palace of Schonbrun in Vienna where his portrait of Emperor Franz Joseph is on display. His painting "*The Horse Market*" is part of the collection at the National Museum in Linz.

Figure 32 - *Arabian Horse, Agha, by Otto Stotz (1805-1873) from The Raswan Archives.*

Figure 33 - *Agil Aga, by Otto Stotz, from The Raswan Archives. This stallion was part of the Brudermann importation and was an important sire at the Babolna Stud in Hungary. From The Raswan Archives*

Figure 34 - *The Stables of Babolna, by Otto Stotz, from the Kunsthistorisce Museum, Vienna, Austria, from Bridgeman Archives.*

Antoine-Louis Barye

Antoine-Louis Barye (1796-1875) was a French sculptor, famous for his work as an animalier. His early training was as a goldsmith under Biennais. However, he became entranced by the animals that he saw at the Jardin des Plantes in Paris, and he developed a lifelong dedication to the art of sculpting animals. He studied sculpting at the Ecole des Beaux-Artes, and he achieved a widespread reputation for works depicting animals on both a large and a small scale. He was elected to the Academie in 1868. Barye made a business from his highly regarded equine bronzes, but poor business acumen led to his bankruptcy. All of his molds were sold to a foundry that mass produced inferior quality reproductions, flooding the market. Barye died penniless and forgotten.

Figure 35 - *The Lion Hunt, by Antoine-Louis Barye "Surtout de table", Walters Art Museum.*

Figure 36 - *Two Arab Horsemen Killing a Lion, by Antoine-Louis Barye, The Walters Art Museum, Creative Commons.*

Figure 37 - *Horse Attacked by a Lion, by Antoine-Louis Barye, The Walters Art Museum, Creative Commons.*

Horace Vernet

Horace Vernet (1789-1863) was the son of Carle Vernet. Horace, whose full name was Emile Jean-Horace Vernet, was, ironically, born in the Louvre Museum in Paris, where his pregnant mother, out of desperation had sought safety and protection from the mobs during the French Revolution. As a young man, he disdained the Classical art style, and worked in oils depicting contemporary scenes. He made his reputation painting battle scenes during the Bourbon Restoration. King Louis-Philipe was his most valued patron. During the Second Empire under Napoleon III he traveled with the French army to the Crimea as a war illustrator. This experience in the Middle East brought him into contact with oriental horsemen, and his later career consisted of works based on his historical interpretations of political events.

Troisième suite de Chevaux (N.32) d'après Carle & Horace Vernet.

Figure 38 - *In The Style of Vernet, from the Victoria and Albert Museum.*

Figure 39 - *Un Mameluck, by Horace Vernet, used with the permission of the Musee de Louvre, Paris.*

Figure 40 - *Mameluck Bridling His Horse, by Horace Vernet, from the Bibliotheque Nationale, Paris.*

Figure 41 - *Mameluck Charging the Enemy, by Horace Vernet, from the Bibliotheque National, Paris.*

Figure 42 - *A Mameluck Loading his Gun, by Horace Vernet, from the Bibliotheque National, Paris.*

Figure 43 - *Mameluck au Trot, by Horace Vernet, from the Bibliotheque National, Paris.*

Figure 44 - *The Lion Hunt, by Horace Vernet, used with the permission of the Wallace Collection, London.*

Figure 45 - *Mohamed Ali, Egyptian Vice-Roi, by Horace Vernet, 1818, image used with the permission of the British Museum.*

Adolph Schreyer

Adolph Schreyer (1828-1899) was born in Frankfurt-am-Main in Germany. His training took place at the studios of Munich. As a young man, he accompanied Maxmillian Karl, Prince of Thurn and Taxis through Asia Minor. He also traveled to Egypt, Syria, and Algiers. In 1862 he settled in Paris, later returning to his home near Frankfurt. His works depicting horses is unmatched. He was admired for his excellent draughtsmanship and his ability to portray natural movement among the seemingly endless stream of oriental horsemen, who were always in motion.

Figure 46 - *Arab Warriors on Horseback, by Adolph Schreyer, undated, oil on canvas, used with the permission of the Dahesh Museum of Art.*

Figure 47 - *Arab Caravan, by Adolph Schreyer, used with the permission of the Dayton Art Institute.*

Figure 48 - *Arab Irregular Troops, by Adolph Schreyer, used with the permission of the Hamburger Kunsthalle, Hamburg, Germany.*

Figure 49 - *Arab Horsemen, by Adolph Schreyer, used with the permission of the Nelson-Atkins Museum of Art.*

Figure 50 - *Arab Horsemen, by Adolph Schreyer, with the permission of the National Gallery of Ireland.*

Figure 51 - *Arabische Reiter, by Adolph Schreyer, image in the public domain.*

Figure 52 - *Arab Horsemen, by Adolph Schreyer, from the Library of Congress.*

Figure 53 - *Two Arab Horsemen, by Adolph Schreyer, used with the permission of the Museum of Fine Art, Boston.*

Figure 54 - *Arab Riders, by Adolph Schreyer, oil on panel, used with the permission of the Frye Art Museum, Seattle Washington.*

Figure 55 - *Escaped, by Adolph Schreyer, from the Library of Congress.*

Figure 56 - *Der Fahnentrage, by Adolph Schreyer, the standard bearer, image in the public domain.*

Figure 57 - *Artillery Shell;War in the Desert, oil on canvas, by Adolph Schreyer, used with the permission of the Santa Barbara Museum of Art.*

Figure 58 - *Arabs in Egypt at Sunrise, by Adolph Schreyer, used with the permission of the Walters Art Museum.*

Figure 59 - *Arabische Krieger, by Adolphe Schreyer, oil on canvas, 1863, image in the public domain.*

Figure 60 - *Arabische Reiter, by Adolph Schreyer, image in the public domain.*

Figure 61 - *Arab Riders, by Adolph Schreyer, from the Library of Congress.*

Figure 62 - *Arab Horsemen, Defeat and Hate, by Adolph Schreyer, 1863, The Heckscher Museum of Art, Huntington, N.Y.*

Figure 63 - *Mounted Horsemen, by Adolph Schreyer, image in the public domain.*

John Alexander Harrington Bird

John Alexander Harrington Bird (1846-1936) was a British painter who spent his youth in Canada where he became a member of the Royal Canadian Academy. Later in life he returned to London, and he made his reputation as a painter of equestrian scenes, many done for the Royal Family, and many depicting Arabian horses. His works were done both in oil and watercolors. He portrayed many scenes of country life in England including fox hunting, coursing, racing, and dressage.

Figure 64 - *Arab Mare and Foal, by John Alexander Harrington Bird, image in the public domain.*

Figure 65 - *Arabian Horses, by John Alexander Harrington Bird, image in the public domain.*

Eugene Fromentin

Eugene Fromentin (1820-1876) was born in La Rochelle in France. He studied art under the landscape painter Louis Cabat. As a young man he visited Algeria, and the impressions that were made there formed the basis for many of the later works. He made a detailed and careful study of the scenery and the habits of the people that he saw in North Africa. His work has been described as ethnographic. Of his art he maintained that "art is the expression of the invisible by means of the visible." His later style was influenced by Delacroix, showing great brilliancy in the use of color and the ability to skillfully compose scenes to suggest a sense of animation and motion.

Figure 66 - *The Simoon, by Eugene Fromentin, image in the public domain.*

Figure 67 - *The Falconer, by Eugene Fromentin, used with the permission of the Chrysler Museum.*

Figure 68 - *Two Arabian Horses, by Eugene Fromentin, used with the permission of the Art Institute of Chicago.*

Figure 69 - *Arab Warriors Returning From a Fantasia, by Eugene Fromentin, used with the permission of the Carnegie Art Museum.*

Figure 70 - *Scene in a Desert, by Eugene Fromentin, 1868, used with the permission of the Hermitage Museum, St. Petersburg, Russia.*

Figure 71 - *La Chasse au Faucon, by Eugene Fromentin, used with the permission of the Musee d'Orsay, Paris.*

Figure 72 - *The Falcon Hunt, by Eugene Fromentin, used with the permission of the National Gallery of Ireland.*

Figure 73 - *Arab Horsemen Watering Their Horses, by Eugene Fromentin, San Diego Art Museum.*

Figure 74 - *Arabs Crossing the Ford, by Eugene Fromentin, used with the permission of the Brooklyn Museum.*

Figure 75 - *Arabs Crossing a Stream, by Eugene Fromentin, from the Metropolitan Museum, public domain.*

Figure 76 - *Moroccan Battle Scene, by Eugene Fromentin, Stanford University Art Museum.*

Figure 77 - *La Chasse au Heron, by Eugene Fromentin, used with the permission of Musee Conde, Paris.*

Figure 78 - *Encampment in the Atlas Mountains, by Eugene Fromentin,Walters Art Museum, Creative Commons.*

Figure 79 - *Arabs Crossing the Ford, by Eugen Fromentin, used with the permission of the National Gallery of Ireland.*

Thomas Bewick

Thomas Bewick (1753-1825) was an English artist and author. He enjoyed considerable popularity and success making wood block engravings to illustrate children's books and an edition of Aesop's Fables. He wrote a book on British birds and was an active animal rights advocate.

Figure 80 - *An Arabian Horse, by Thomas Bewick, 1780, used with the permission of the British Museum, London.*

David Dalby

David Dalby (1780-1849) was from a large Yorkshire family, many of whom were artists. His works were mainly depictions of British Thoroughbreds, horse racing, common carriage horses, and a variety of coaches and carriages of the day.

Figure 81 - *Signal, a Grey Arab, With a Groom in the Desert, by David Dalby of York, 1780-1849, from the Yale Centre for British Art, Yale University.*

Alfred De Dreux

Alfred De Dreux (1810-1860) was a French painter of historical subjects and portraits. He studied in Paris in the studio of Leon Cogniet. He was a fashionable man who painted glamorous and romantic pictures. The work of Theodore Gericault exerted a significant influence on DeDreux. He painted equestrian portraits for the exiled Emperor Napoleon III while in England. De Dreux was killed in a duel with Comte Fleury, Napoleon's principal aide-de-camp.

Figure 82 - *Wild Horse Fighting, by Alfred De Dreux, used with the permission of the Chrysler Museum.*

Figure 83 - *The Pasha's Pride, by Alfred De Dreux, from a private collection, copyright by Christie's Image, from Bridgeman Images.*

Figure 84 - *An Arabian Horse, by Alfred De Dreux, used with the permission of Bridgeman Images, from a private collection, copyright Bonhams, London, UK.*

Figure 85 - *Nubian Horseman at the Gallop, by Alfred De Dreux, used with the permission of Bridgeman Images, from a private collection.*

Figure 86 - *The Horses of Abd el Kader, by Alfred De Dreux.*

Figure 87 - *The Horses of Abd el Kader, by Alfred De Dreux.*

Theodore Gericault

Theodore Gericault (1791-1824) was born in Rouen France and was trained by the strict classicist Pierre-Narcisse Guerin. While studying in Paris he became attracted to equine subjects, and he had free access to the horse breeding stables at Versailles. There he learned horse anatomy and carefully studied the movement and behavior of horses. Some of his finest works featured Arabian horses. He died at a young age from injuries sustained during a riding accident.

Figure 88 - *The Bride of Abydos, by Theodore Gericaut, from the Philadelphia Art Museum.*

Figure 89 - *The Giaour, by Theodore Gericault, used with the permission of the J. Paul Getty Museum.*

Figure 90 - *Mameluke of the Imperial Guard, by Theodore Gericaut, Stanford Art Museum.*

Figure 91 - *Mameluck Retenant Son Cheval, by Theodore Gericault, used with the permission of Musee de Louvre, Paris.*

Figure 92 - *Cheval Arabe de Profil Vers la Gauche, by Theodore Gericault, used with the permission of the Musee de Louvre, Paris.*

Figure 93 - *Cheval Arab, by Theodore Gericault, used with the permission of the Musee Conde, Chatilly.*

Figure 94 - *Cheval Dont on ne Voit Que la Tete, le Poitrail et les Jambs, by Theodore Gericault, used with the permission of Musee Bonnat, Bayonne, France.*

Figure 95 - *Arab Stallion, by Theodore Gericault.*

Figure 96 - *Arabian Horse, by Theodore Gericault, National Gallery of Art, Washington.*

Charles Turner

Charles Turner (1774-1875) was an English artist best known for his mastery of the mezzotint technique of engraving and printing. In his youth he studied at the Royal Academy in London. He first came to public attention in 1801 when he issued a well received print of Napoleon Bonaparte. Turner was an ambitious man, and he was able to rapidly produce high quality images for the changing interests of the public. He was quick to capitalize on the art market, and in his lifetime produced 921 prints. He worked in close association with the Swiss artist J.L. Agasse and the British painter J.M.W. Turner. In 1812 he was appointed "Mezzotint Engraver in Ordinary to His Majesty", and in 1828 he was elected to the Royal Academy. The Wellesley Arabian depicted in this mezzotint was in fact one of two horses, one grey and one chestnut, imported to England from India in 1803 by the brother of the Marquis of Wellesley. The horses are registered in the General Stud Book where they are said to be "not Arabians, but evidently Gulf Arabs, or Persians".

Figure 97 - *The Wellesley Arabian, by Charles Turner, used with the permission of the Yale Centre for British Art, Yale University.*

Antoine-Jean Gros

Baron Antoine-Jean Gros (1771-1835) was born in Paris to an artistic family. The young Gros showed early promised and entered the studio of Jacques-Louis David. Events conspired to bring the young painter to the attention of Bonaparte, who appointed him inspecteur aux revues. His career was made by his many portrayals of the heroic deeds of his benefactor. Sadly, his patron was soon out of favor, and Gros' career faded. He was bypassed by the rising tide of Romanticism. He died, disillusioned and desponded. His body was found drowned in the Seine with a note saying that he was "tired of life, and betrayed by lost faculties which rendered it unbearable".

Figure 98 - *Napoleon at the Pyramids, by Antoine-Jean Gros, image in the public domain.*

Figure 99 - *Mameluck Warrior, by Antoine Gros, used with the permission of the British Museum.*

Figure 100 - *Cheval Arab, by Baron Antoine Gros, used with the permission of the Musee des Beaux-Arts, Valenciennes.*

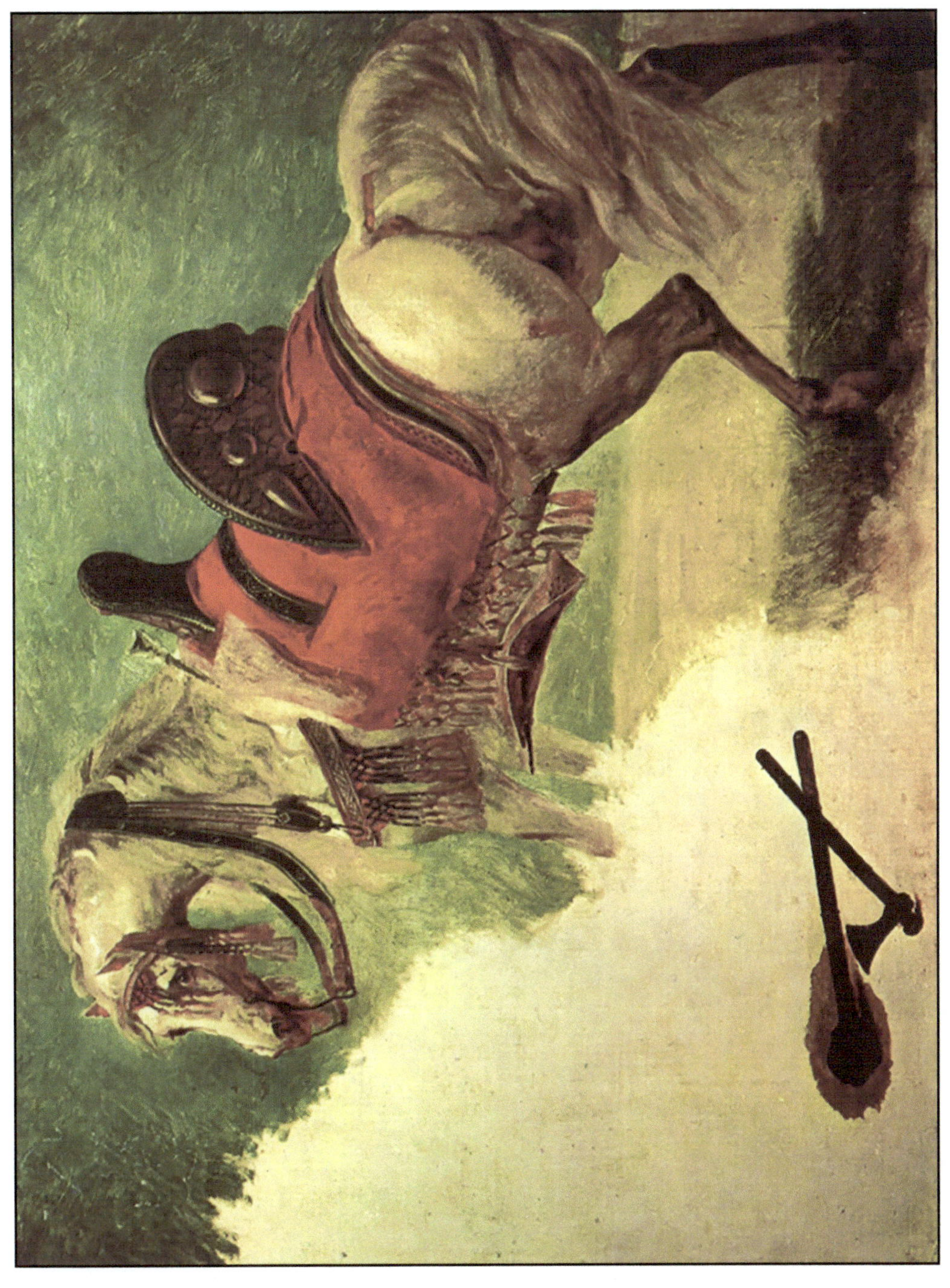

Figure 101 - *Arabian Horse, by Baron Antoine Gros, image in the public domain.*

Figure 102 - *Arab of the Desert, by Baron Antoine Gros, National Gallery of Art, Washington.*

George Stubbs

George Stubbs (1724-1806) was an English artist, and he was largely self-taught. As a young man he became interested in depicting horses, and undertook a study of dissecting horses. He published The Anatomy of the Horse in 1766. Through his paintings, he came to the notice of the nobility whose love of all things equine led them to commission works from him. He did paintings for the Duke of Richmond; his works were so true to life, superior to earlier painters, that he quickly achieved a stellar reputation among his many patrons. Sixteen of his works are in the Royal Collection. The famous *Whistlejacket* hangs in the National Gallery, London.

Figure 103 - *Warren Hastings Arabian, by George Stubbs, used with the permission of the British Museum.*

Figure 104 - *Arab Belonging to Lord Grosvenor, by George Stubbs, used with the permission of the British Library, London.*

Figure 105 - *Mr. Ward's Black Arabian, by George Stubbs, 1770, used with the permission of the British Museum, London.*

John Herring

John Frederick Herring (1795-1865) was a British painter in the Victorian tradition. His first work came when he moved to Doncaster, the Thoroughbred heartland of England. He was employed to paint signs for inns and stables, as well as coach insignias and coats of arms. In his spare time he drove coaches for the wealthy. After moving to London, his serious work began. He was appointed Animal Painter to HRH the Duchess of Kent and was also given commissions from Queen Victoria. He was in no sense an Orientalist.

Figure 106 - *Arab Caravan Approaching Giza, by John Herring, image in the public domain.*

Figure 107 - *The Pharoah's Horses, by John Herring, image in the public domain.*

Thomas Smith

Smith, an English painter, was known primarily for his landscape paintings and hunting scenes. His best known works were views of Chatsworth House and the Lake District.

Figure 108 - *The Cullen Arabian, by Thomas Smith, used with the permission of the British Museum, London.*

Martin Theodore Ward

Martin Theodore Ward (1799-1874) was a British painter who trained under the master Edwin Landseer. Ward was the nephew of the noted equine artist James Ward, but his career was not a successful one. He was of an uneven temperament, given to drink, and he gambled. His career was at first successful, with exhibits at the Royal Academy in the 1820s. But he was an eccentric man, always at odds with the establishment. He was, after accumulating a series of unpaid debts, incarcerated in York Castle Prison as a debtor. In prison, his gift for painting dogs was discovered. He excelled in painting dogs. After his release, he painted constantly, but his dissipative lifestyle continued. He died penniless. His collection of paintings of dogs resides in the York Art Gallery.

Figure 109 - *Head of A Grey Arabian Horse, by Martin Theodore Ward, used with the permission of the Yale Centre for British Art.*

Pierre Jules Mene

Pierre-Jules Mene (1810-1879) was a French animalier. As a successful pioneer in the field, he opened his own foundry to cast reproductions of his most popular works. His success among the bourgeois was so great that his reputation as a serious artist was tarnished, and forgeries of his works abound. Despite this, he was considered the master of the lost wax method of casting and his regard by scholars was not surpassed until the works of Auguste Rodin appeared, such as *The Thinker* in 1902.

Figure 110 - *The Falconer, by Pierre Jules Mene, from the author's collection.*

Figure 111 - *Arab Horseman, by Pierre Jules Mene, from the author's collection.*

Emile Prisse

Emile Prisse d'Avennes (1807-1879) was a French-born archaeologist whose art work was an extension of his primary interests in Egyptology. He was born in France to an old English family of the nobility, and as a youth trained at Chalons in engineering and drafting. His early fascination with the orient led him to a lifetime of work in Egypt, documenting the artwork of the temples and monuments of the Nile delta. Working under the patronage of Egyptian Viceroy Mohammad Ali, he was instrumental in creating the discipline of Egyptian art preservation. Over a span of 40 years he obsessively catalogued Islamic art from the Middle East. The results were a series of art history book still in print today. His volume L'Art Arabe is considered a masterpiece of antiquarian documentation of murals and friezes which are now lost to time and the elements. It is only through his tireless efforts at art conservation that we have any idea today of the magnificence that was once present in the Egyptian tombs and monuments of Egypt.

Figure 112 - *An Arabian Horse, by Emile Prisse, from the collection Oriental Album, published by George Lloyd, London.*

David Roberts

David Roberts (1796-1864) was a Scottish painter well known as an orientalist illustrator. His early training was limited, and he worked in a series of jobs as a designer and painter of theatre sets in London. His genius did not become evident until he began to travel in the orient and the Middle East. He toured Egypt, Nubia the Sinai, Palestine, Jordan and Lebanon, making sketches for later use. Back in England, he began an association with the lithographer Louis Haghe, and began to produce finished works of the sights that he had seen in the East. The project was financed by subscriptions, and the first patron of over 400 was Queen Victoria. Roberts died suddenly, at the peak of his popularity, of an apoplectic seizure.

Figure 113 - *Sebaste, by David Roberts, detail from the author's collection.*

Figure 114 - *Sebaste, by David Roberts, detail from the author's collection.*

Figure 115 - *Sebaste, by David Roberts, detail from the author's collection.*

Figure 116 - *Sebaste, by David Roberts, detail from the author's collection.*

Figure 117 - *Sebaste, by David Roberts, detail from the author's collection.*

Figure 118 - *Sebaste, by David Roberts, detail from the author's collection.*

Emil Volkers

Emil Volkers (1831-1905) was a German painter who specialized in equine painting. He was born in Birkenfeld, and trained at the Art Academy in Dresden and later at the Munich Art Academy. His study of equine form and function was derived from his many visits to the state studs of Trakehnen, the State Stud Celle, and the Krupp Stables in Essen. He came to the attention of regal patrons and enjoyed the support of Prince Charles of Romania and Grand Duke Friedrich August of Oldenburg. A consummate draughtsman, his critics dismissed his works as "too accurate and realistic". His portrayals of horses in scenes of Romanian gypsy life were particularly vivid.

Figure 119 - *Zarife, by Emil Volkers, from The Raswan Archives.*

Honzes Ali

Figure 120 - *Hengst Von Schubra, by Honzes Ali, used with the permission of the Olms Archives.*

Carle Vernet

Carle Vernet (1758-1836) was born in Bordeaux France. He studied with Lepicie and found that he had a natural affinity for portraying horses. He studied equine anatomy in the stables and riding schools of Paris. His work as a lithographer of his own works made his reputation. Vernet began to paint battle scenes on commission from Napoleon I, and for his achievements was awarded the Legion of Honor. He was himself an avid rider, often seen on horseback in advanced old age.

Figure 121 - *Wild Horses Fighting, by Carle Vernet, used with the permission of the Art Institute of Chicago.*

Figure 122 - *Hussard Striking a Mameluck, by Carle Vernet, from the National Gallery of Art, Washington.*

Figure 123 - *Mameluck Leading His Horse, by Carle Vernet, image in the public domain.*

Figure 124 - *Mameluck in Repose, by Carle Vernet, National Gallery of Art, Washington.*

Figure 125 - *Mameluck Gravissant les Montagnes, by CarleVernet, image in thepublic domain.*

Figure 126 - *Gazal, by Carle Vernet, used with the permission of the British Museum. London.*

Figure 127 - *Chief Mameluck, by Carle Vernet, used with the permission of the Art Institute of Chicago.*

Figure 128 - *Cheval Arab, by Carle Vernet, used with the permission of Harvard University.*

Figure 129 - *An Arabian Horse, by Carle Vernet.*

Figure 130 - *Arabian Horse, by Carle Vernet, used with the permission of Harvard University.*

Figure 131 - *Arab Family, by Carle Vernet, National Gallery of Art, Washington.*

Figure 132 - *Arab and Horse, by Carle Vernet, Stanford Art Museum.*

Figure 133 - *An Arabian Horse, by Carle Vernet.*

Figure 134 - *A Mameluck Horseman, by Carle Vernet, Victoria and Albert Museum.*

Abbas Al-Musavi

Figure 135 - *Battle of Karbala, by the Persian artist Abbas Al-Musavi, late 19th century, image in the public domain.*

Alexandre Bida

Alexandre Bida (1813-1895) was a Frenchman, trained by Eugene Delacroix. In his youth he traveled in Egypt and Palestine, and was in his later years was considered an Orientalist. He worked primarily on textured cream paper with a white pigment ground. His drafting technique was executed with black crayon over ink washes.

An annual ten day celebration of the birth of Mohammad was practiced in Cairo until the 19th century. It was considered by the faithful to be a miraculous event. The Dosseh was a display of fanaticism which legend states was initiated by an illustrious 10th century Muslim saint. The legend held that the dervische's horse, being magical and sanctified, would not injure the truly devout Muslim who lay in its path. The ceremony was subsequently taken up and performed by the dervisches of Cairo, practitioners of the mystical branch of Islam, Sufism. They were the seers and miracle workers of the Islamic faith. The Chief Imam, Sheikh of the Dervisches, rode a wildly prancing, nobly bred, sacred Arabian horse from his mosque to the home of the Superior Dervische El Bekri. Along the course of the ceremony, pious and faithful Muslims laid down in the path of the horse, forming a human carpet on which the Sheikh would ride. All of Cairo's members of the mystical brotherhood were present, performing zikr (spinning, whirling dances, and all the while uttering invocations to Allah) all along the course of the procession. As the ordeal proceeded, the horse was gradually urged on by the Sheikh to trot faster, and he began to advance more restively as the procession continued.

Those persons who were injured were considered by the crowd to possess insufficient piety. Those who were killed were thought to be fortunate, for they had gone directly to Paradise. The practice was emblematic for the Islamic faithful, as it demonstrated their religious creed; "Are we not all in the hands of fate?"

The practice was abolished by the Egyptian Khedive Tewfik Pasha in 1880.

Figure 136 - *The Ceremony of Dosseh, by Alexandre Bida, 1855, used with the permission of the Walters Art Museum.*

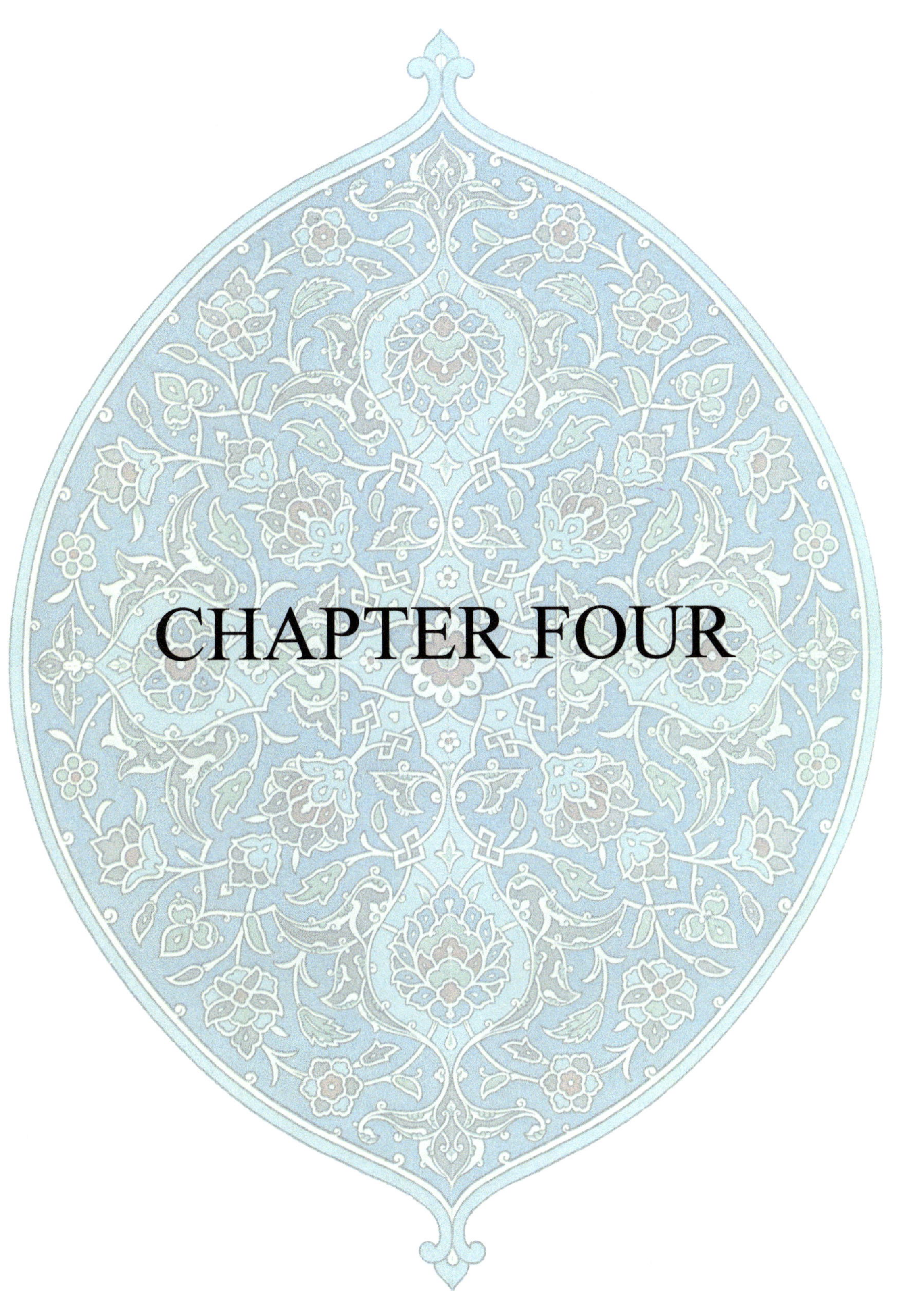

CHAPTER FOUR

CHAPTER FOUR

THE ARABIAN PENINSULA DESERT SOURCES OF THE EGYPTIAN ARABIAN HORSE

PART ONE

Since the purpose of this book is to examine the facts that are known about the matrilines of the Egyptian Agricultural Organization, an examination of the sources of these horses from the Arabian Peninsula is useful. However, no such definitive records exist. Lacking this information, some insights can be found by surveying the methods of Bedouin horse breeding, record keeping, and observations of the horses, the people, and their environs as made by Western explorers during that most critical of centuries, the 19th.

The purpose of this chapter is threefold. First, a review of the centuries-old methods of Arabian horse pedigree recordation on the peninsula is presented. Secondly, an overview is given of the geographic regions of the Arabian Peninsula in which horse breeding was historically prominent. Thirdly the observations of Western explorers will be presented which tells much about the state of asil horse breeding in Arabia.

Figure 1 - *A Mameluck on Horseback, by Carle Vernet.*

Truth, it is said, never dies, but it does appear to live a tortured and wretched life. Arabian horse author and specialist Carl Raswan reflected on this fact throughout his career, observing that, "The truth in horse breeding comes first. I must advocate this and not conceal anything." The facts concerning the components of Egyptian Arabian horse pedigrees remain a subject of debate among Arabian breeders around the world today. The truth has remained elusive. As Raswan wrote, "The Arabian horse (as an archetype) retains what we significantly call his distinctive Arabian characteristics so long as no other foreign blood is added to his own PURE (original, authentic blood)."

Figure 2 - *The Wellesley Arabian Stallion, after Benjamin Marshall, 1820, from the collection of the British Museum.*

Prior to the advent of DNA identification of horses, there was little factual basis for debate. Discussions of this topic have been largely based on conjecture, hearsay, innuendo, prejudice, and opinion. The majority of the debates are based almost entirely on the self interest of the parties involved, rendering any conclusions suspect. Modern DNA research, which is discussed later in this book, is adding factual information to the debate.

Figure 3 - *The French "pie de gru" or "crane's foot".*

The evaluation of any Egyptian Arabian horse begins with the analysis of its pedigree. The English term "pedigree" is derived from an old English word taken from the French "Pie de gru" or "crane's foot", an apt visual description, portraying the many branches of the pedigree.

For Arabian horse breeders, the pedigree of a horse is a useful predictive tool that guides breeding decisions. The pedigree is an essential tool, but it is only useful to the extent that it is accurate. A pedigree that contains inaccurate information is the horse breeder's greatest impediment. An incorrect pedigree is often worse than no pedigree at all. The problem of pedigree inaccuracy has plagued breeders for centuries.

The desert horse was essential to the survival and prosperity of the desert nomads. The Bedouin treasured his horse. Possessing the breeder's natural instincts, he was aware of the need to protect the breed from the deleterious effects of indiscriminate crossbreeding with non-Arabian stock. Toward this end, legend maintains that pedigrees were systematically retained in memory, both by individuals and entire tribes and transmitted orally from one generation to the next. The Bedouin insisted on purity in breeding horses.

According to Max von Oppenheim (1860-1946), the renowned German archaeologist, Islamist, and an expert on the "Bedu".

> "The Bedouin is immeasurably proud of his ancestry, which he guards and treasures. For him, only the Bedouin are asil. The belief in the uniting and binding power of the blood is rooted deeply in his nature… the animals bred by the Bedouin are quite naturally also termed asil, a sign of the strong bond between man and beast in the harsh struggle for survival in the desert."

Figure 4 - *Max von Oppenheim (1860-1946) studied and lived with the Syrian Bedouin during 1892. In 1896 he was assigned to a German government position in Cairo and operated from there for the next thirteen years. His books documenting the life of the Bedouin were widely read and regarded as authoritative.*

Figure 5 - *A Bedouin encampment, photograph by Gerald de Gaury in the 1930s, photo used with the permission of the Royal Geographic Society, London.*

Arabian horse provenance research begins with an analysis of the birth certificate. The written pedigree of an Arabian Desert horse, the equivalent of a birth certificate, was known in the desert as a *hujjah*, and the Bedouin presented these when required for horse sales to foreigners and outsiders. The Western horse buyers of the 18th and 19th centuries were well aware of the importance of obtaining only those horses derived from pure desert stock, uncontaminated by alien blood. This eagerness for the pure blood was based on the generally acknowledged animal breeding principle that "sooner or later, genetic contamination in an otherwise *purebred* population will lead to the emergence of undesirable traits." Verbal representations regarding genetic purity were viewed with increasing skepticism by the horse-buying public, and it soon became standard practice for a Middle Eastern horse seller to produce a written and witnessed document of pedigree. The Western horse buyer required it.

To the unlettered desert Bedouin, the concept of a written pedigree was alien, but this is precisely what Western buyers and horse dealers insisted upon. Western buyers often bought horses not from the tribe that bred them but through intermediaries and agents. This led to unscrupulous dealing and pedigree fabrications.

The English Arabian horse enthusiast Lady Anne Blunt learned from her direct experience in the horse markets and deserts of Syria and Arabia that some horse dealers were not averse to fabricating written pedigrees to facilitate the sale of a horse. In

1898, Blunt sold a number of horses from the Sheykh Obeyd stud in Egypt because of errors discovered in the documentation of their provenance. What she called the "first attempt" had to be abandoned, and all but one were given away because their pedigrees proved to be insupportable. For Blunt, blood purity was essential to her project, and the specific qualities of the individual horse came second. For Blunt, purity was the issue. She found that lacunes in the pedigrees given to her in the desert could not be passed over casually and that dealing with the horse sellers of Egypt, Syria or Arabia required a good deal of circumspection and skepticism. She is remembered today primarily because of the horses that she obtained and bred in the "The New Venture".

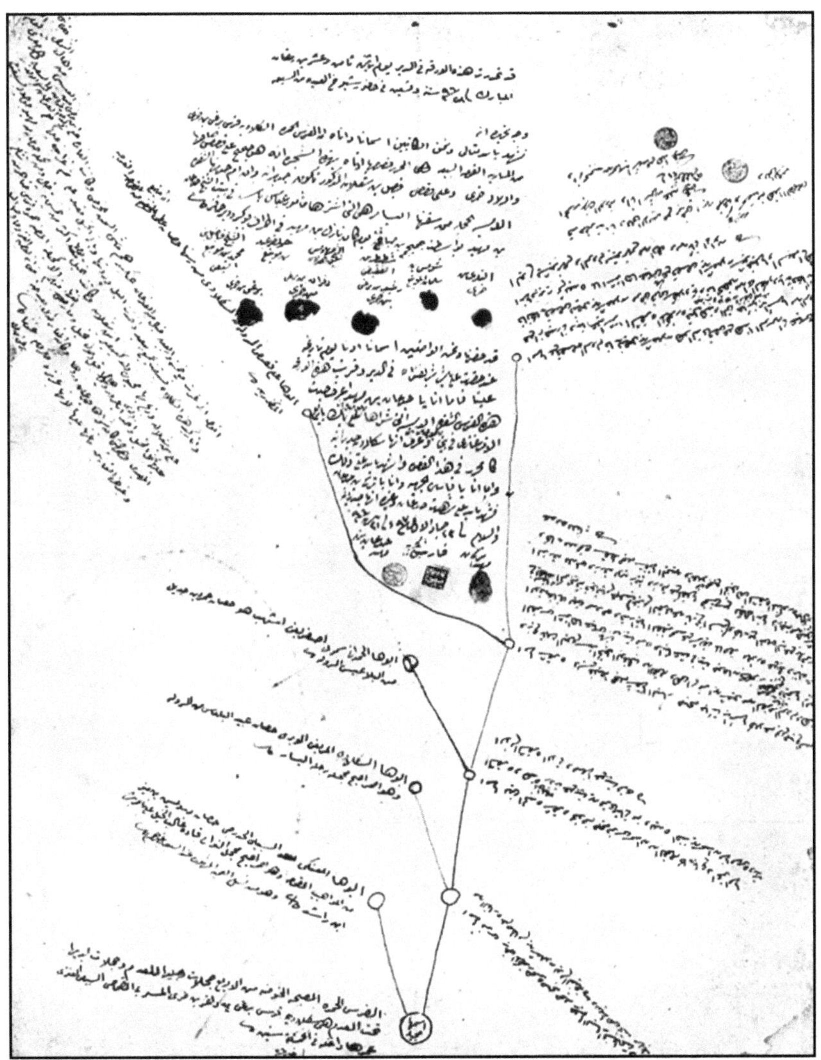

Figure 6 - *Hujje of the Blunt horse Meshura, from* <u>*The Authentic Arabian Horse*</u> *by Lady Wentworth.*

In 1898, after culling the horses that she considered to be impurely bred Arabians, she set out to continue breeding with horses that were from the bloodlines of Abbas Pasha or Ali Pasha Sherif. The only exception that she made was to keep the bay mare Yamama who was born in 1885 in the stud of Ibn Khalifeh of Bahrein. The mare was a Jellabi, from the Kehilet Ajuz line. She also kept the Yamama offspring that had been foaled at Sheykh Obeyd. The Yamama daughter, a grey mare named Yashmak was born in 1893. The stallion Ibn Yashmakwas sent to the Blunt's farm in the U.K. in 1904. Bint Yamama, born in 1904, was given as a gift and exported to Greece.

Meshura was one of the horses that the Blunts bought from a desert tribe during their 1881 trip through northern Arabia as they bought horses in the "First Attempt". Raswan states that the mare named Meshura that was bought by the Blunts was an impure Arabian and was sold to the Blunts by a Turk, who acquired her from her Bedouin breeder because she was contaminated with Mu'niqi blood. Raswan attributes this to an accidental breeding. Despite the apparent authenticity of the *hujje*, Meshura was not kept for the Blunt's "New Venture".

The Egyptian horse breeding records were also at times found to have errors and misstatements. Researchers have attempted to trace the extended pedigrees of horses that belonged to such highly regarded Arabian horse breeders as Abbas Pasha and Ali Pashas Sherif with little success.

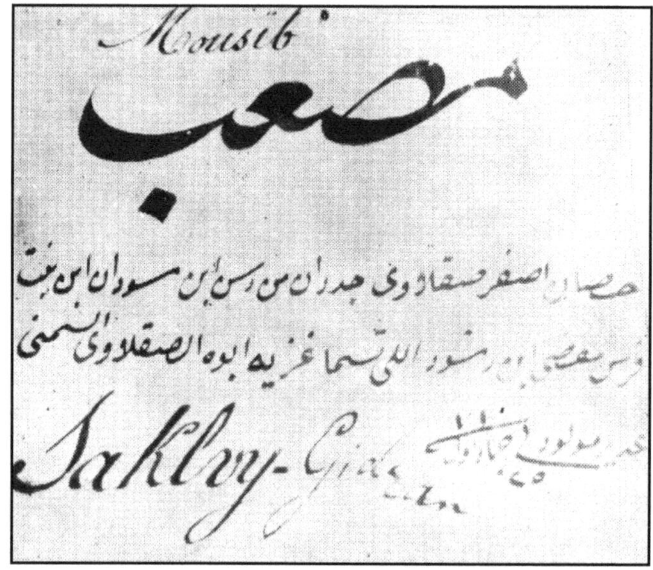

Figure 7 - *The stud book entry of Ali Pasha Sherif's stallion Mousib, a horse that was purchased from the stud of Al Hami Pasha, the son of Abbas Pasha, in 1861, image from The Raswan Index.*

Opinions on the matter of pedigree validity varied. HRH Prince Mohammed Ali of Egypt expressed regret that Bedouin pedigrees were not kept with the same rigor exhibited by European studs. "The tribes," he said, "were mostly occupied in wars and quarrels; secondly the tribes were not friendly with each other. Order was never a strong point with Orientals."

Figure 8 - *Prince Mohammad Ali Tewfik.*

The Prince's observation was noteworthy. Even in his own era, with its earnest adoption of European culture and a high degree of education and literacy among the wealthy, the matter of pedigrees was for most horsemen less important than the animal itself. Records were kept by many of the nobility of the time, but they were often incomplete or filled with errors. Most significantly, the documents themselves were often hastily written. Many stud records were lost, as one generation succeeded the next. Very few Arab horse owners and breeders in the Middle East kept records with an eye to posterity.

Ali Pasha Sherif was said to have kept extensive records on the herd of Arabians that he bred in Cairo for over 36 years. His manuscripts were inherited by his son Huseyn Bey Sherif, and later the records came into the possession of the Egyptian King Fouad. For many years they were considered lost.

In 1957, Carl Raswan published a series of hand written Arabic documents in The Raswan Index, which were said to be records from the stud of Ali Pasha Sherif. The original documents were photographed by Raswan in Cairo in the study of Mohammad Ali Tewfik at Manial Palace on the Isle of Roda in the early 20th century. The images were then reproduced, untranslated, in The Raswan Index.

age (B1) (belongs to the beginning of Book (B)).

Figure 9 - *A page from The Raswan Index, a photostatic copy of 19th century documents, with the translation following.*

These badly faded and nearly illegible pages were translated during the research for this book by a professional Arabic translation service. The results of the following translation illustrate some of the many problems and difficulties that are encountered by researchers attempting to analyze the documentation of Egyptian Arabian horse pedigrees.

A Book that Lists Thoroughbred Male Horses raised by Abandi [Afandi] [illegible] 1895 in a stable

Horses found until [illegible]

1-A horse [illegible] with an [illegible] named Al-Hamdan al-Simri. His father is [illegible] and his mother is [illegible]. Was sold in 1896.

2-An azraq Qouti horse called [illegible].

3-An azraq Qouti horse called Al-Saqlawi.

4-An azraq Qouti horse named Kahaylan Al-Jilabi [illegible].

5-An ahmar limping horse called Kishan Al-Daas [the one that treads] because [illegible]. Was sent [illegible].

6-An azraq Ethiopian horse called Abu Junub.

7-Azraq Qouti horse called Mohamadani al-Simri- came from Mohamed [illegible]. Was sent for sale [illegible].

8-Ahmar horse named Obayan – His father is [illegible].

9-Ashqar horse named Najdani al-Simri Al-Saghirah [From a stable- son of [illegible] Al-Simri Al-Kabir. Was gifted to [illegible] Agha in 99.

10-Azraq mare [illegible] named Kahaylan Al-Saghir son of Kahaylan Al-Jilabi [illegible]. Was sent out for sale [illegible].

11-Azraq horse [illegible] named Kahaylan Al-Ajouz son of ashqar Kahaylan al-Ajouz [illegible]. Was gifted to Mohamed [illegible]. Was born in a stable in Ghaza in 1893.

12-Azraq Qouti horse called Abu Janoub Al-Saghir. His mother is Kahayla al-Ajouz. He was born in a stable in 1890. Was given as a gift to [illegible].

13-Azraq Ethiopian horse named Al-Saqlawi. He was given as a gift to [illegible].

14-Azraq qouti horse named Hadyan [illegible words] – [illegible].

15-Ahmar horse named Nafousani al-Alqami – [illegible].

16-Azraq horse named Obayan.

17-Hadidi azraq horse named [illegible]. Was given as a gift to [illegible].

18-Azraq Qurushi horse called Obayan. His father is Obayan and his mother is the English horse. He was born in [illegible], and was given as a gift to Hussain [illegible].

19-Ahmar horse called Kahaylan al-Ajouz [illegible]. Was sold to [illegible] on [illegible].

20-25-Faded [illegible].

Page 2:

Riding Horses consisting of thoroughbred horses and offspring mares His Excellency Dawlat Afandim Abdel Halim Pasha Bashir [illegible]. Continued: Horses found until 1864

26-A [illegible] horse named Abu Janoub son of his father Janub Al Azraq and his mother is Najibiyah. Born [illegible]. Was gifted to Ibrahim [illegible] Rushdi in 1865.

27-An ahmar horse named Obayan al-Sharaki. Was given as a gift to Shawqi in [illegible] palace in 1865.

28-White horse called Hadban. Was given as a gift to Shukri Bek in 1865.
29- Azraq horse named Kuhaylan al-Ajouz. Was sent out for sale in 1866.

30-Ashqar horse named Dadaman. Was sent to Kafr El-Sheikh in 1865 [illegible].

31-Ashqar horse named Al-Saqlawi. Was given to His Excellency Farid in 1865.

32-Azraq Ethiopian horse called Obayan al-Sharaki. His mother is Jawharah. Was given as a gift to His Excellency Khalil Pasha in 1895.

33-Azraq horse named [illegible]. Was sent out for sale in [illegible].

Horses brought in 1895:

Horses that were brought from Baghdad with the knowledge of Khawaja [illegible]:

34-Ahmar horse with a small [illegible]. Died in 99.

35-Azraq Ethiopian horse. It was given as a gift to [illegible].

36-Azraq white-footed horse . It was given as a gift to His Excellency Fairouz Agha [illegible].

37-A horse with the same description above. Was sent out for sale [illegible].

38-Azraq Ethiopian horse. It was given as a gift to the supervisor of the stable in Abbasiyah [illegible].

39-Azraq qouti horse. It was sold in 99.

40-Ahmar [illegible] horse with a small [illegible].

41-Black white-footed horse. It was given as a gift to the physician of [illegible].

42-Mawazdi azraq horse. It was sent out for sale.

43-Azraq qouti horse.

44-Grey horse.

45-Black horse with a small [illegible]. It was given as a gift to the inspector of [illegible] of Kafr al-Sheikh in 99.

Horses coming from the shaykh of the Arabs Maari Ibn [Samir or Taymiyah]:

46- An ashqar horse called Lebanon. It was sent out for sale in 99.

47-An azraq Mawazdi horse called Al-Saqlawi. It was sent out for sale in 99.

48-An azraq horse named [illegible]. It was sold [illegible].

Page 3: Continue: Thoroughbred Male Horses

Product of 1895:

49-An ashqar pony. His mother is the English horse named "The [illegible]" and his father is Kahaylan al-Ajouz Al-Ashqar, born in 1895. Sold [illegible].

50-An ashqar pony [illegible] white-footed (left side). His mother is an English horse called "The Rose". The father is Najdani Al-Simiri. Born in Stable 209 in 95. Sold [illegible].

51-An ashqar pony. His mother is called [illegible]. She is an English horse. His father is Dahimdan and was born in stable 76 in 1895.

52-An ashqar pony. His mother is English and is called "Julie Jana"[or "Johnny Jana"]. Son of [illegible]. Born in stable 244 in 92.

Product of 1896:

54- An ashqar pony. His father is the ashqar Horse Najdani al-Simiri and [his mother is] the English female horse [illegible]. Was born in [illegible].

55- An ashqar pony, his father is Najdani al-Simiri and his mother is the English horse mentioned [illegible narration about the horse]. It was given as a gift to [illegible].

PAGE 4:

Rishah Dahmah Ibn Bassis

Continued . . . Mares [illegible]

At this point the text became entirely illegible and the translation was abandoned. However the following points are noted:

1. These are indeed stable records from the 19th century, but there is no evidence that they were from the stud of Ali Pasha Sherif. There is no indication that the records refer to stables in Cairo. The records appear to describe studs belonging to two different men or stable owners, perhaps noblemen, and apparently wealthy men, who kept "*purebreds*" as riding horses. The records appear to be from the 1860s through to the 1890s. The definition of the term "thoroughbred" as it is used in these pages is not defined. Mohammad Ali Tewfik was born in 1875.

2. The physical deterioration of the original records, compounded by loss of image resolution due to the photo-reproduction process itself, have rendered most of the document illegible.

3. There is no pedigree information given for any of the horses, beyond the names of the sire and dam.

4. The horses' colors are described, and their names were often given simply as "strain names", terms that will be familiar to students of the history of the horses of Arabia. Ahmar is a term that denotes a coat color that is "red", ashqar translates as "pale, blonde, or yellow" and azraq is "steel-gray".

5. The presence of English horses in some of the pedigrees is noted.

Figure 10 - *Arabian Mares, from the collection of Prince Mohammad Ali Tewfik.*

Prince Mohammad Ali Tewfik wrote Breeding of Pure Bred Arab Horses in 1935, to provide future breeders with a clear picture of the characteristics of the Arabian horse, "and, as all thoroughbreds are descended from him, it is only right that he should be looked upon with the veneration which he deserves." The prince is remembered today for his involvement in the case of the "two Yemamas" discussed later in this book.

Other early Egyptian horse breeders took a more scholarly approach to the selection of *purebred Arabian horses*. From its beginning in 1898 the directors of the Royal Agricultural Society adopted a very rigid approach to the purity of the horses that they selected for its foundation. They selected horses that they believed, without reservation, the finest and purest examples of desert breeding, drawing at first from the available horses descended from stock of Abbas Pasha and Ali Pasha Sherif. Genetic Purity was their guiding principle.

Figure 11 - *Prince Mohammad Ali Tewfik.*

From very early times the Arabs have been reputed for the preservation of the pedigrees of horses as a precaution against any intermingling of other strains. The pedigrees were conveyed by word of mouth from old to young, and the horses were even immortalized, together with their qualities, in poems and verses until the advent of the era of recording and writing.

(Dr. Abdel Ashoub, RAS Studbook Vol. 1, 1947)

The Egyptian Arabian horse breeder is interested in pedigree research because of the value of the information contained in the pedigree with respect to horse breeding of the present day. Any pedigree information that can be obtained from sources in the Nejd region of Arabia is particularly important because of the purported links between the horses of the Nejd during the 19th century, the acquisitions of Mohammad Ali and his descendants, and the subsequent selection of foundation stock by the RAS.

Carl Raswan expressed the opinion that the preservation of the *purebred Arabian* possessing a classically pure and superior pedigree was essential. The antique blood obtained directly from the desert tribes would be needed in the future to regenerate the part-bred horses which will have lost Arab qualities. Writing in the 1950s Raswan was certain that "uncontaminated Arabians of doubtless pure desert ancestry are still available in sufficient numbers (outside of Arabia itself) in Egypt, Europe, and America."

The reality of breed degeneration can be seen today in the English Thoroughbred. By breeding horses for profit from a small pool of famous and well known sires, the industry has produced poorly conformed animals that are unable to meet the demands of competitive race jumping. During the 2006 Cheltenham Festival, 11 horses died from leg injuries and broken necks. There are 8,200 betting shops in Britain.

Regrettably, most all documentary connections between Nejd horses of the past and living horses of the present day have been lost. Despite this fact, a study of the traditional desert practices regarding the maintenance of pedigrees remains illuminating.

The *hujje* is the birth certificate of the foal. It is not a written document drawn up by horse dealers at the time of the sale of an adult horse to European buyers. The construction of a *hujje* was based on a conventional formula, but the matter of assessing the validity of the *hujjah* (also transliterated as *hojja*) is in fact quite complicated. In the West, the concept of a certificate of authentication carries the connotation of a business document. In the Muslim world, the word itself carries a significant religious connotation and it is used by the Arab with a feeling of gravity and seriousness. In Islamic theology, *hujja* means "proof" and is used to refer to a Mohammedan saint or holy man whose piousness and sanctity embody the "proof of God" to humanity. The Arabs took a *hujje* very seriously. Western attempts to study these documents, however, have been hampered by the fact that the way of life of the inhabitants of the interior of the Arabian Peninsula had remained unknown until recent centuries.

Opinions have varied over the years concerning the usefulness and validity of the *hujjah*. Pedigree skeptics opine that no written document originating from the desert can be accepted with certainty, due in large part to widespread illiteracy among the Bedu. Pedigree purists are more likely to accept the documents as valid based on the Bedouin's reputation for an attitude of gravity and seriousness when attesting to a horse's pedigree. As it turns out, neither is necessarily correct.

As with all legal documents, some of these documents are probably true and correct, some are in all likelihood fabricated, and many are a conglomeration of fact and fiction. Since all legal documents are based on a series of assumptions and assertions, and since the truthfulness of these assumptions is dependent on the truthfulness of the people who make them, even the well-constructed document may contain information which is simply not true. Much has been written by pedigree purists regarding the honesty of the Bedouin, claiming that a Bedouin would not give false information on the sale of a horse because of the dishonor and reprisals that would result from his own people. Purists emphasize this point in order to support the case for accepting the written pedigree records from the desert as factual.

Figure 12 - *Cheval Arabe, by Carle Vernet, from the British Museum.*

> Compromise is unknown to the Bedouin, who cannot tolerate the least uncertainty. The quantity of alien blood was not important to him, only its quality. He esteemed his own breeding to be better than all others, and even the smallest drop of alien blood would in his eyes only worsen, never improve. As to his own breed, he had no wish to improve it as he already had the best.
>
> (Erika Schiele, The Arab Horse in Europe, 1970)

Apart from mendacity and a host of human frailties, pedigree analysis and verification is plagued by the more mundane sources of inaccuracy. There is the matter of inadvertent and unavoidable human error. *Hujjaj* analysis is full of examples of mis-copied names, mistranslated terms and illegible handwriting so cryptic that a definitive reading is not possible. And there was at times frank deception by the authors of the documents.

The English journalist Samuel Johnson described the universal problem of human probity:

> *"We are all prompted by the same motives, all deceived by the same fallacies, all animated by hope, obstructed by danger, entangled by desire, and seduced by pleasure."*
>
> Samuel Johnson - Rambler No. 60

THE ELEMENTS AND CONSTRUCTION OF THE HUJJE

The form of the *hujje* in the Middle East was standardized, consisting of a series of conventional elements that most of the documents had in common.

In the final analysis, however, there is no proof that any particular *hujja* was a document of fact. The reliability of a given instrument was ultimately based on the truthfulness of the original breeder. It has been noted that the Arabs as a people have been regarded since antiquity as a group deeply rooted in the virtues of nobility, honor and fidelity.

Figure 13 - *Bedouin portrait, Library of Congress collection.*

Herodotus (born 484 BCE) wrote in Book Three of the Histories:

> No nation regards the sanctity of the pledge more seriously than the Arabs. When two men wish to make a solemn compact, they get the service of a third, who stands between them, and with a sharp stone cuts the palms of their hands near the base of the thumb; then he takes a little tuft of wool from their clothes, dips it in the blood, and smears the blood on seven stones which lay between them, invoking as he does so the names of Dionysus and Urania; then the person who is giving the pledge, commends the stranger – or fellow citizen, as the case maybe – to his friends, who in their turn consider themselves equally bound to honor it.

A number of firsthand accounts of pedigree documentation have been preserved by the earliest Europeans traveling in Arabia. One of the most authoritative of these is the compilation Historical Reports on Arab Horse Breeding and the Arabian Horse, by Karl Wilhelm Ammon. Published in Nurnberg Germany in 1834 and available in reproduction from Olms Press, this collection gives a lively and vivid picture of the horses of the Bedouin of the 17th and 18th centuries, written at a time long before the corrupting influence of Europeans became dominant in Arabia. The remarks are notably varied and at times contradictory. However, these reports warrant the close attention of the student of the desert horse because a remarkably clear and detailed understanding of the authentic Arabian horse emerges. After the infiltration of Western influences reached the interior of Arabia, the horses of antiquity appear to have changed.

Laurent D'Arvieux was born in Marseilles in 1635. His family was descended from the nobility of Florence, Italy. He entered the French diplomatic corps and, with a natural inclination for languages and travel, was assigned to a post in the Levant and Syria where he learned Arabic, absorbing the culture and manners of the inhabitants. He served in this position for 12 years. Later in life he was given the position of French Consul to Syria where he distinguished himself for six years. He was regarded in his time as an accomplished Arabist and a reliable author.

The Chevalier D'Arvieux in Sitten der Beduinen and Travels in Arabia wrote:

> "The Arabs are much less concerned over details of their wives' birth certificates than over those of the strain registers of their horses, the latter are dealt with most attentively. They often produce documents tracing them back 400 or 500 years."
>
> "The Arabs know by long custom the Race of all the horses they or their neighbors have; they know the name, the surname, the coat, and marks of every horse and mare in particular; and when they have no noble horses (stallions) of their own they borrow some from their neighbors, paying so much money, to cover their mares and that before witnesses whose attest it under their Hand and seal before the Emir's secretary or other public person, where the whole creation, together with the Names of the Creatures, is set down."
>
> "As the Arabs have only a tent for a house it serves then too for a stable; the mare, the colt, the man, the wife, and the children retire together, and all pig together. There you'll see little children asleep upon the mare's belly, upon her and the colt's neck, without the least harm."
>
> "They usually ride on mares as is properest for their business, experience has taught 'em that they bear fatigue, hunger and thirst better than horses, and bring them every year a colt, which they presently sell. Mares never neigh, which is very convenient for'em in their ambushcades to surprise passengers."

"An Arab would not be reckoned an honest man if he had not a mare to bestride."

"Mares leap rivulets and ditches as nimbly as stags, and if the rider happens to fall whilst they are leaping or upon full speed, they instantly stop and give him time to get up and mount."

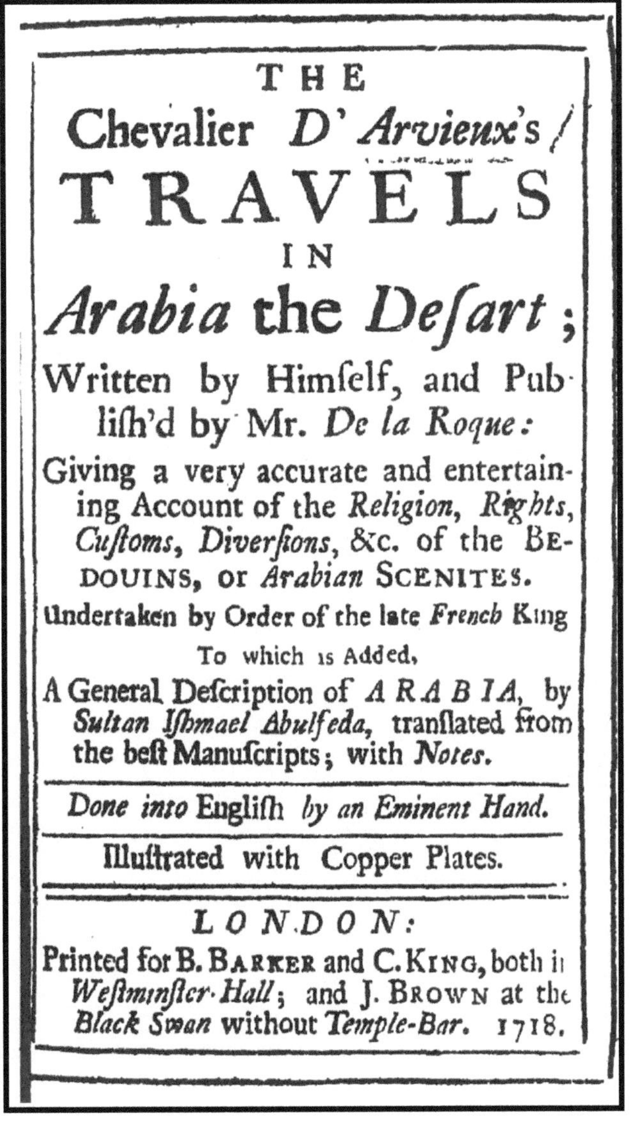

THE
Chevalier *D' Arvieux's*
TRAVELS
IN
Arabia the *Deſart*;
Written by Himſelf, and Publiſh'd by Mr. *De la Roque*:
Giving a very accurate and entertaining Account of the *Religion*, *Rights*, *Cuſtoms*, *Diverſions*, &c. of the BEDOUINS, or *Arabian* SCENITES.
Undertaken by Order of the late *French* King
To which is Added,
A General Deſcription of *ARABIA*, by *Sultan Iſhmael Abulfeda*, tranſlated from the beſt Manuſcripts; with *Notes*.

Done into Engliſh *by an Eminent Hand.*

Illuſtrated with Copper Plates.

LONDON:
Printed for B. BARKER and C. KING, both in *Weſtminſter-Hall*; and J. BROWN at the *Black Swan* without *Temple-Bar*. 1718.

Figure 14 - *Title Page from d'Arvieux's Travels in Arabia the Desart.*

Jean de la Rocque (1661-1745) was a French explorer and journalist. He visited the Levant in 1689, and on subsequent trips he explored Yemen. He was the first Westerner to write an account of the coffee trade in the region.

De La Rocque (Le Voyage de Mr. de la Rocque), Fouche, and Mariti report that the natives kept regular lists of their horses' strains, but generally only for those horses of great importance.

Figure 15 - *Carsten Neibuhr.*

Carsten Niebuhr (1773-1815) was a German explorer who recorded carefully written reports of the desert horse. He was also a surveyor, and was chosen by Frederick V, King of Denmark, to join an expedition of six men to map the Nile River and conduct scientific surveys of the areas of Arabia that they found accessible; Mecca, Mocha, and Saana. Their journey lasted five years, and Neibuhr was the only man to survive.

Niebuhr and von Rosettti state flatly that the Bedouin have no written records of the pedigrees or strains of their horses. Niebuhr wrote: "In Arabian cities, many common people can read and write." But he says that the inhabitants of the inner deserts are illiterate; only the occasional sheikh can read and write.

Niebuhr writes; "Although there are many Arabs with no qualms about perjury, we evidently have no knowledge of an Arab ever signing false records on the birth of a foal because they firmly believe that their whole family would be wiped out if they do not tell the truth in this matter."

The intrepid explorer Count von Veltheim agrees with the travelers who find the idea nonsensical that the pedigrees of the noble horses of the peninsula are in any way recorded.

"The incongruence in the argument that a nomadic people constantly wandering about in the immense deserts, should put down in writing the exact lists of origin of its horses for more than 12 centuries, and should have preserved them without any established archives, despite their constant raids, is all too obvious."

Burckhardt observed that the *hujje* or birth certificate was written down at the time of foaling by a scribe or literate chieftain. The genealogic information written down, however, contained only the names of the parents. It was assumed that no more than this was necessary since all members of the tribe would have known the grandparents of the foal personally.

Burckhardt has left a wealth of scholarly information regarding horses in the Middle East. After a university education at the University of Gottingen and at Cambridge University in England, Burckhardt accepted an offer from the Africa Association to discover the origin of the Niger River. He first went to Syria to perfect his disguise as an Arab doctor, acquiring the language and the cultural sensitivity to pass through Africa as an Arab, posing as a learned doctor of Islamic law. He adopted the name, Sheikh Ibrahim Ibn Abdullah. He made subsequent journeys into central Arabia, all of which he carefully documented. Burckhardt was never able to make his journey to seek the source of the Niger. He died in Cairo from dysentery in 1817.

Burckhardt states that while among the tribes of the Arabian interior there is some attempt to write down specifics of a foal's birth, the situation is very different when these tribesmen take a horse to one of the major trading cities for sale. In Mecca and Medina, a written pedigree is provided to the buyer of the horse. "That is the only time when a Bedouin is in possession of a written pedigree of his horse; in the desert, however, he would just laugh if asked for his mare's strain register."

Figure 16 - *Johann Ludwig Burckhardt (1784 to 1817), born in Leipzig, Germany.*

The legends concerning the extensive documentation of Arabian horse pedigrees in the desert appear to stem from an actual practice that did occur at times, but not always, and not everywhere. d' Arvieux wrote:

> A mare is always covered in the presence of several witnesses, who

> then issue a document which is signed and sealed by the Emir's secretary. This document contains the entire strain register, and the names and colors and markings of both parents. When the mare is foaled, witnesses are again called in, and then another document is set up in which the sex, the conformation, the color and markings of the newborn foal, and in addition the time of its birth, are recorded exactly.

This practice existed among the Arab nobility but was far from common among the rank and file of Arab society. Its usage was restricted to the breeding of stallions of superior quality and fame. Niebuhr confirms this and adds that after the mare is covered, a man is assigned to stay with her for twenty days to make certain that no subsequent matings take place. Similar accounts are found in De La Rocque. Niebuhr says that "the genealogic tables, called *hujje*, never go back to the granddam, since by general assumption every Arab in the tribe is already familiar with the tradition of the whole breed's pure lineage." He points out that strictly speaking, a *hujje* is a birth certificate. This term should not be applied to the documents drawn up on the spot for the sale of a mature horse.

Deep in the interior of the Nejd, far from any towns or settlements, matters were very different. There the tribes were truly nomadic and were constantly on the move. There were no scribes, no secretaries, and no learned sheikhs. These Bedouin relied entirely on memory to relate the pedigrees of their horses. They had no documents. In the 17th century, there was little trade between the inhabitants of central Nedj and the outside world. If a horse was to be sold, the owner took it to the horse markets in Mecca, Medina, Baghdad or Basra. There, knowing that a written pedigree would be required for a sale to occur, the owner gave testimony to a local judge or kadi who wrote the document for him. Witnesses were required, and if none were available the owner would be forced to hire some from the local populace. Burckhardt wrote that although the horse traders in Syria were unscrupulous, the Nejd Bedouin were "not familiar with the deceitful ways of the European horse dealer who is very skilled at cheating the buyer. Without worrying one may accept a horse from them on the first trial, or after having seen the animal for the first time, without the fear of being cheated."

Ammon quotes Russell, a long time resident of Allepo on this matter:

> *"In the interior parts of Arabia the people are not too corrupt and*

> *therefore still regard an oath with respect. However along the borders of the desert where Europeans have settled, the spirit of avarice reigns, and the Arab's former integrity has been transformed into the horse dealer's cheap cunning."*

Von Rosetti repeats the common thread of the earlier writers, but takes care to qualify his opinion:

> *"At the birth of the foal the Bedouins of Arabia and Syria occasionally bring in several witnesses in order to record in their presence the names and breed (noble strain) of the foal's sire and dam, as well as the characteristic markings of it color. This however, is no common usage, and this type of birth certificate never refers to the grandsire or granddam...Now if the numbers of horses in the desert were greater it would make it more difficult for the Arabs to keep the genealogy of each horse in his mind...however..it is sufficient for an Arab to have seen a mare just once, in order to remember all details for her genealogy and her life's history."*

The early accounts from the interior of Arabia all agree on the point that the Bedouin horse breeders of the Nedjean highlands were, without doubt, the true and original guardians of the *purebred* authentic desert horse. In 1809, the Rualla Bedouin of Nejd engaged in a battle with the Ottoman Pasha and his forces from Baghdad, capturing 500 of the Pasha's horses. Distrustful of the purity of the equine war booty, they did not keep a single one of the horses, selling them instead to horse traders in Yemen.

Very few original documents from this period have survived. Several translations of *hujjaj* have been preserved in the records of early European adventurers. Most of these, however, are, strictly speaking, not *hujjaj* but are bills of sale, executed at the time of the sale of a mature horse to a foreign buyer. The *hujje* itself, strictly speaking, was a birth certificate written at the time of the birth of a foal.

Figure 17 - *Sebaste, Ancient Samaria, by David Roberts. From the collection of the author, photograph copyright Dr.William Hudson.*

Ammon has translated several *hujjaj*, such as the following from Pennant's Britische Zoologie:

> The writing of this document, or instrument was occasioned by the following incident. Thomas Usgate, the English consul, appeared at the court in Acca in the house of the kadi, the legally recognized judge, and with him Sheikh Morad ebn el Hadgi Abdolla, Sheikh of the land of Safad. The aforesaid consul requested from the mentioned sheikh the attestation of the grey horse's origin which he was buying from him. He was given the assurance that it was a Manaki Shadahi. However, not satisfied with only this information, he demanded the testimony by the Arabs who raised the horse and would know how it came into the possession of Sheik Morad. Hereupon some reputable Arabs arrived, whose names will be mentioned below; they testified and

declared that the grey horse, which the consul had formally acquired from Sheikh Morad, was a Manaki Shadahi, a horse of pure breeding, purer than milk, and that its history went as follows: Sheikh Saleh, Sheikh of Alsabal, bought it from the Arabs of the El Mohammadat tribe, and Sheikh Saleh sold it to Sheikh Morad ebn el Hadgi Abdollah, Sheikh of Safad, and Sheikh Morad sells it to the aforementioned Consul. This matter has come before us, and we are well acquainted with the issue. Since the mentioned gentleman requests a certificate and the signatures of witnesses, we have therefore set up this pertinent certificate, which he shall take as the proof hereof.

Dated Friday the 28th of the second Rabia in the year 1135 (that is the 29th of January 1722)

The following translated *hujje* is supplied by Ammon, who has taken it from Fundgruben des Orients, written by the French consul in Aleppo, Rousseau. The formula is typical:

The invocation of Allah, the statements of the facts, the attestations and assurances, and the names of those present. Seals were used for those who could not sign their name.

In the name of God the All-Bountiful and All-Merciful, to whom we look for guidance and assistance. The Prophet said: My people will never gather to give false testimony.

We, the undersigned, declare before God the most high, we assert and testify, swearing by our luck and our destiny, that the bay mare with the white forehead, with a white front and hind foot, descends from noble ancestors through three direct and uninterrupted generations on the sire's side as well as on the dam's side; that she is

> truly the daughter of a mare of the Seglawi breed of Nedj, and of the stallion Choueyman-Elsebbah breed; and that in her are blended the characteristics of those mare of which the prophet says: Their womb is a treasure chamber, and their backs the seat of honor.
>
> Based on the testimony of our forebears, we attest by our luck and destiny, that the mentioned mare is of noble origin and as pure as milk; that her gallop is distinguished by fleetness and speed; that she patiently suffers thirst; and is accustomed to the hardships of long journeys. We have set up this present instrument for the certification of the above according to what we have seen with our own eyes, and what we know. God is our first witness. Follow the signatures and seals of the testificators.

This example is undated, and the location of the transaction is not stated. There is no indication that the seller is actually the breeder of the horse. This document is typical of those which were in use in the early 1800s.

The next example illustrates how brief these bills of sale can at times be:

> *"By the present document testimony is given that Bahyan was born out of a mare that had been covered by Aesrak, who sprung from Abian, the son of Abid, one from the great illustrious family. The mare had been covered in the presence of the noble Sheikh Soliman, whereupon the present document has been issued."*
>
> Damascus, in the month of Dulkadah in the year 1204 (1790).

There were no signatures, and no seals. The names of the seller and the buyer are not given.

Figure 18 - *An Arabian Horse, from the London Illustrated News, 19th century.*

Ammon relates the following *hujje* which he translated from Berenger's Geschichte des Reitens:

> The brief report of this strain register and the reason for the sale are the following: I, Fakir Mohammed, son of Hagdi Chalil, son of sheikh Suleiman, Sheikh of the market-town of Alchadar which borders on the ridge of Mount Sighangan, have today sold my brown skew-bald horse Bik, which is an authentic Arabian, son of the bay mare Alkahila, sired by Nif, the Gailf, a bay with black eyelids, a noble Arabian. The sire's dam was Hussein Bey's mare. He is of pure noble blood. I Fakir Mohammed, to whom the highest God-praise be to him - may be merciful, son of Hagdi Chalil, I myself have sold my above mentioned horse, which is still among my horses and in my

paddock. It is a brown skewbald with black eyelids. The witnesses mentioned below will testify to its breeding and its strain. On the last of the month of Safar in the year 1173 (1760 by Western calendars). At the very same time the aforementioned horse has been sold by way of an express messenger, following the drawing up of this document, to Mr. N., son of N., a commander of the British Company of Frankish Merchants in the British Agency, situated on the fringes of the desert at Aleppo. I made this contract with him, and have received payment amounting to the full sale price in valid and correct currency.

Mohammed, son of Hagdi Chalil

Hagdi Isa, dervish

Hussein, aga Suleiman

Hagdi Mohammed, dervish

Seid Ibrahimi, head aga of the khan at Toman Sid Abdallah Alynas

The following certificate of origin was recorded by Ammon, who had obtained it from a royal Prussian equerry named Ehrenpfort at the time of the acquisition of the stallion Nishty. The document was found hanging about the neck of the stallion by a thin metal chain. It consisted of a small tin paper that had been folded into the shape of a triangle, in which various herbs and roots, presumably a talisman, were enclosed.

In the name of God the All Merciful; Praise be to the one God,

Purpose and occasion for the writing of this document is the noble foal of unmixed pure origin of the Nedsjedi family, out of a noble mare belonging to Obeid-eltemen-o-awy, and by the sire Elzekhlauwy, who again descends from one of Mohammed Elokheidy's noble mares; and which foal Sheikh Abdelkaadir, Sheikh of Jebel-ann, has sold to Mr. Tschaky, consul of the Nether-lands community at Haleb and one of its wealthy and respected members, under attestation by the reputable and prosperous citizen Hossein Eltshjemal, Ishmael Elaswed and the Sheikh Hossein; that the same is truly a Koheilan, purer than pure sweet milk, and with luck and blessings no fault will ever be found on it; also that the same is a straightforward Koheyl and, possessing a characteristic praised chiefly of them, rises bravely against an enemy army. To preserve such noble offspring whose lineage has been guarded carefully, firmly and uninterruptedly in consideration of all points just referred to: that is the object of this instrument.

Dated, the 25th of the month of Shawaal in the year 1203 (July 17th, 1789)

"The Poor One"

"Sheikh Abdelkaadir"

"Sheikh of Jebel Ali"

"With the help of God"

This is to record:

We whose signatures and seals are below, SHIYUKHS of the Suwaylimat a branch of the Anazeh, do testify, by Allah, and by Muhammed son of Abdullah, truly, without compulsion, in respect to the horse of Ma'ashil Hashai of the Suwaylimat: and he is a bay, with a mark like the new moon on his forehead; by our stars and fortune, his dam was of the strain Wadna-Khirsan; and his sire Kuhaylan Abu Junub - the well known strain. He is a horse used as a sire. It is also known to us that his price has stood Khider, the Uqayl, in 550 *ghazis* (about 440 gold pounds). According to our knowledge and information we have written this certificate. (Sealed and signed by twelve witnesses - the owner of the horse and Shiyukhs of the Amarat tribe.)

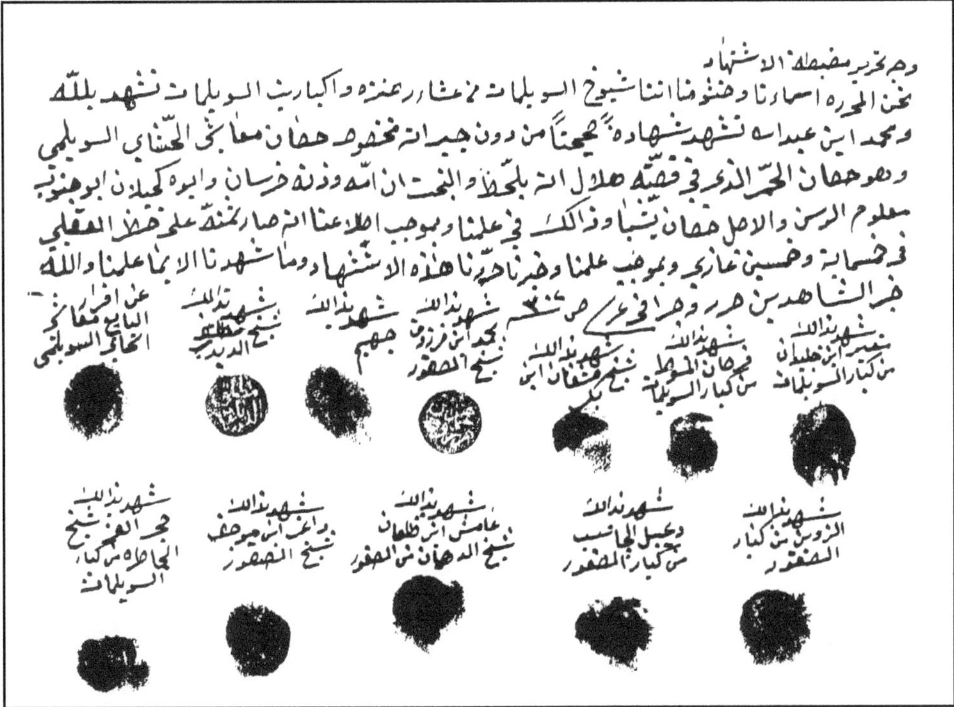

Figure 19 - *A facsimile of an original hujje preserved in* <u>*The Raswan Index*</u>*. Most of the marks identifying the testifying sheikhs are thumb prints.*

In some Bedouin tribes, the birth of a foal was documented by writing the names of the sire and dam on a scrap of paper which was then placed inside a brass ball that was hung around the foal's neck with a leather strap. This device, called a *kojet*, was never removed from the animals and proved useful at the time of a subsequent sale. In some cases, the genealogic information was written down at the time of sale of an adult horse. The document was often folded into the shape of a triangle, filled with flowers and herbs, and fastened around the neck of the horse with a thin iron chain. One surviving document of this type was written on one thin piece of tissue paper, one foot long and four inches high. About 43 lines of very small Arabic characters presented the details of the parentage of the horse for the prospective buyer.

Figure 20 - *Bedouin warrior, early 20th century, from the collection of the Library of Congress.*

In summary, the most reliable certificates of origin came from the Bedouin in the central highlands of Nejd. Certificates written by horse traders and dealers in the Syrian Desert and in the vicinity of Allepo were less reliable, since the horses for sale had by that point been through several hands, and the corrupting influence of European traders was strong, exposing the natives to the temptation of committing fraud and abuse. All observers of the 18th century agreed that the value of any Arabian horse pedigree was uncertain. Paradoxically, a questionable pedigree attested to by an original horse breeder in the Nejd was considered superior to any other pedigree attested to outside Nejd. A Nedji may sell a horse with an inaccurate pedigree, but the horse was still irrefragably asil: the Nedjis had nothing but asil horses to sell.

Figure 21 - *Cheval Arabe, by Theodore Gericault, from the British Museum.*

Figure 22 - *This drawing from the Lukomski treatise is a chest decoration that accompanied a mare imported to Poland from Arabia in the 19th century. The pedigree of the horse was often enclosed in a pocket of a chest decoration.*

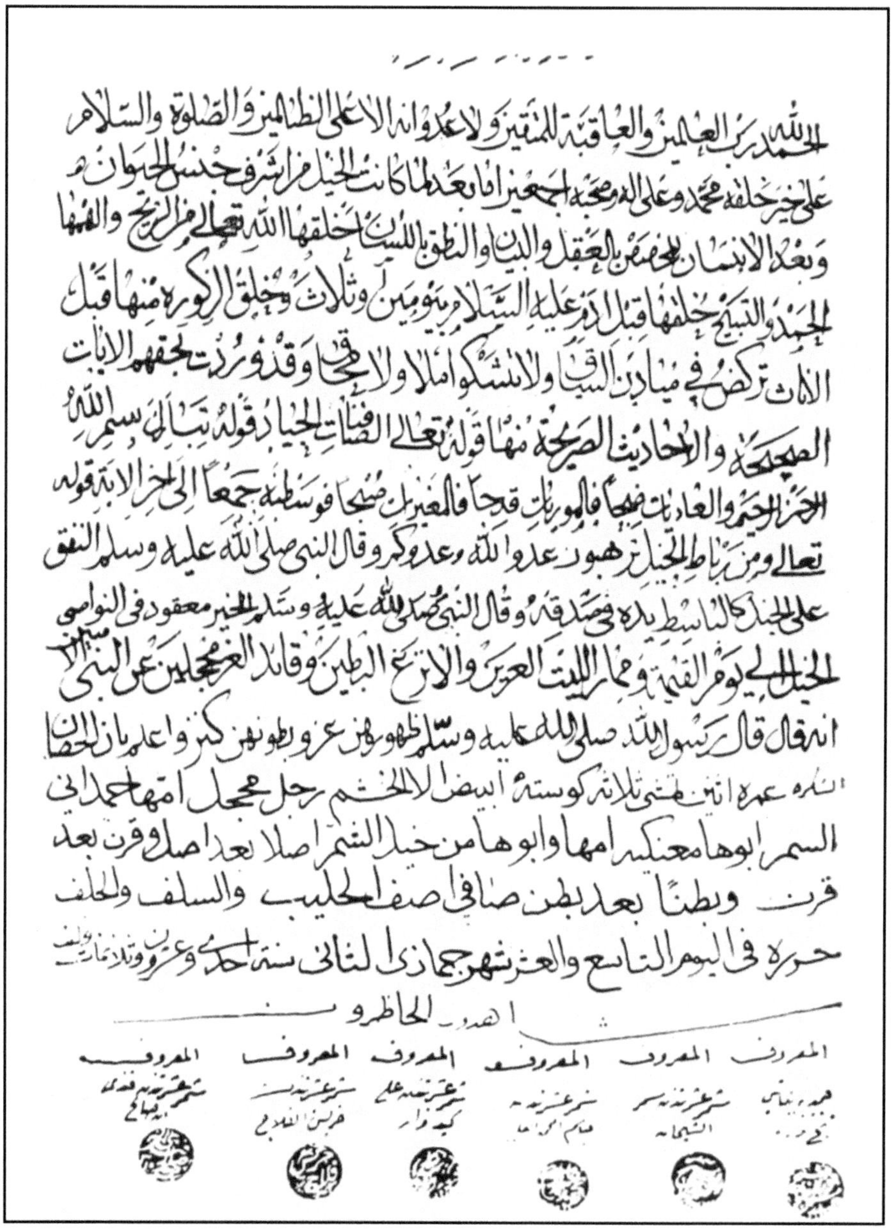

الحمد لله رب العالمين والعاقبة للمتقين ولا عدوان الا على الظالمين والصلوة والسلام
على خير خلقه محمد وعلى آله وصحبه اجمعين اما بعد لما كانت الخيل من اشرف جنس الحيوان
وبعد الانسان المخصص بالعقل والبيان والنطق باللسان خلقها الله تعالى من الريح [illegible]
[illegible] والتسبيح خلقها قبل آدم عليه السلام بيومين وثلاث وخلق الذكورة منها قبل
الاناث تركض في ميادين السباق ولا [illegible] ولا [illegible] وقد وردت بحقهم الايات
الصحيحة والاحاديث الصريحة منها قوله تعالى الصافنات الجياد قوله تبارك بسم الله
الرحمن الرحيم والعاديات ضبحا فالموريات قدحا فالمغيرات صبحا فوسطن به جمعا الى اخر الاية وقوله
تعالى ومن رباط الخيل ترهبون به عدو الله وعدوكم وقال النبي صلى الله عليه وسلم المنفق
على الخيل كالباسط يده بصدقة وقال النبي صلى الله عليه وسلم الخير معقود في نواصي
الخيل الى يوم القيامة ومنها الليث العريض [illegible] البطين وقائد الغر المحجلين عن النبي
انه قال رسول الله صلى الله عليه وسلم ظهورهن عز وبطونهن كنز واعلموا ان الحصان
[illegible] عمره [illegible] ابيض [illegible] رجل محجل امها حمداني
السمر ابوها [illegible] امها وابوها من خيل الشمر اصلا بعد اصل وقرن بعد
قرن وبطنا بعد بطن صافي [illegible] الحليب والسلف والخلف
حرره في اليوم التاسع والعشر شهر جمادى الثاني سنة [illegible]

الحاظرون

المعروف المعروف المعروف المعروف المعروف المعروف

Figure 23 - *The hujje of an Eastern horse imported into Spain from the desert, image from the Spanish Stud, as reproduced in* The Arabian Horse of Europe*, by Erika Schiele. The document was witnessed by six sheikhs, each of whom appended a seal.*

GEOGRAPHY, THE HUJJE, AND THE QUESTION OF GENETIC PURITY

Arabia was a closed book to the West until recent centuries. As Western commerce became more interested in the usefulness of the peninsula, explorers began to collect geographic data on the largely unknown land. Knowledge of the geography of the Arabian Peninsula was a long and protracted process as Western cartographers slowly added details of the cities and land formations over a period of several centuries. The following series of maps illustrates the progress.

Figure 24 - *The Arabian heartland was isolated from the wider world, as late as the 17th century. This image shows a 1675 East India Company map of coastal ports from the southern extremity of Africa to Eastern Africa and India. This was an active maritime trade route for Europeans who traded for spices and silks. When the centuries-old Spice Road was closed by the Ottomans in 1493, a sea route to India was needed. The Portuguese navigator Vasco da Gama reached India by sailing from Europe around the southern tip of Africa and from there to the East. Following his discovery, trade with the Far East resumed. Port cities along the route were important to the vessels on the sea routes as a means of obtaining provisions and as an escape from pirates and inhospitable weather conditions. While the names of the port cities of the Arabian Peninsula were well known and commonly visited by sailing vessels in the 17th century, the interior had yet to be explored. The inhabitants and their civilization were to remain completely unknown to those on the outside world for another century.*

Figure 25 - *Landscape Map of the world in 1138 from* *Tabula Rogeriana*. *This map was prepared by the Arab geographer Muhammad al-Idrisi for the Norman King of Sicily Roger II. Each dot represents a town or city, and the Arabian Peninsula contains many of these notations. To this Arab geographer, the interior of Arabia was not a complete blank; it was well known to him. He indicated the location of towns as well as major topographic features. The map appears to have disappeared into obscurity in the Middle Ages. Subsequent European cartographers were unaware of this remarkable work for many centuries. Today only three copies are still in existence.*

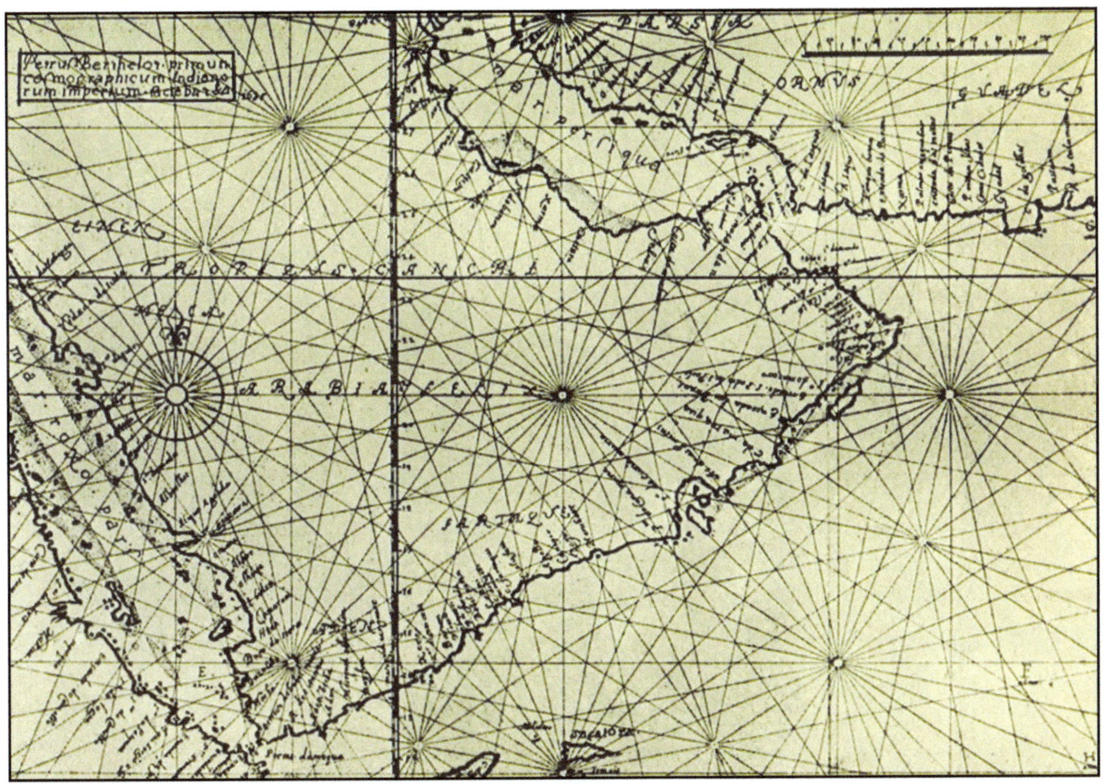

Figure 26 - *Bertelot's map of the Arabian Peninsula, 1635. While coastal ports are well represented in this map, the towns of the interior are not noted. This map was primarily for nautical use in the lucrative trade between India and Europe. Specific longitudes and latitudes were expertly determined and drawn. Compass roses are noted and the numerous radiating rhumb lines, useful to navigators are seen.*

Figure 27 - *A Map of Arabia by the Frenchman Nicolas Sanson, 1654. Sanson's map greatly overestimates the boundaries of "Arabia Felix" or, "Arabie Heureuse" omitting the Empty Quarter altogether.*

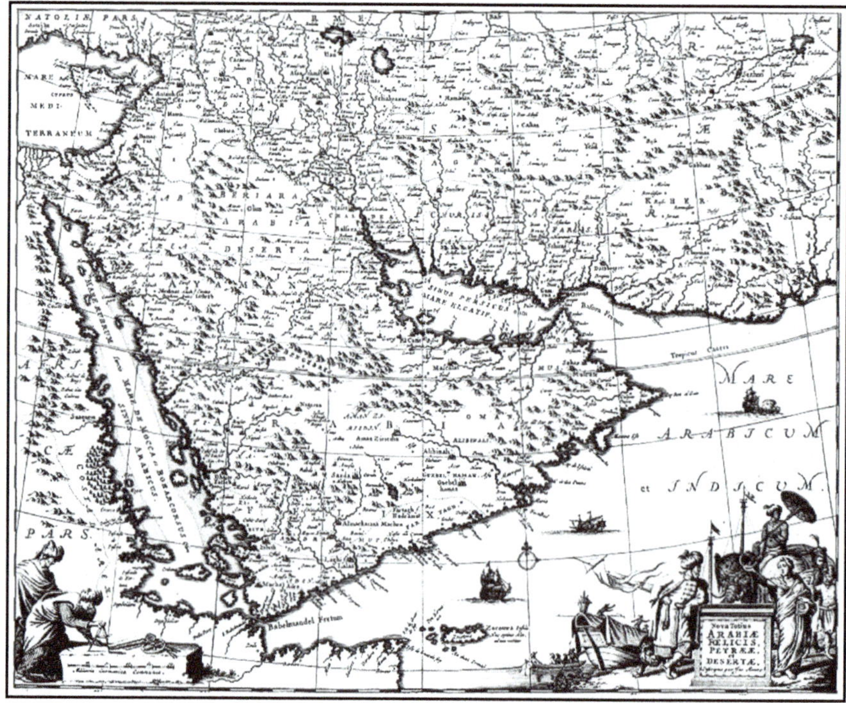

Figure 28 - *1680 map of Arabia by the Dutch cartographer Jac van Meurs. The geographic details of the interior of Arabia are shown accurately.*

Figure 29 - *Laroque's map of the southern Arabian Peninsula from 1716. Medina, Mecca, Yemen and Muscat are accurately depicted. The region named Naged is noted in the central peninsula.*

Horse breeding and horse trading in the Middle East during the 17th and 18th centuries consisted of a network of individuals and tribes, some of whom genuinely valued and protected the asil horse, and some of whom sought to make a commercial success from the breeding, racing, and sale of part-bred Arabian horses. The degree of honesty and accuracy in horse pedigrees was intimately connected with the places where traders and breeders lived and carried out their operations.

When studying the perplexing issue of the Egyptian Arabian horse pedigree, much can be inferred from an appreciation of these geographic patterns and the inconsistent and contradictory manner in which place names were often used. Geography determined the likelihood that the pedigree of a given Arabian horse was valid, and also determined the likelihood that a given horse that was said to be asil was actually asil.

History and tradition relate that the authentic genetically pure blooded and original desert Arabian horse was bred by the Bedouin in Nejd, the central highland region of Arabia. There is no scientific data to support this contention. It is an assumption that is based on the historical reports from reliable sources and has been the foundation of the received wisdom concerning the Egyptian Arabian breed.

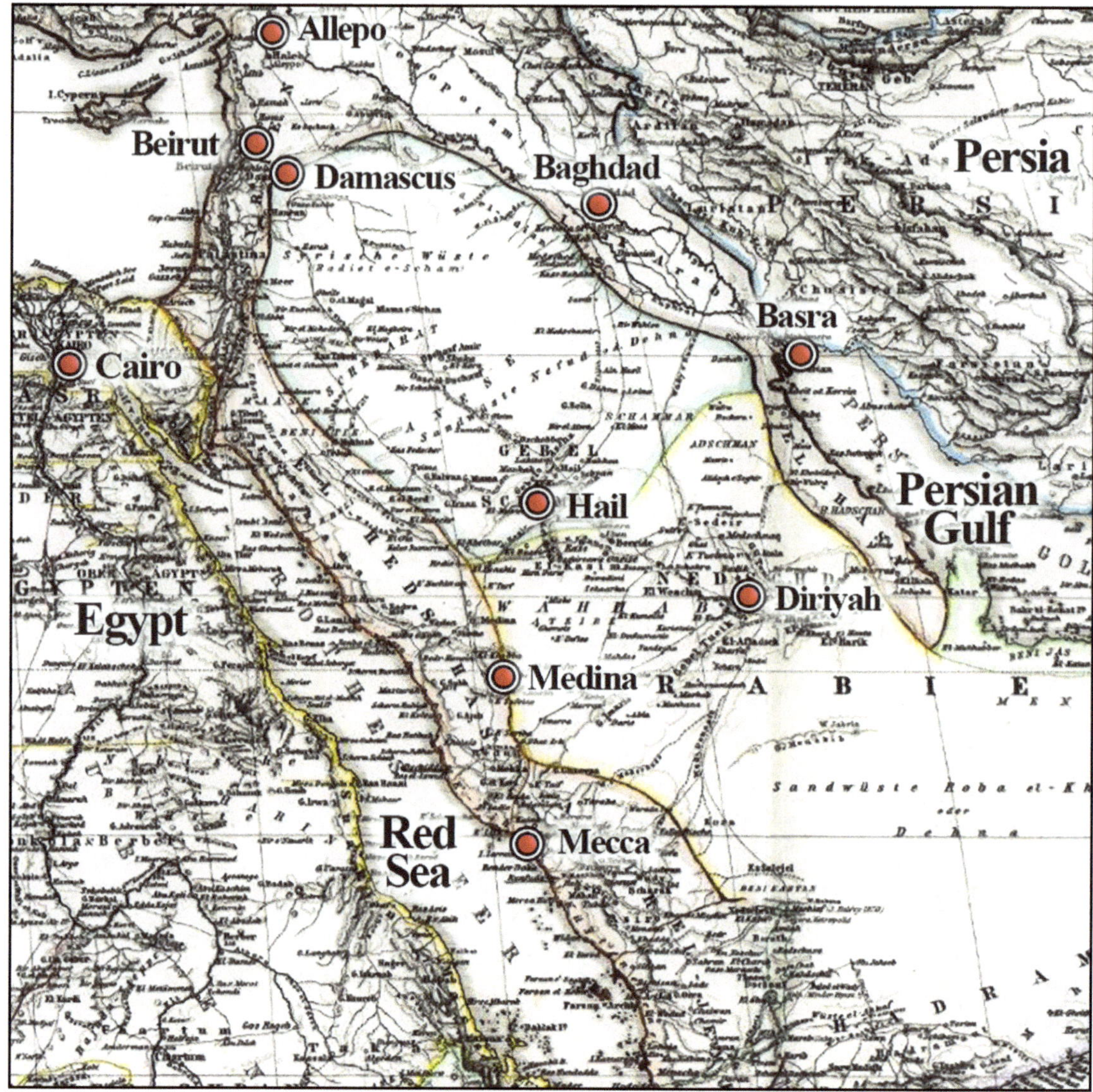

Figure 30 - *Middle eastern cities involved in the trade in Arabian horses during the 18th and early 19th centuries.*

Carl Raswan wrote:

> Nedjean horses are those bred in Nedj (Central Arabia) by the princes of Hayil and/or the camel-breeding tribes of Nejd. Desert Arabia and the Bedouin sustained the PURE blood of Arabian horses for thousands of years. It is with this PURE blood (kept uncontaminated like the

> original blood of wild animals) that horse-breeders all over the world were able to create (with their native breeds) new types of horses and later were able to regenerate them with pure Arabian blood. (Part-bred Arabians have to be regenerated from time to time with PURE Arabian sires). Hence the importance to keep some of our Arabian stud farms absolutely uncontaminated from any outside blood.

The Nejd is a topographical island, surrounded on three sides by desert and on the fourth side by the Hejaz and the Red Sea. The reports of Europeans emphasize this island-like isolation, and collectively these reports are in agreement on the question of the authenticity of the Bedouin horses from this region. The isolation resulting from its geography played a key role in the preservation of the Nejd pure blooded Arabian horse for centuries. The Bedouin were a fierce and independent people. They resisted the attempts of the Ottoman government to lure them out of the desert and domesticate them. They resisted accepting the use of Ottoman customs, language and modes of dress. Their physical isolation helped them to preserve the Nedjean culture from contamination by the outside. The study of surviving *hujjaj* appears to confirm this impression.

The relative validity of any *hujja* can be assessed by considering three principle factors: the geographical location in the Middle East where it was written, the particular time in history when it was written, and the identity of the persons involved in its preparation and execution. For example, a *hujja* written in the 16th century, in or near central Nejd, prepared by a scribe for a nomadic Bedouin breeder who actually bred the horse in question, is far more likely to be valid than one written in the 19th or 20th century, in the horse markets of Baghdad or Aleppo, executed by a horse trading intermediary as part of the sale of a horse to a European buyer.

In the evaluation of the validity of any particular *hujje*, geography was destiny. This is due to the fact that each of the Bedouin tribes possessed their own hereditary regions on the peninsula. Each tribe had its own reputation for truthfulness in dealing with the sale of horses to foreigners.

While the northern tribes occupied a place of prominence as the principle source of *purebred Arabian horses*, nomadic tribes existed in the south as well. There may even have been wild horses present on the peninsula as late as the 19th century.

Figure 31 - *The Great Rualla tribe of Nejd, from The Raswan Archives.*

The asil horse of Arabia, legend has it, came originally from the southern part of the peninsula. Lady Anne Blunt's diary of 1880 contains a note regarding the possible presence of *purebred* desert horses still living in feral herds in southern Arabia.

> "Wilfred has met a man who came from Sana and told him that some distance from Sana in the interior there are the Beni Husyan, Mohammed Bedouin who catch wild horses. They live in the district called Jofr el Yemen and are very 'adroit' in riding."

An appreciation of the location and boundaries of three particular geographic regions in Arabia and the Middle East (Nejd, the Levant, and the Syrian Desert) are useful in understanding the historical dislocations of the Arabian horse. These terms were used by early importers of Arabian horses from the Middle East to the West to describe the locations from which many horses were purchased. The imprecise boundaries of these three regions led to much confusion and misunderstanding about the geographic areas referred to in the old records by authors such as Burckhardt, the Blunts, and Palgrave.

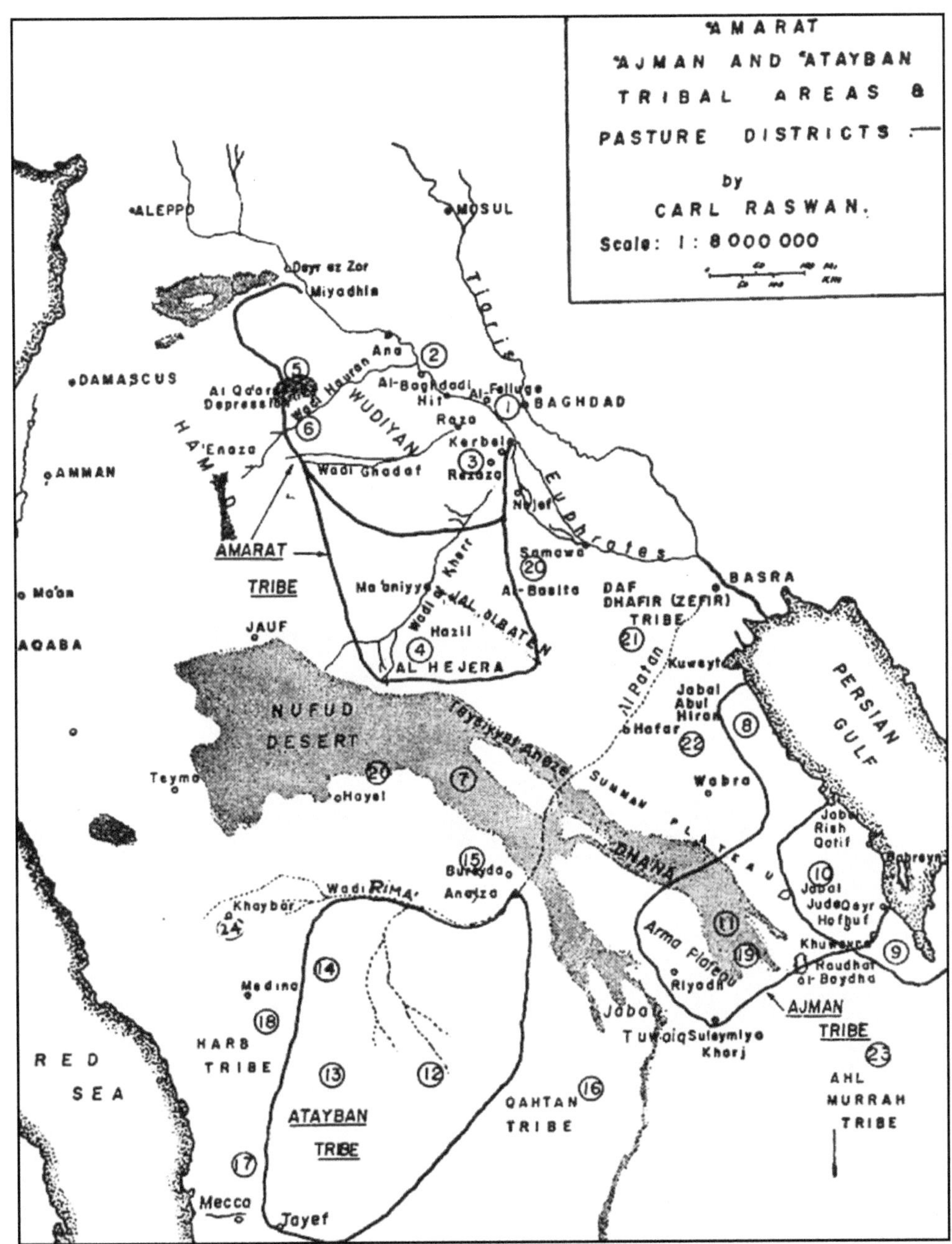

Figure 32 - *Tribal migration regions of the Amarat, Ajman, and the Atayban, from* The Raswan Index. *The numbers indicate details of specific tribal information, found in* The Raswan Index *following page 25, Volume 1. A knowledge of the boundaries of these districts, along with the seasonal nomadic migrations, provided European horse buyers with information necessary to locate the major horse breeding tribes.*

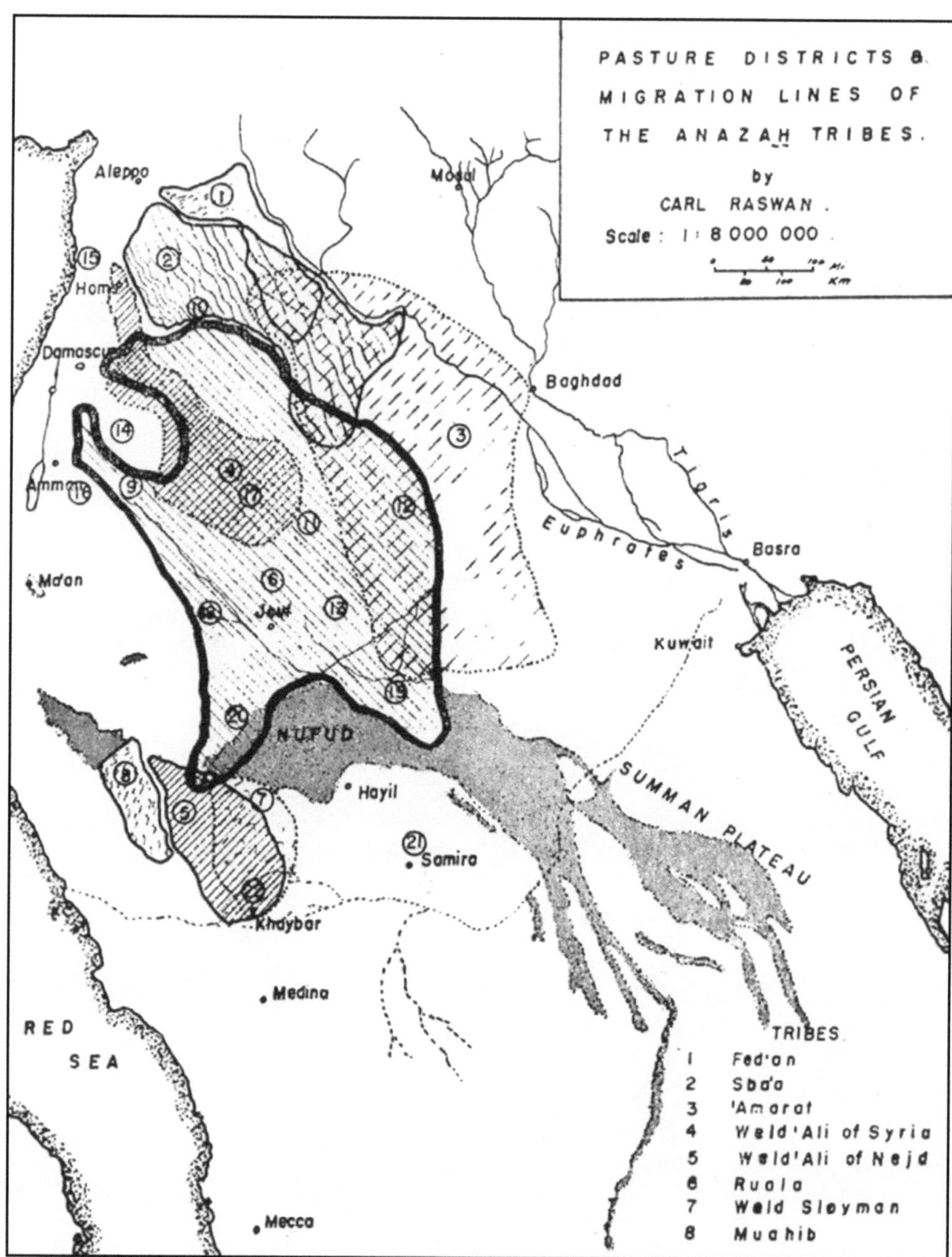

Figure 33 - *Tribal migration areas of the Anazah and allied tribes, from The Raswan Index. The numbers indicate details of specific tribes, and are described in The Raswan Index, Volume 1, following page 48.*

When reading the texts of explorers from the 19th century regarding the source of the *purebred Arabian horse*, it is important for the student of the pedigree to recognize that countries in the Middle East as they exist today did not exist then. The nations that constitute the Near East (Turkey, Syria, and Lebanon) did not exist until well into the 20th century. Until 1918, most of the Near East was a part of the Ottoman Empire, and national boundaries were vaguely defined or non-existent. The national borders that exist today were created for purely political reasons having nothing to do with geography.

As the geography of the central deserts became known, the tribal districts and annual migration pattern also became known to Westerners. This information provided useful information for horse buyers, especially those whose interest was primarily the acquisition of *purebred Arabian horses*.

At the end of World War I, the Axis powers were defeated, the Ottoman government collapsed, the Treaty of Sevres was signed, and Constantinople was captured. The Western Allies, primarily France and Great Britain, assumed control of the region. The process of portioning the region among the victors began with the implementation of the controversial British and French Sykes-Picot Agreement. During the War, secret arrangements between Sherif Hussein of Mecca and the Western Allies proved essential in the defeat of the Ottomans.

Following the end of the war, strong nationalistic sentiment in Anatolia led to the Turkish War of Independence and the formation of the Republic of Turkey in 1923. Early on, Yemen was acknowledged as an independent state by the Treaty of Sevres. Syria and Lebanon became French protectorates and their political borders were established. The regions that became known as Palestine and Iraq were created by the British government and the political boundaries of these newly formed nations were drawn.

Backed by the British, Hussein's son, Faisal, was appointed King of Iraq. A subdivision of the land of Palestine, newly named Transjordan, was created in order to give another of Hussein's sons, Abdullah, a place of authority in the new Arabic world order. Palestine and Jerusalem itself remained under direct British control. The Sherif's eldest son, Ali bin Hussein, became King of Hejaz and Grand Sharif of Mecca in 1924. His rule over the Hejaz would be brief, terminated by the invasion of the Al Saud tribe and their Wahabbi warriors.

Nejd, Arabic for "upland", comprises the central highland plateau of Arabia. Nejd is highest in the west with an elevation of 1360 meters, and lowest in the east at 750 meters. The heart of the Nejd is Jabal Tuwayq, a long low ridge. The tallest mountain is Jebel Shammar, the ancestral home of the Shammar tribe. It is bordered on the north by the Nefud and Syrian Desert, on the west by the mountains of the Hejaz and Yemen, in the south by the Rub al Khali, and to the east by the ad Dahna, "the river of sand". This region is a rocky highland with scattered sand deserts and mountain peaks, most notably the Jabal Shammar. Numerous wadis, ancient river beds, drain in a pattern from the highlands to the east towards the Persian Gulf. The largest are Wadi Hanifa and Wadi ad-Dawasir. Scattered artesian oases are found there as well as *sabkahs*, or salt marshes. Burckhardt reported that "Nejd is famous all over Arabia for its magnificent pastures, where after a rainfall even the desert turns green."

The satellite photograph showing the Arabian Peninsula (Figure 35) illustrates the geographic isolation of the highlands of Nejd, (1), which is a virtual island in a sea of sand. Before the time of mechanization, travel by foot into and out of the Nejd entailed certain hardships and hazards. Arabia is in some ways an island, cut off from the rest of the Middle East by the Nafud Desert. Nejd is an island within the island; the central highlands are surrounded by desert and mountains. The relationship between the An Nafud Desert, (2) al Dhana, (3) and the Rub al Khali (4) is shown with the three great deserts forming an isolating curvilinear pattern of sand around the plateau of Nejd. The An Nafud Desert is in the northwest and the Rub al Khali is in the southeast. The Al Dhana is comprised of an arcuate belt of parallel sand dunes known locally as "irqs". The Al Dhana connects the Nefud and the Rub al Khali, creating a massive continuous desert ecosystem. The western border of Nejd is demarcated by the hills and mountains of Asir and the Hejaz which can be seen on the southwest coast as dark brown/green areas. The Hejaz completes the geographic circle of isolation. The Asir is named from the Arabic word for "the difficult", referring to the hardships of travel. The Hejaz takes its name from the Arabic word for "barrier", a term that reinforces the idea of the isolated geography of Nejd. The Nejd highlands rise up from the mountains and deserts, reaching an elevation in the eastern region of 2000 ft above sea level in Riyadh, and rising to 3,488 feet above sea level at Afif in the west.

European narratives of horse buying in the desert often refer to the Levant. The term Levant is a collective and imprecise European term used geographically to describe the lands of the Eastern Mediterranean, including parts of Lebanon, Syria, Jordan, and Iraq. It is a useful term as a shorthand way of referring to that region of the Near East where European influence was dominant. In general, those areas of the Near East where European influence was strongest were also the source of many of the horses with questionable or incomplete pedigrees.

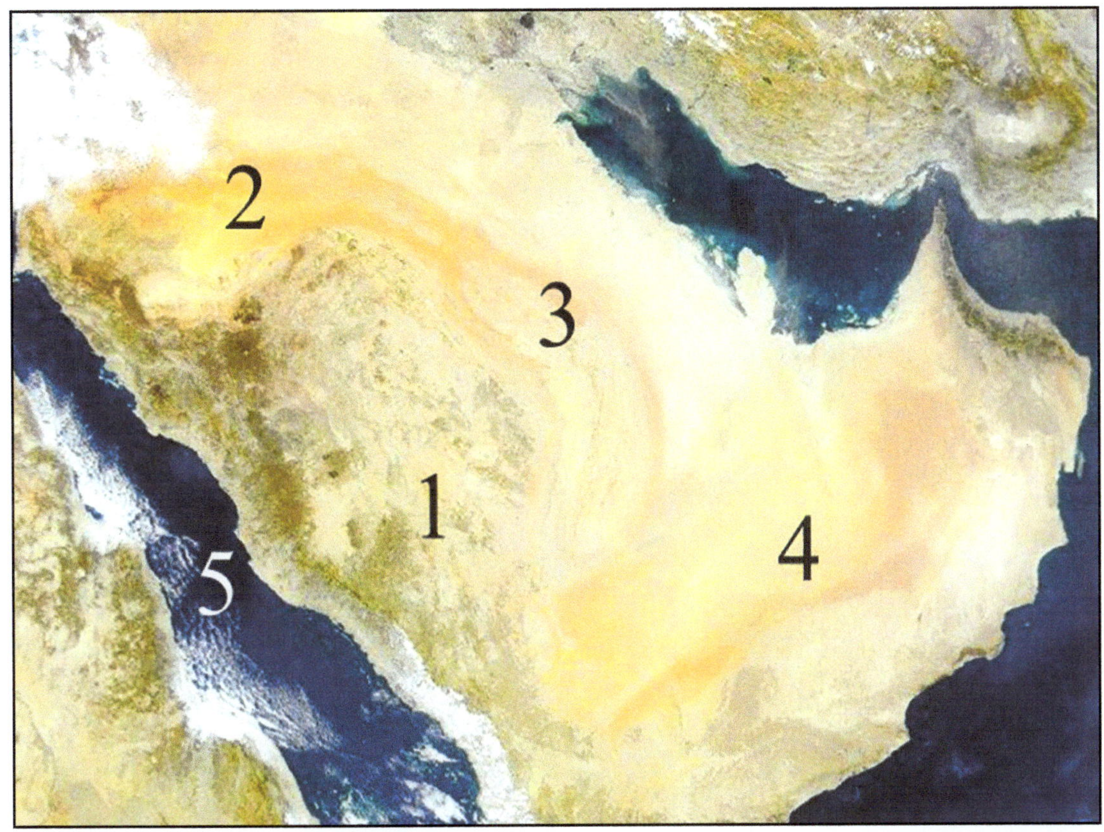

Figure 34 - *The Arabian Peninsula, NASA photograph.*

Figure 35 - *Sabkha, or salt flats, in the region of Nejd.*

The Levant is bounded on the north by the Taurus Mountains, on the south by Arabia, and on the east by the Zagros Mountains. The Levant is an important region in the history of the Arabian horse because of the long-standing and pervasive influence of Europeans; it has been a cosmopolitan region since the mid-19th century, heavily influenced by French business and commercial interests.

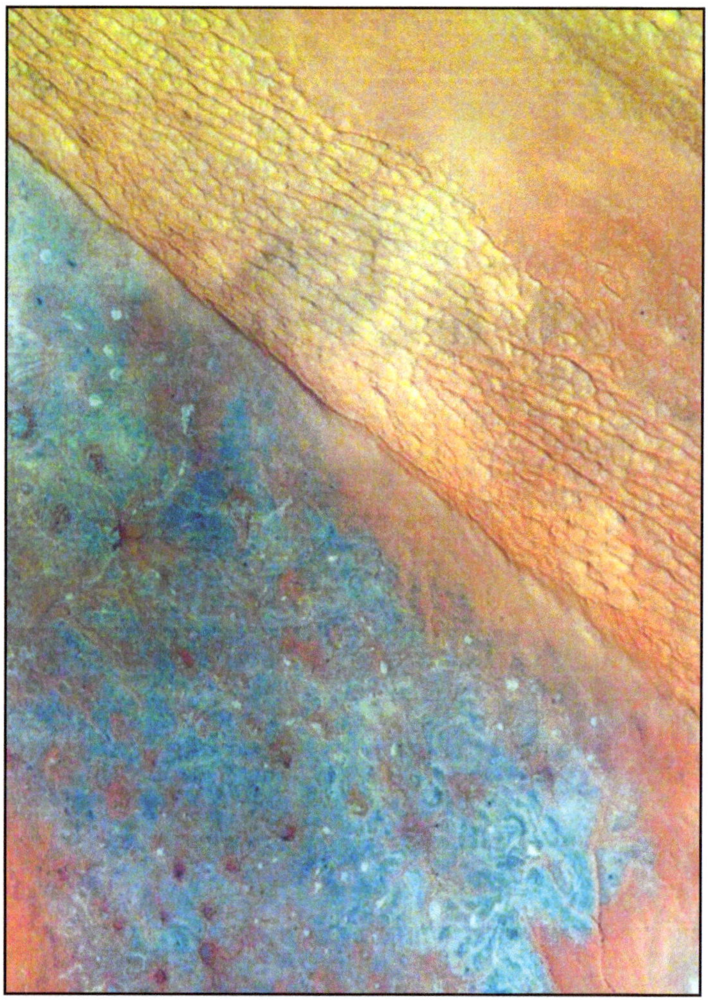

Figure 36 - *Northeast Nejd, as it rises up from the Ad Dahna sand cordon, photo by NASA, oriented with magnetic north at the top of the photograph. The steep cliff-like border of the desert and the plateau, running from the top left to the bottom right, casts a dark shadow. The edge of the Nedjean highlands, with rock outcroppings and wadis, dried ancient river beds, is lower left. The sands of the Ad Dahna (upper right) consist of both linear organized dunes and flat featureless sand sheets.*

Figure 37 - *The Arabian highland region of Nejd.*

Beirut has been a highly Europeanized city since French, British, and American commercial developers began investing in port construction, rail systems, and trade in the 19th century. By 1911, Beirut's population was made up of 80% foreigners. After 1918, free from Ottoman restrictions, Lebanon was a flourishing French protectorate and prosperity was sustained in the European manner for many decades. European manners, style, commerce, and culture were predominant, including the consuming interest in horse racing. Track racing, and Western-style betting, previously unknown in the region, were introduced by Europeans and the practice was avidly adopted by many Arabs. The associated financial incentive to win races increased. Beirut was not a center of horse breeding but was rather a center of Arab horse racing, with several busy tracks and, even today; it has over 400 registered racehorses.

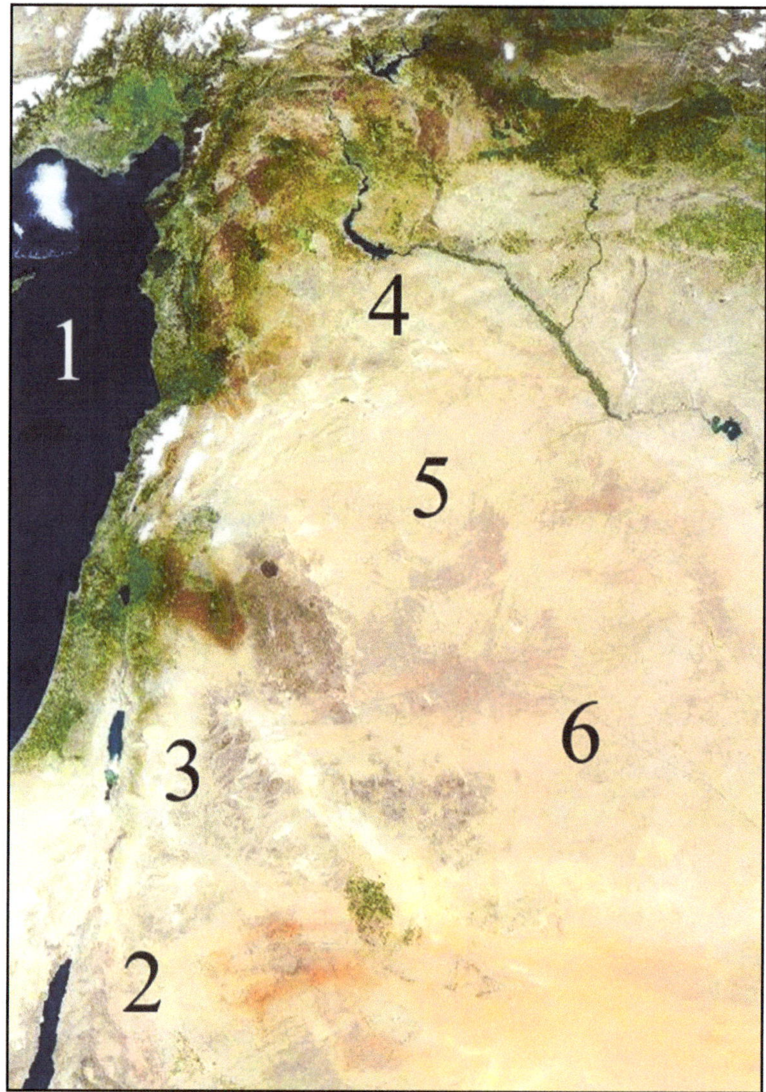

Figure 38 - *The Levant, NASA photograph. A satellite image of the Levant, with the Eastern Mediterranean Sea (1), the Gulf of Aqaba (2) and the Dead Sea (3). The head waters of the Euphrates River and Lake Assad (4) are in the upper right portion of the photo. The snow-capped Lebanese Mountains in the upper left demarcate the coastal city of Beirut, Lebanon from the inland Syrian city of Damascus, which is east of Beirut east of the Anti-Lebanese Mountains. The modern political boundaries are faintly visible as dark lines on the map. The demarcation between "the desert and the sown" along the Mediterranean Sea is visible as the green belt of vegetation merges into the sands of the Syrian Desert. The unbroken continuity of the Nafud Desert of Saudi Arabia (6) and the Syrian Desert (5) is noted. The Syrian Desert, or Al-Hamad, is bordered on the south by the Arabian Desert after crossing both Jordan and Iraq, but the line of demarcation between the two regions is arbitrary and politically defined. The two deserts are in reality one continuous sand expanse. The Syrian Desert is bounded on the north by the fertile terrain of Syria, on the west by the Orontes River, and on the east by the Euphrates River.*

Figure 39 - *The Isle of Graia, in the gulf of Aqaba, lithograph by David Roberts.*

Figure 40 - *Bedouin horse racing, early 20th century, image from the Library of Congress.*

For the Arabian horse pedigree researcher, an appreciation of the turf activities in and around Beirut is particularly useful. Here the Western passion for turf sport was at its zenith during the latter half of the 19th century. Westernization of the French-allied city brought with it all of the vices of the West. While distance racing was a well known part of the Bedouin way of life, track racing was not. The Beirut tracks were flooded with English Thoroughbreds. The tracks ran races for English horses, Anglo-Arabs, and Arabian horses. The pure Arabs simply could not compete on the short oval course against the Thoroughbreds. But, since the time of Charles II, the English Thoroughbred was bred for only one purpose, and that purpose was superior sprint performance. The English Thoroughbred illustrates the horseman's adage "a defect here creates an excess there, and an excess there creates a defect here."

> "It is not surprising that he (the English Thoroughbred) should have acquired peculiarities of form and temper that render him undesirable for the more sober and steady uses of everyday life… he is too nervous and excitable."
>
> National Livestock Journal 1868

This was just the type of horse that the Europeans loved for their own use on the track. And it was this passion for the race, the addiction of gambling, and its excitable, agitated animals that led horse breeders to cross English Thoroughbreds with an Arabian horse in order to produce horses that looked Arabian but ran like Thoroughbreds. Deceit was not unknown at the tracks.

In contrast to Beirut, other cities of the Levant were not overwhelmed by westernization; Damascus, Allepo, and Amman retained their Arabic culture and character.

Figure 41 - *The extent of the Syrian Desert, interpreted in a liberal sense. The term as used by Western writers of the 18th and 19th centuries is indicated by the pale area above.*

The vast desert expanses southeast of the border of the Syrian Desert blend imperceptibly into the Arabian Desert. The Arabian Desert likewise has no precise geographic boundaries. Syria and Saudi Arabia do not share a political border, so technically the Syrian Desert extends across western Jordan and eastern Iraq before merging with the Arabian Deserts proper. Since there are no geographic or political definitions of the Syrian Desert or the Arabian Desert, the term is used by various individuals and governments in variable ways. Some older travel writers and explorers also use the term without clear definition. Local and regional residents of Damascus and environs also use the term without precision, of necessity, since there is no generally agreed upon extent of the Syrian Desert. Little useful geographic information was conveyed when early importers of horses from the Middle East cited the point of purchase as "The Syrian Desert".

Geography was indeed destiny with regard to the breeding of the genetically unsullied Arabian horse. The horse breeding practices of the people inhabiting the regions outside of the Nejd did not always adhere to the time-honored principles of pure breeding but bred horses for commercial purposes. During the 17th century, Bedouin of the Hejaz bought generic mares from pilgrims traveling from Egypt during the annual Hajj to Mecca and bred them to high caste Nejd stallions. The foals resulting from this practice were then sold to Yemen with the assurance that they were *purebred.* This practice was made possible by the fact that while purchasing a pure Nedji mare was beyond the means of the average Bedouin, the price of service to a noble Nejdi stallion was very modest. During this era, the Bedouin of Syria followed a similar practice, as the quality of the native mares of Syria was considered to be below average. Bedouin of the Persian Gulf, and especially Baghdadis, also bred horses in this same manner.

The Bedouin of the Nejd preserved the ancient tradition of pure breeding and were horrified at the misalliances produced by their coastal brethren. The purist horse breeders of the central highlands despised the crossbreds, and adamantly refused them access to their tents.

Arabian horses that were sold in the towns were almost without reflection considered by a Bedouin to be part-bred. Burckhardt found no horses among the wealthy inhabitants of Medina. In general, he found the Hejaz was nearly devoid of horses. Mules, asses, and dromedaries were commonly seen. Medina, noted Burckhardt, specifically forbid dogs, based on the fear that the inhabitants that the dogs would desanctify the local mosques.

The German traveler and author Karl Wilhelm Ammon records that by the early 1800s the Arab horse markets of the Middle East were well-established, each with its own source of horses to sell and each with its own clients to sell to. Arabs sold horses among themselves and they sold horses at the markets of Baghdad, Basra, Damascus, and Aleppo. Stallions were primarily sold, as the Bedouin had little use for them. Gelding was unknown in the desert at this time. Ammon estimates that one-third of the horses sold in these markets were *purebreds*, foaled in the desert by Bedouins. The rest were part-bred.

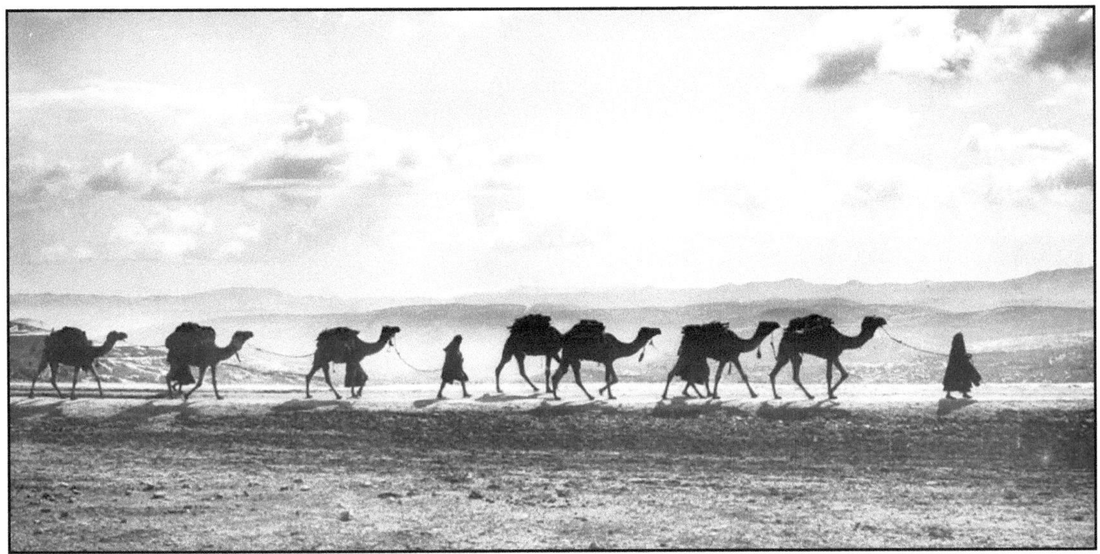

Figure 42 - *A Bedouin caravan in the early 20th century, from the collection of the Library of Congress, Washington.*

The Baghdad and Basra markets were supplied by the Montefik tribe, who were known as indiscriminate breeders, and sold mainly to customers and agents who bought for export to India and the East Indies. The horses were of mediocre quality. Ammon calls these "half-bred" or "improved" horses. Desert horses from Nejd of true nobility were seen very rarely, but when brought by the Bedu to market due to pressing economic need, these horses sold for very high prices. The Bedouin of the Syrian Desert supplied most of the markets of Damascus and Aleppo. The Bedu used horse traders as intermediaries, and these horses were sold to customers from Egypt, Libya, Turkey and Europe. The quality was generally poor but some excellent specimens were seen.

Louis Damoiseau, born in France, 1775, was sent by the French government to the Levant for the purpose of obtaining *purebred Arabian horses* (about 40 stallions) for use in the government studs. The Imperial French Studs, once filled with fine specimens from Napolean's short-lived adventure in the Middle East, had been depleted by invading armies. Damoiseau's training as a veterinarian made him well suited to observe and report on their findings of the expedition.

Damoiseau was in Damascus in 1819 and witnessed an auction in the horse market in which one of the bidders was Emir Akhor, equerry to Mehmet Ali Pasha of Egypt, who bought several of the choice mares for his master but was forced to pay ten times the normal amount for them. In order to stimulate bidding during the auction, the "horse dealers were chasing around on them". The Aleppo market was held every

Thursday and consisted of Arabians, Turkoman, Kurdish and Syrian horses. Ammon says that the Arabs were usually stallions, of poor quality, and not pure blooded. Foreign buyers, not fluent in Arabic, were often deceived in the pedigrees of the horses that they bought.

Ammon suggested that the best way to obtain full-blooded horse of known pedigree at fair prices was to travel to the interior of the Arabian Desert and buy from the Bedouin of Nejd who bred them. He adds "Alas, since these Bedouin are fierce robbers and mercilessly plunder any stranger who is not their guest, nay sometimes kill them, one cannot possibly envisage such an undertaking." The best alternative, he said, was to visit the Bedouin camps in the spring when they drew near to the margins of Syria and Palestine. The camps were so numerous that they stretched in an unbroken line from Allepo to Damascus - an eight-day journey.

Ammon writes that the Anazeh, who camped in the Hawran district in Syria, possessed the finest horses. While they migrated to Syria in the spring to sell horses, their ancestral home was Nejd, and it was in the Nejd that they had bred pure Arabians for centuries. Here the chances of deception are diminished. The prices were higher for horses with convincing and written certificates of pedigree, and for those horses descended from the more famous noble strain.

Figure 43 - *Image courtesy of Saudi Aramco.*

THE ARABIAN PENINSULAR DESERT SOURCES OF THE EGYPTIAN ARABIAN HORSE

PART TWO

ما لا يعلم كله لا يترك كله فان العلم
بالبعض خير من الجهل بالكل

Figure 44 - *"All knowledge of that which is not wholly known is not to be abandoned; knowledge of part is better than ignorance of the whole." Historian and geographer Abu Al-fida 'Ishma'il Ibn ali Ibn Mahmud Al-Malik Al-mu ayyad'imad Ad-Din of Damascus (1273-1331) from Upton, Gleanings from the Desert of Arabia, Olms Presse.*

This book is centered around an investigation into the genetic identities of the RAS/EAO Arabian foundation mares and their matrilines. Very little is known about the extended pedigrees of most of these mares. It is generally assumed, though not proven, that they were derived from *purebred* equine stock bred by the Arabian Desert Bedouin in the heart of Arabia, in a region known as the Nejd. To many Arabian horse breeders, this point is significant. For centuries, breeders have held that horses authentically and purely bred by the Bedouin tribes of Nejd possess certain special qualities that were valuable in breeding Arabian horses. Further knowledge of the genetic identities of these mares would bolster the claim that they are *purebred* horses.

Since this element of uncertainty will probably never be fully resolved, the student of the Egyptian Arabian horses will find the following section of some value. This section chronicles reports written by 19th century Europeans travelers who describe the desert milieu of the Arabian Peninsular interior, the region from which these historically important matrilines were likely derived. Taken as a whole, these reports give the Arabian horse enthusiast a fairly clear and coherent idea of the nature of the Bedouin and his horse in the interior of the Arabian Peninsula during this seminal century.

The 19th century is especially important to the Arabian horse pedigree researcher because it was a time prior to the despoliation that resulted from the infiltration of Western interests. In addition, the 19th century produced a great deal of reliable, first hand, well-written reports from Western travelers concerning the Bedu and horse breeding in the Arabian interior. It was also the century of endless war and bloodshed between the Ottoman/Egyptian armies and the Bedouin. The loss of *purebred Arabian horses* was the unfortunate result.

Not to be forgotten, this was the century in which most of the Arabian horses of the Egyptian Governors Mohammad Ali, Ibrahim Pasha, and Abbas Pasha were obtained from their desert owners. It is also the period during which the emissaries of the Egyptian Khedive Abbas Pasha traveled throughout the country, interviewing the tribal leaders and elders (the graybeards) who knew and had owned these *purebred* horses and their predecessors. The horses of the family Ali formed the core of the RAS at its inception.

It is likely that a large number of Arabic language texts and pedigrees of desert-bred Arabian horses from this period exist, but remain unknown to Western researchers. New information remains to be discovered.

During the 19th century, the horses of northern Syria and Mesopotamia (the fertile crescent, controlled by the Ottomans) were mostly of mixed breeding, combining equine types from Persia, Turkmenistan, and Arabia. These were common "grade" horses; they were inexpensive to buy and were in common usage. There were half-bred horses also, the result of mating a *purebred* stallion to a common mixed breed mare. These were also inexpensive to produce, but could be sold to naïve foreigners at high prices as "*purebred* horses".

All authorities agree on the consensus observation that horses of these two types were commonly found in the Hejaz and southern Arabia (Hadramut, Oman, and the regions bordering the Persian Gulf).

However, things were different in the heart of the Arabian Peninsula, where *purebred* horse breeding was, at least according to Bedouin testimony, carried out with exactitude. To be clear, researchers that have written on the topic of the *purebred Arab horse* have always had to deal with one central problem; the absence of a generally accepted definition of the *purebred Arabian horse*. Most experts, like Ammon, writing in 1834, define the *purebred Arabian horse* in this way:

> "...Noble horses...belong to an ancient breed which the conscientiousness of the Arabs have always kept pure and unmixed on the sires' and on the dams' side. This pure breeding has been a habit observed without interruption for a period of more than one thousand years; hence horses have acquired a certain uniqueness of conformation and build which is as unmistakable as it is indelible, and which definitely distinguishes them from the other horses of Arabia and of the whole Orient...one calls these horses *Koheylans* or *Nejdis*."

The desert tribes who possessed *purebred Arabian horses* had ample opportunity to breed their mares to impurely bred horses if they wished to do so. It is generally accepted that, prior to Western infiltration of the Middle East, they did not wish to do so. They had no economic motive to do so. In fact, they had powerful economic interests not to do so.

The notion of Western infiltration means, of course, the introduction of the often unacknowledged introduction of the blood of the English Thoroughbred into the breeding population of the horses of the peninsula of Arabia.

But then, little is known on this subject with absolute certainty. The Eastern mind has never demonstrated that relentless obsession with certainty that has preoccupied the mind of Western man.

The Arabian horse has been known to European horse breeders since the 10th century. The first imported horses from the desert, Arundel and Truncefice, arrived in London from Damascus in 957 CE. The Royal Stud Book of England of 1624 lists an Arabian colt in its possession, and states that the colt was used there for breeding. Beginning in 1657 the Levant Company, a British trading venture, began importing large numbers of Arabian horses from Allepo to Europe. During the reign of Charles II of England (1660-1685) the systemic use of Arabian stallions in the studs of the

King and the nobility was the norm.

The Byerley Turk, the Godolphin Arabian, the Darley Arabian and the Alcock's Arabian were the cornerstones of the English Thoroughbred horse. Their value as breeding stock, their capability of regenerating local horse quality, was immediately recognized. However, these stallions were bought by horse dealers operating on the fringes of the Ottoman Empire and the Middle East. Little was known about their actual provenance or extended pedigrees.

During this era, horses were purchased by agents who were dispatched by the nobility of Europe to obtain fine quality Arabian horseflesh for their own private studs, but they were constrained by circumstances to buy horses only at the coastal horse markets along the Mediterranean; Algeria, Tunisia, Morocco, Beirut, Allepo, and Constantinople. Baghdad was the only inland center in the Middle East that horse traders from Europe could enter. Traders could not penetrate into the interior of the Arabian Peninsula. The stallions and mares of the English King's Royal Stables, for example, had passed through several hands before reaching the King.

The Arabian blood was not uniformly praised by horse breeders in Europe. There were vocal detractors. There were skeptics in Europe who claimed that Arabian blood was ruining the Thoroughbred, causing the introduction of serious leg conformational abnormalities. One such cynic, the English horse breeder J. Justinus wrote in 1815 "How many of the horses which have been brought to Europe as Arabians ever saw Arabia?" This sentiment, time would reveal, was neither impertinent, isolated, nor entirely incorrect.

At the dawn of the 19th century, the horsemen of Europe were well aware of the value of the blood of the *purebred Arabian horse*.

The genetic qualities of the Arabian horse were found to improve the quality of virtually any equine stock on which it was used. Horsemen wanted access to that high-quality bloodstock, and they wanted to obtain it directly from the heart of the desert, straight from the Bedu. The reports that reached Europe indicated that those Bedu who bred these horses lived in a region known as the Nejd.

The Bedouin tribes who originally bred the *purebred Arabian horse* were known to the outside world by way of scattered and unsubstantiated reports; at the time such information was simply not available outside of Arabia. Rumors circulated that the true wellspring of these remarkable animals was in the very heart of the Arabian

Peninsula itself, a vast region of endless scorched burning desert, the details of which were unknown to the outside world. It was said that any man who entered the desert would never return. The central desert, it was said, was a fiery furnace. Stories and tales of life in the desert and the Bedouin theories of horse breeding trickled into the European psyche, but few facts were known with certainty. Fables of the mysterious Arabian Desert and its quixotic inhabitants, its poets, and its warriors, were widely circulated.

Figure 45 - *A Horse Market in Baghdad, 19th century watercolor by Lady Anne Blunt.*

Until the 19th century, however, no horses had been purchased directly from their original desert breeders. No factual details concerning the source of the remarkable animals were known. This began to change as serious-minded, intrepid and, at times, foolhardy European horsemen, scientists, academics, and researchers ventured into the heart of the desert itself. Most of these early explorers were highly educated and well-trained people, whose observations bear the stamp of authenticity.

A few of the early explorers to Arabia were ethnographers. Some of the Europeans were military men seeking geographic and political insights that could be politically useful to their governments. Some of these adventurers were European horsemen who were interested primarily in learning about the state of pure Arabian horse breeding in its homeland. Some of them came to the peninsula specifically to buy horses for their respective governments. Some of them came to the interior purely seeking adventure. One of them came entirely by accident, and, what is more, against his will.

Figure 46 - *An Arabian horse, by Carle Vernet.*

Decade after decade, new explorers came, and they began to publish books describing the fabric of native life in the interior of the peninsula. Scientific advances of the late 18th century, particularly the development of accurate and portable geographic positioning systems, led to an upsurge in explorations and mapmaking; scientists were eager to learn about the flora and fauna of Arabia. Cartographers sought geographic data and local information in order to produce updated maps.

But there were substantial barriers to exploration.

The dangers posed by political and military events that occurred in Arabia during the 19th century were largely hidden from the Western eye and only came to the notice of the outer world through the works and reports of European diplomats, writers, travelers and diarists of the era. Even the geography, land routes and cities of Arabia were poorly documented. Although the coastal geography of the peninsula was well known to Western maritime traders as early as the 17th century, the geography of the interior remained unknown. Reports on horse breeding among the tribes were scant and of questionable reliability. All of this began to change in 1819 when the first European explorer crossed the Arabian Peninsula.

Arabia was closed to the West until the first tentative European explorations of the 17th and 18th century began to shed light on the activities of the natives in the interior of the peninsula. Travelers who ventured into Arabia during the 17th and 18th centuries were confronted with a vast uncharted desert. 19th century adventurers would be armed with more scientific instruments and knowledge, but they would face the additional dangers associated with the emergence of a radical form of Islam, Wahhabism.

The greatest deterrent to 19th century European exploration was the hostility shown toward Western visitors by the Wahhabis. This radical sect of revolutionary Islam regarded any foreigner as an infidel. When detected, the infidels were summarily executed. Compounding matters, the native Arabians were constantly alert for the presence of spies and secret agent, not only those from Europe but also those from Constantinople. They guarded their isolation jealously. The Bedouin were wary; they had become accustomed to a life punctuated by raids and thievery. They were a suspicious people, and with good reason. Any stranger was a threat.

The Arabian Peninsula was a dangerous place for a wandering European traveler. Bands of Arab brigands were a common sight, and theft and murder were commonplace. Any traveler suspected of being an infidel was cornered and questioned by the Wahhabi Bedu, challenged to repeat the standard allegiance to the Prophet. Those who failed or refused the challenge were executed on the spot. At one time the rotting corpses of these unfortunates were hung on the gates of the Nedji town of Hail, displayed there by the Rashidis as a deterrent to foreign intrigues.

The safest means for a European exploring Arabia during the 19th century was through the use of subterfuge. Many Western travelers were able to move about the country disguised as native Bedouin. Many European explorers were darkly complected, capable of learning the rudiments of Arab culture and dress, and most importantly spoke fluent colloquial Arabic. Many of the first travelers adopted professional credentials, presenting themselves as Arab scholars, physicians, and clerics. Without the use of some kind of artifice, a European infidel would be detected immediately and, in Wahabbi manner, assassinated. In the latter half of the 19th century, the Wahhabi fanaticism against foreigners had begun to abate, and disguise was no longer essential.

The increasing presence of foreigners in the desert was especially disturbing to the mightiest of the desert princes, jealous of their power. The English, one Emir remarked, were like ants. When one found something of value in Arabia, hordes of English were sure to follow. Apart from real concerns about the intentions of the British and the French scouting out the interior, the Bedouin were by their very nature xenophobic in the extreme. They had little knowledge of the outside world and had no wish to have their way of life disturbed.

K.W. Ammon, Arabian horse historian, wrote in his 1834 book Historical Reports on Arab Horse Breeding "How marvelous if one day a true expert on horses, who also understood the Arabic language, should travel through Arabia (particularly through its interior) and give us thorough first-hand information on the Arab's horse world."

Ammon was lamenting the complete absence of reliable authentic insight into the origin of those remarkable animals and how they came to be in the possession of isolated desert nomads in the first place.

Ammon discounts the *purebred Arabian horse* buying and *purebred* breeding efforts of Ehrenfort, Deporter, Rzewusky, Gertsinger, and others who had made trips into the Syrian Deserts prior to 1834 to buy Arab horses. None of these horsemen advanced more than a short distance into the northern Syrian Desert, buying horses from the Bedouin tribes that summered near the cities. None of these horsemen even reached the Arabian Desert, much less the Nejd. It is not likely, says Ammon, that they were able to purchase the truly first class animals. It is not likely that they were shown or offered first-class animals.

Ammon was well aware of the reasons that men were reticent to travel into the interior. It was a dangerous place for foreigners. By 1865 the Wahabbis had gain control of a large part of the peninsula, and their hatred of infidels within their borders led to violent conditions for travelers. The Bedouin were, by reputation, barbarous and bloodthirsty. It was said that they only sold their poor quality animals to foreigners, who they considered to be fools with regard to the knowledge of horseflesh and legitimate targets for deception. Besides, the Bedouin rarely sold mares.

To compound matters, the climate of Arabia posed a serious obstacle to exploration in the interior. The lack of water in the central desert had killed more than one European adventurer. The mysterious reports of poisonous spring water were often repeated. The sun was said to shrivel the skin, requiring travelers to wear several layers of thick clothing to avoid being burned. Face scarves were required to avoid the dangerous effects of directly breathing the hot air. The "*Samum*", or deadly sandstorms, were lethal.

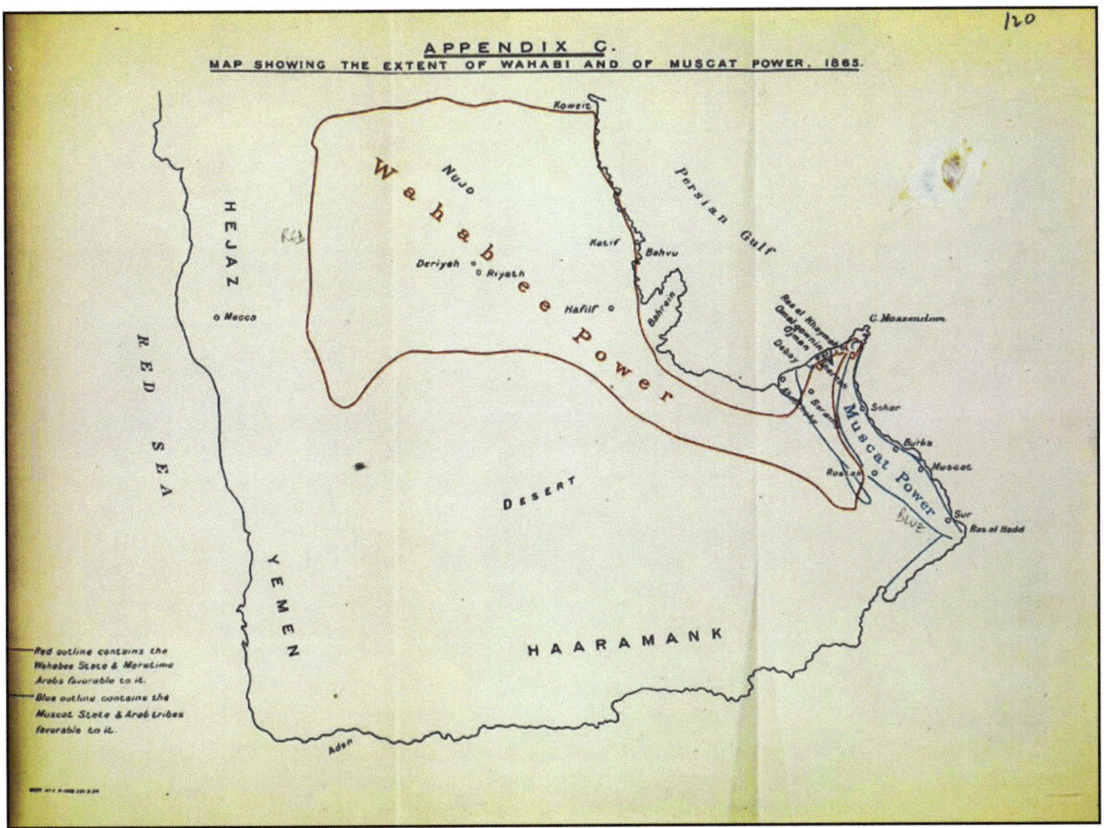

Figure 47 *The extent ofWahabbi and Muscat power in 1865, image from the Qatar Digital Library.*

The following tables illustrate the climatic condition of Arabia for Riyadh, Buraidah and Ha'il.

Month	Jan	Feb	Mar	Apr	May	Jun	Jul	Aug	Sep	Oct	Nov	Dec	Year
Record high °C (°F)	25 (77)	24.8 (76.6)	32 (90)	37 (99)	46.1 (115)	47 (117)	48 (118)	47 (117)	44.5 (112.1)	42 (108)	39 (102)	33 (91)	48 (118)
Average high °C (°F)	20.3 (68.5)	23.3 (73.9)	28.3 (82.9)	32.6 (90.7)	38.6 (101.5)	41.6 (106.9)	42.2 (108)	41.3 (106.3)	40 (104)	34.5 (94.1)	27.3 (81.1)	22 (72)	32.67 (90.83)
Daily mean °C (°F)	14.3 (57.7)	16.8 (62.2)	21.4 (70.5)	25.7 (78.3)	31.1 (88)	33.6 (92.5)	34.7 (94.5)	32.6 (90.7)	31.8 (89.2)	26.6 (79.9)	20.6 (69.1)	15.7 (60.3)	25.41 (77.74)
Average low °C (°F)	8.4 (47.1)	10.4 (50.7)	14.5 (58.1)	18.8 (65.8)	23.7 (74.7)	25.7 (78.3)	26.9 (80.4)	23.6 (74.5)	23 (73)	18.7 (65.7)	13.5 (56.3)	9.5 (49.1)	18.06 (64.48)
Record low °C (°F)	−2 (28)	0.5 (32.9)	4.5 (40.1)	11 (52)	18 (64)	21 (70)	23.6 (74.5)	22.7 (72.9)	16.1 (61)	13 (55)	7 (45)	1.4 (34.5)	−2 (28)
Rainfall mm (inches)	11.7 (0.461)	10.5 (0.413)	24.7 (0.972)	23.3 (0.917)	2.6 (0.102)	0 (0)	0 (0)	0 (0)	0 (0)	0.7 (0.028)	6.9 (0.272)	13 (0.51)	94.6 (3.724)
Avg. precipitation days	5.8	4.8	9.8	9.0	3.5	0.0	0.0	0.0	0.0	1.2	3.4	6.3	45.2
% humidity	47	38	34	28	17	11	10	12	14	21	36	47	26

Figure 48 - *Climate data for Riyadh: Due to its lower elevation at 2008 feet above sea level, Riyadh in southern Nejd is among the most environmentally inhospitable areas of the peninsula.*

Month	Jan	Feb	Mar	Apr	May	Jun	Jul	Aug	Sep	Oct	Nov	Dec	Year
Average high °C (°F)	19.2 (66.6)	22.3 (72.1)	26.5 (79.7)	32.4 (90.3)	38.5 (101.3)	42.1 (107.8)	43.2 (109.8)	43.4 (110.1)	40.9 (105.6)	35.3 (95.5)	26.7 (80.1)	21.3 (70.3)	32.65 (90.77)
Daily mean °C (°F)	12.8 (55)	15.2 (59.4)	19.2 (66.6)	24.9 (76.8)	30.6 (87.1)	33.5 (92.3)	34.5 (94.1)	34.8 (94.6)	32.1 (89.8)	26.9 (80.4)	19.8 (67.6)	14.7 (58.5)	24.92 (76.85)
Average low °C (°F)	6.3 (43.3)	8.1 (46.6)	11.9 (53.4)	17.4 (63.3)	22.7 (72.9)	24.9 (76.8)	25.7 (78.3)	26.1 (79)	23.3 (73.9)	18.5 (65.3)	12.8 (55)	8.0 (46.4)	17.14 (62.85)
Rainfall mm (inches)	21.2 (0.835)	10.8 (0.425)	26.7 (1.051)	28.8 (1.134)	13.9 (0.547)	0 (0)	0 (0)	0.1 (0.004)	0.1 (0.004)	4.3 (0.169)	24.2 (0.953)	15.6 (0.614)	145.7 (5.736)
Avg. precipitation days	6.2	3.0	6.3	8.1	3.5	0.0	0.0	0.1	0.1	1.4	5.3	4.5	38.5

Figure 49 - *Climate data for Buraidah. Located at an elevation of 2,126 feet above sea level, Buraidah is an oasis, located in the heart of the Nejd. It has a similar climate to Riyadh, but has twice the annual rainfall.*

Month	Jan	Feb	Mar	Apr	May	Jun	Jul	Aug	Sep	Oct	Nov	Dec	Year
Record high °C (°F)	27 (81)	29 (84)	34 (93)	37 (99)	42 (108)	44 (111)	44 (111)	45 (113)	42 (108)	38 (100)	31 (88)	28 (82)	45 (113)
Average high °C (°F)	17 (63)	19 (66)	24 (75)	30 (86)	35 (95)	38 (100)	38 (100)	38 (100)	37 (99)	32 (90)	24 (75)	18 (64)	29.2 (84.4)
Average low °C (°F)	3 (37)	5 (41)	9 (48)	14 (57)	19 (66)	22 (72)	23 (73)	22 (72)	16 (61)	12 (54)	10 (50)	5 (41)	13.3 (56)
Record low °C (°F)	−10 (14)	−7 (19)	−4 (25)	0 (32)	10 (50)	16 (61)	17 (63)	16 (61)	14 (57)	5 (41)	0 (32)	−6 (21)	−10 (14)
Precipitation mm (inches)	17.8 (0.701)	19.1 (0.752)	12.4 (0.488)	10.6 (0.417)	5.5 (0.217)	0 (0)	0 (0)	0 (0)	5.4 (0.213)	8.9 (0.35)	12.6 (0.496)	08.3 (0.327)	100.6 (3.961)

Figure 50 - *Climate Data for Ha'il: At 3,255 feet above sea level, Ha'il, in northern Nejd, has a more moderate climate, but little rain.*

Figure 51 - *Arab Horses, by Carle Vernet.*

There was a desert saying among the Bedouin; "After enduring a sandstorm for three days, even murder is excused ." The desert was infernally hot, dry, and dangerous.

Ammon's lament would soon be answered by a string of intrepid Western explorers.

As with the preceding section on the Arabian horse *hujjaj*, the following reports on the Bedu and his horse are also subject to the caveat that the information presented may be incorrect, incomplete, confused, misleading, or patently untrue. Only a Bedouin will ever know the Bedouin mind or the Bedouin spirit.

The first report of foreign contact with the peninsula is found in the 20th century BCE records of the ancient Egyptians. During the Second Egyptian dynasty of Mentuhotep, an Egyptian explorer named Hannu traveled by boat, exploring south in the Red Sea, reaching Punt (Ethiopia) and southern Arabian, where he collected myrrh, precious metals, rare stones, and exotic wood.

Figure 52 - *Simoon, by Eugene Fromentin, from a private collection.*

Figure 53 - *The Oasis, by Adolph Schreyer used with the permission of the Joslyn Art Museum.*

Military interest in the peninsula first began during the period of Hellenic expansion of Alexander the Macedon in the 4th century BCE. He sent Androsthenes of Thasos and Archias of Pella on expeditions down the Euphrates to explore the Persian Gulf. They reached as far south as Bahrain. They found nothing of interest to report.

Modern Western explorers first crossed the Arabian Peninsula in 1819, and their accounts began to gradually provide clear and reliable information available concerning the history and activity of asil horse breeding in the interior. The most comprehensive account of the opening of Arabia is found in The Penetration of Arabia by David Hogarth, published in London in 1904.

Before discussing the landmark 1819 expedition, there are two peculiar cases to mention.

Among the first accounts by a European of the desert horses came from a European explorer of the 16th century. The Italian Lodovico di Varthema of Rome journeyed into the desert, passing for a Syrian Arab. He took the name "Yunis" Arabic for Jonah. His motivation for the adventure is unclear, but his time in Arabia provided the material for a book The Travels of Ludovico di Varthema, published in 1510. He claims to have traveled from Medina to Mecca, then moving on to Sanaa and the Yemeni coast.

Of the desert Arabs and their horses he wrote :

> "They ride on horses covered only with a loose cloth or mat, and wearing nothing but a petticoat. For weapons they use a long dart made of reed, of the length of ten or twelve cubites, tipped with iron after the manner of the javelins and fringed with silk. They march in order and are despicable and of little stature, of color between yellow and black, which some call olivastro. They have the voices of women and long black hair. They are of greater numbers than a man would believe and are continually at strife and war among themselves."

His claim that he saw 40,000 stallions and 100,000 thousand mares in one Bedouin herd may be regarded as hyperbole. His work as a whole demonstrates a tendency toward exaggeration and fantasy. He was, it would seem, a tireless self-promoter.

No discussion of early European travelers to Arabia is complete without mentioning the peculiar case of the 17-year-old British cabin boy Joseph Pitts (1663-1735). He was captured by Algerian natives when he was shipwrecked on the Spanish coast was sold into slavery. His third owner was a Muslim, and Pitts converted to Islam and made the Hajj. He eventually escaped, returned to England and wrote a book describing his experiences in Arabia. His description of the Bayt Allah, the Ka'ba, was so vivid that it can be assumed that his narrative was authentic. His book, A Faithful Account of the Religion and the Manners of the Mahometans was published in 1704. He did not mention horses.

The first academically trained geographer, Western explorer, and competent Arabist to penetrate into the core of the Arabian Desert was Johann Ludwig Burckhardt. He was a resident of Syria for two years, and made a number of trips into the interior of Arabia, from 1812 to 1815.

Figure 54 - *The portrait of Varthema on the title page of his book more or less speaks for itself.*

A fellow European happened to encountered Burckhardt in Cairo and described his appearance:

> "He wore a blue cotton blouse covering a coarse shirt, loose white trousers and wearing a common calico turban. He had a full beard, was without stockings, only the slipshod slippers of the country, and he looked completely like an Arab."

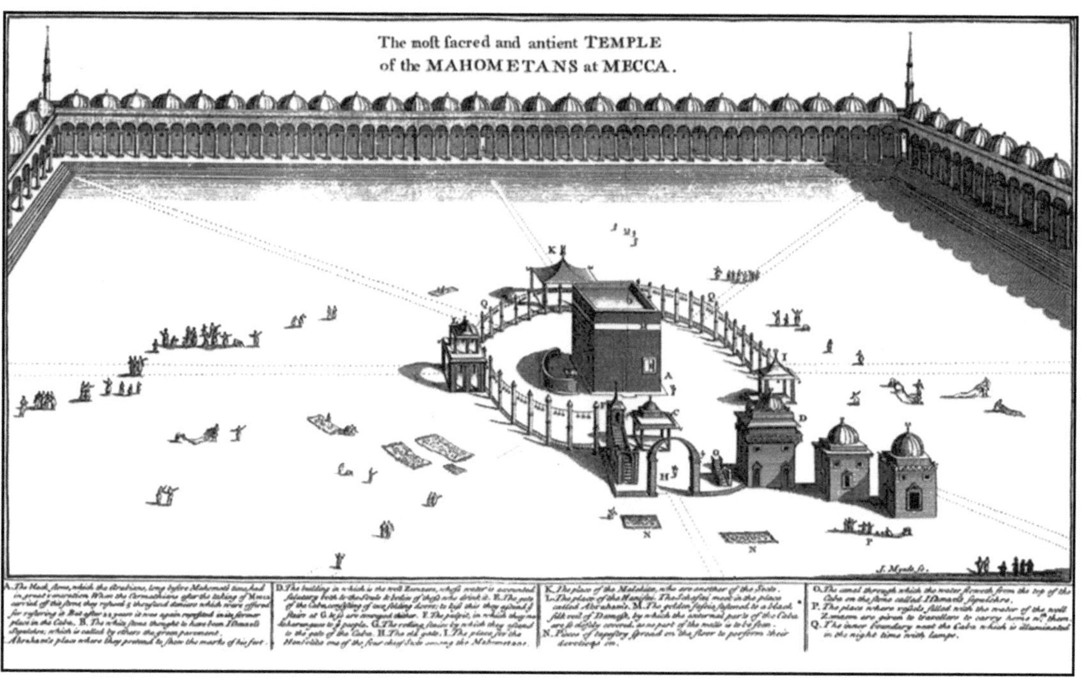

Figure 55 - *The Ka'ba at Mecca.*

He died in 1817 at age 32 in Cairo from dysentery. His observations, the earliest reports made of Arabia by a trained academician, are known to the Western world from the regular packets of reports that he had sent to England. They have been preserved at Cambridge University.

Burckhardt wrote that the Anazeh tribe, living on the Syrian border, possessed about 10,000 horses. The Montifik tribe living near the Euphrates Valley had about 8,000 horses. The Dhofur and Beni Shamar had large numbers of *purebred* horses.

He also reported that:

> "Under the rule of the Wahabbi leader horses become rarer every year among his Arabs. Their owners sold them to foreign buyers who took them to Yemen, Syria, and Basra, and the latter town supplied the Indian market with Arabian horses. The reason for this is that the Bedouin were afraid that Saud, their leader, or his successor, would take their horses from them, for it was quite a common occurrence that at the slightest disobedience or unlawful behavior a Bedouin's mare would be confiscated in favor of the funds. Every Bedouin owning a horse was obliged always to be ready to accompany the leader in his wars. Therefore many Arabs found it preferable not even to own a horse."

Burckhardt was readily received by the Egyptian Viceroy Mohammad Ali who was in residence at the Viceroy's summer home at Taif. The Viceroy was interested in the foreigner's assessment of his troops and interrogated Burckhardt with keen enthusiasm, eager to hear news of Napoleon, the French government, and the British military intentions in the Middle East. The Viceroy was well aware of the dangers faced by his government from foreign governments, and he was especially interested in the information that he could glean from this educated and intelligent adventurer. Ali was so impressed with Burckhardt's erudition that he gave him money and other forms of material assistance in his continuing researches on the peninsula. He even allowed the foreigner the privilege of visiting his wife who was also in Arabia, traveling this great distance to visit her son Tousson. This was at the height of the Wahabbi wars.

Burckhardt was, posthumously, praised in England for his vivid and accurate portrayals of the sights of Arabia as well as his thorough portrayals of the nomadic inhabitants.

His later biographer Alan Morehead wrote:

> "He is, as it were, a grammarian of exploration, a pedant turned explorer, and a classicist in the wilderness."

Burckhardt's book Travels in Arabia was published in London by his friends at the African Association in 1829.

The outcome of the first attempts by foreign academicians to make scientific surveys of Arabia did not bode well. In 1810, the German Natural Scientist Ulrich Jaspar Seetzen made the journey to Medina and Mecca disguised as a dervische. His intention was to observe, catalogue and collect zoologic and botanical specimens from Arabia. He concocted the name Hajji Musa and as an accomplished Arabist, he spoke the language without difficulty. He began his journey on the west coast of Arabia. After exploring the Hejaz, he went to Aden and then to Sana'a, where he was murdered. It was alleged that he was poisoned by his guides on orders from the Imam of Sana'a.

Seetzen kept a journal, and he made some notes on horses that he saw in Arabia in the early 19th century. Seetzen states that there were very few Arabian horses of high quality in the regions he was familiar with; The Hadramaut, Yemen, Asir and the Hejaz. Horses were found only in the possession of wealthy traders or sheikhs, and even then they were seen to have only one or two mares. Seetzen never saw a Bedouin tribe with horses during his entire time on the peninsula. It was his impression that the cost of feeding a horse in the desert was so great that only wealthy men could afford to own them. Many tribes had no horses, and those that did had only a few, belonging to the Sheikh of the tribe. The Harb tribe formerly had no horses, but after their conversion to Wahabbism, they received gift horses from the Wahabbi chief. The Sherif of Mecca, a very wealthy man, kept 50 or 60 horses at his country estate outside Mecca. In Yemen, the animal was as rare as it was in the Hejaz. Oman had no horses. El Bahrein was the home of a few horses but the sandy soil and lack of water seemed to Seetzen to be the explanation for this fact.

The province of Nejd, however, was different. Although he did not see Nejd, he was told by the Bedouin throughout his travels that the region abounded in horses. The high quality and nobility of the horses of Nejd were well known to the Arabs of the peninsula.

A contemporary geographer of Seetzen by the name of Pedro Nunnes traveled in the interior more extensively and reported that the Arabian horse was so scarce that Ibn Saud himself had only two or three hundred horses among his 45,000 Wahabbi warriors.

Figure 56 - *Ulrich Jaspar Seetzen (1767-1811).*

The first European to cross the Arabian Peninsula was a British military infantryman by the name of George Foster Sadleir. He was sent by the British Foreign Office to glean information about the Wahhabi uprising as well as the Turkish response to the desert unrest. To the British, he was a soldier on reconnaissance. To the Arabs, he was a spy. The British government was constantly vigilant with regard to any factor that might pose a threat to the health of the British East India trade. Since the overland trade routes to India often involved way stations in Arabia, the British government was keenly interested in military and political events that might hamper their success. They were prepared to take any steps necessary to keep their trading routes open and safe, especially in the crucial crossroad of Arabia. To enhance his access to Ibrahim, he was to deliver a ceremonial sword to the Pasha as a token of tribute from Queen Victoria.

The British army officer Captain George Forster Sadleir (1789-1859) was an infantryman in the British army, the 47th Regiment of Foot, posted to India. He was instructed by the Governor of British India Lord Hastings to go to Arabia to investigate the positions and the condition of the troops of Ibrahim Pasha in his campaigns against the Wahabbis. The British were concerned about the long-term intentions of Al Saud and needed information on the potential impact that this might have on British trade. He was instructed by his government to propose a treaty of alliance between the Egyptian Ibrahim and the British government in order to place strict controls on the power of the Saudis, the Wahabbis, and the Ottoman Turks. These forces tended to threaten British interest in the area, and the British government wanted to see them suppressed.

Figure 57 - *George Foster Sadleir, 1789-1859.*

Sadleir began his daunting task from a port in the Persian Gulf. He was at first frustrated in his attempts to find Ibrahim, whose course across the desert in pursuit of the Al Saud rebels was circuitous and unpredictable. Sadlier, attired in full military regalia throughout his trek, was eventually received in audience by Ibrahim in Medina. Sadleir presented the magnificent sword as well as the proposal of the British government. The Pasha was not interested in a treaty with the British, or anyone else for that matter. Having accomplished his mission, Sadleir then took the safest route to safety, Yenbo. From there he obtained passage by ship to India, to resume his post.

In the course of his wanderings across Arabia in 1819, Sadleir became, quite by accident, the first European to cross the peninsula from coast to coast. He traveled by camel. His route took him from Qatif on the Persian Gulf to Yenbo on the coast of the Red Sea. He published his findings in The Diary of a Journey Across Arabia. Sadleir noted the position and strength of the Ibrahimic forces that he saw, consisting of 200 Turkish cavalry, 200 Maghribi horsemen, and 900 infantry. He gave a fairly complete and lucid account of the Cairene's travails in the desert. His descriptions are extensive, his orthography precise, and his tone toward the natives was typically European.

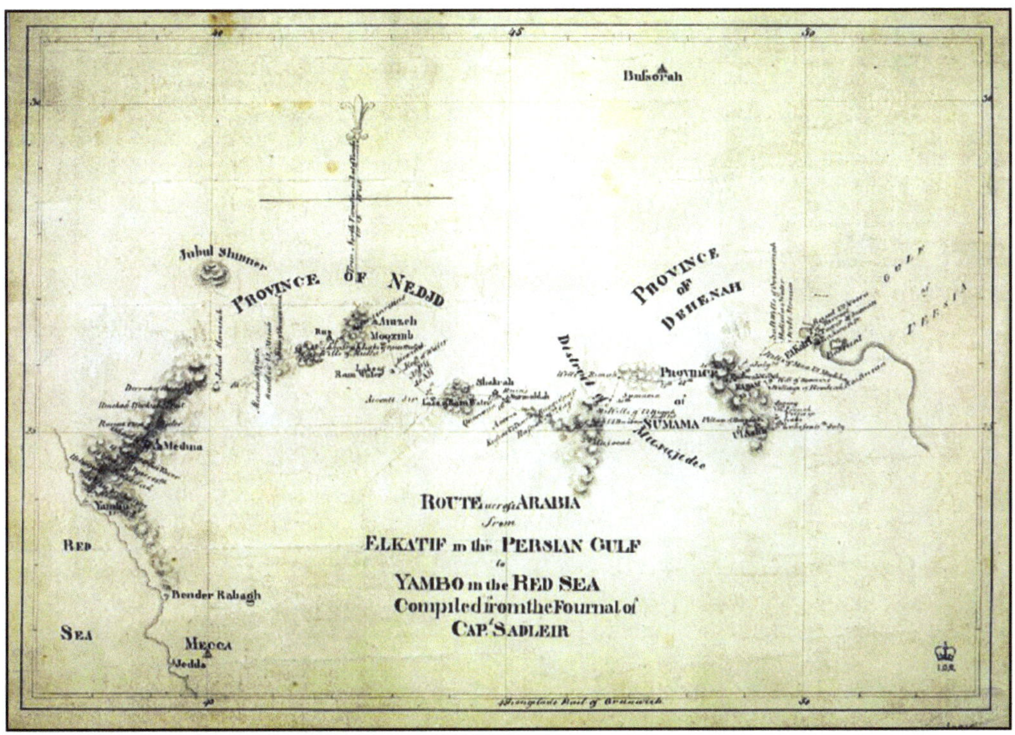

Figure 58 - *Route of Captain Sadleir from the Persian Gulf to the Red Sea, image from the Qatar Digital Library.*

Of the Egyptians, he wrote that their methods in destroying the Saudis were harsh and draconian. Ibrahim often tied his captives head to the open barrel of a cannon, firing the cannon and thus decapitating the victim. He wantonly destroyed trees, gardens, and fields of any type, to undermine the will of the people to resist his forces. Sadleir called the Egyptians "turbulent barbarians". He was outraged by the gratuitous cruelty of the Egyptians.

The 19th century exploration of Arabia had a distinct political dimension. Many of these early Arabian explorers may have been secretly obtaining information for their various governments. No one was more aware of the Middle East aspirations of European powers than was Mohammad Ali of Egypt. Napoleon's failed invasion of Egypt of 1798-1801 alerted the region's leaders to the threat posed by the French. During the 19th century, France's Napoleon III continued to pursue France's quest for empire. It was his intention to expand his empire to include the Middle East. The international influence of Napoleon III led to the French being awarded the contract to build the Suez Canal, completed in 1869, and operated under French control. Once the Canal was completed, the British interest in the Middle East increased exponentially. For the British, India was always the jewel of the empire that must be protected and remain under British control.

Figure 59 - *Napoleon III.*

Intrigue abounded.

Britain wanted to keep Arabia safe for trade routes and commerce with India. In addition, the Russians were intent on obtaining lands with seaports that bordered on the Black Sea. Since all Russian ports were frozen for half of the year, Russia was landlocked in winter, and a warm water port was essential for Russian imperial aspirations. Russia especially wanted to possess Constantinople. Without it, Black Sea control was impossible. The Russians were also keen to obtain a presence in the Arabian Peninsula because of the value of its trade routes.

Figure 60 - *Coffee Hills of Yemen.*

Louis Du Courbet (1812-1867) was the son of a French army colonel. In 1836 he served in the army of the Egyptian Viceroy Mohammad Ali. Du Courbet fought at the Battle of Nezib in Syria in 1839. He converted to Islam, took the name Abd al-Hamid Bey and performed the hajj. During 1844 and 1845 he journeyed through the Arabian heartland, beginning in Yemen and ending in Oman.

His findings were published in Life in the Desert in 1860.

"The true Arab has a special passion for horses, to satisfy his fancy for which he will sacrifice thousands of dollars."

"At Mareb horses are scarce, but among the Beni Schiddad and Beni Nauf, on the contrary, great numbers of them are raised, the original stock of which came from Nedjed. These animals which are so docile as to be manageable with a single halter, are never shod."

"Horses of superior class are also raised here, but not in great numbers, as the country is not adapted for them."

"The Arabs seldom, if ever, put clothing on their horses, which, nevertheless, exposed as they constantly are to the heat of the day and the chilly dews of the night, are quite exempt from distempers, retaining their vigor and elasticity to a great age."

"Instead of tethering the horses they sometimes hobble them with chains. They water them but once a day, foddering them all year round upon barley and raw meat, with green herbage for a change, and hay which latter is twisted up in the form of a rope and untwisted according as wanted."

"These horses are never shorn or trimmed in any way. Scissors are unknown to them. The Arab grooms curry and rub down their horses as we do ours."

"The Marebeys and Hadramites use the Mameluke saddle, while the Arabs on the western side of the peninsula prefer the kind in fashion in Cairo. Both these saddles are set on without cruppers."

Louis de Courbet

Figure 61 - *George Augustus Wallin, portrait from the University of Helsingfors.*

The desert sheikhs were asked by Du Courbet to describe the ideal asil desert horse.

> "He should be broad in the forehead, chest, croup and limbs. His chest should be deep also, his forearms, belly, and haunches long, his loins, pasterns, ears, and tail short. He ought to stir up the mud before drink-

> ing of the water. He must not start at the rustling of a standard in the wind, nor object to drag a dead body."

He was followed in the exploration of the desert by the Arabist scholar George Augustus Wallin (1811-1852), a Finnish academic who came to Arabia in 1845. In his youth, he became interested in the East and learned Arabic. Wallin traveled across the Arabian Peninsula in the guise of a doctor of law, using the name Abd Al-Wali. He was the first European to view the land and the people with the eye of a scientist. His actual work was that of an anthropologist and a geographer. He had studied the works of Arabian geography that were written by ancient Arab scholars, written in the Arabic language: he came to central Arabia with a full knowledge of those authentic Arab scientists who had gone before him.

Unconfirmed sources claim that he was sent to Arabia by Mohammad Ali, Viceroy of Egypt, to gather intelligence concerning the activities of the Wahhabi radicals, and Ibn Saud in particular. By 1845 the Al Saud family had been reconstituted at its new capital in Riyadh, Wahhabism re-emerged, and the Second Saudi state was a reality. This development was a cause for concern in both Cairo and Constantinople.

Wallin's disguise as an Arab was complete. Beginning in Cairo, he engaged two Bedu as his guides. He crossed Sinai and traveled on to Ma'an. South of there, he began to encounter warring desert tribes. He witnessed the constantly prowling bands of Arab horsemen lurking about in the distance. He was welcomed in the camp of the Hegaia, a tribe rich in horses and mighty in battle. He witnessed an attack on the Umran by the Eneze. Several days before his arrival, there had been a vicious battle between the Huweitat and the Ruwala fought over grazing rights.

His prose descriptions were elegant. Offered safe passage across the Shera chain to Ma'an by a sympathetic Sheikh of the Umran, he wrote "I could not but regard it as an advantage to avoid as much as possible every communication with people settled in towns and villages, as I had already been taught the maxim of the Bedawies, 'always keep with them,' and therefore, I readily accepted his proposal."

He was admitted to Hail which he described as a clean and well built city. In the surrounding fields, Wahhabites tended crops of corn and dates. The crops were irrigated by groundwater. When asked what items the natives need from the outside world they answered, "Gunpowder, lead, and weapons".

He learned from the natives of the Shammar that the Egyptian Viceroy especially favored the horses of the Shammar, sending a caravan of slaves annually from Alkahira to buy horses for the Pasha's collection.

Wallin saw the wealth of horses owned by the Shammar tribes and concluded that they "fully deserve the credit of being the finest and swiftest of this noble race". The stud of Adb Allah alone contained 200 horses. The Emir of the Shammar sent a few horses annually to the rulers of Medina, Mecca, and Baghdad as tribute. He occasionally sold some horses to the Al Saud family.

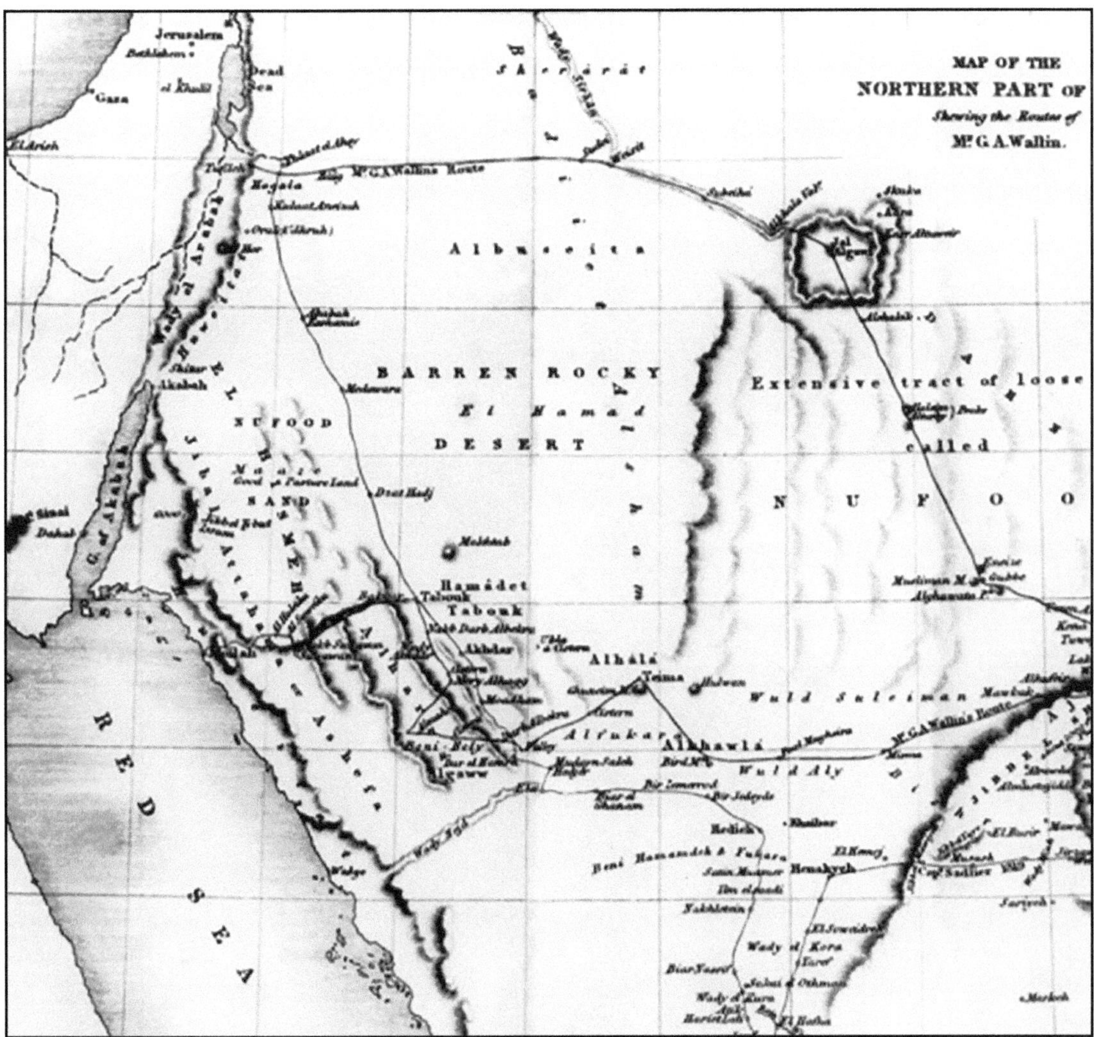

Figure 62 - *Wallin's Route through Arabia, part one, from the Qatar Digital Library.*

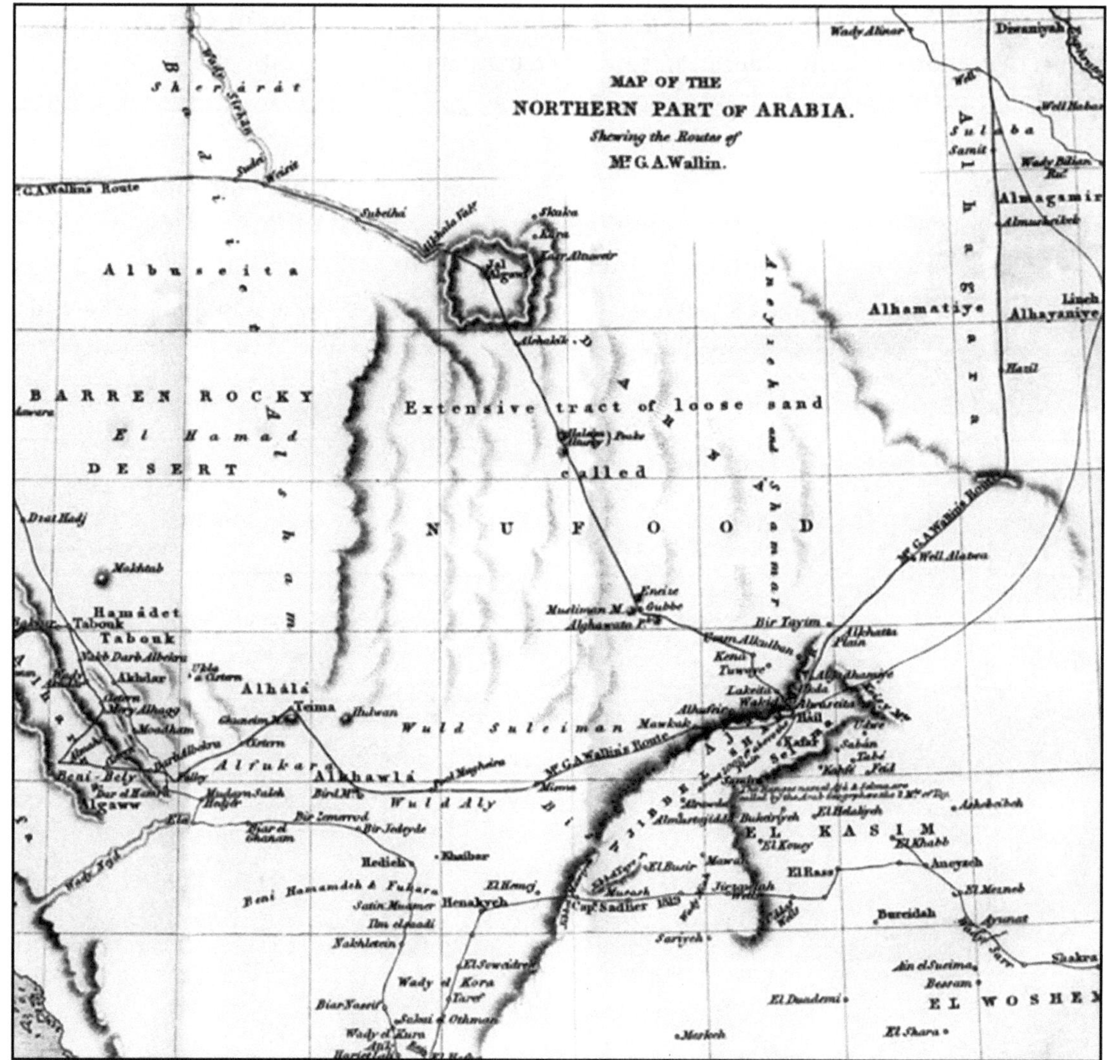

Figure 63 - *Wallin's Route through Arabia, part two, from the Qatar Digital Library.*

After leaving Hail, Wallin visited Mecca in 1845, making the Haj with other pilgrims.

Between 1846 and 1848 he traveled to Palestine and Persia. He died soon after his return from Arabia, but he was able to finish his doctoral thesis. He died at 41 years of age.

He left only one report that was read before the Royal Geographic Society in 1850. The Royal Geographic Society published these notes of his journey as Notes

Taken During a Journey Through Part of Northern Arabia. His works were formally published by his friends in 1854 as Notes of a Journey from Cairo to Medina to Mecca, by Suez, Araba, Tawila, al Jauf, Jubbe, Hail, and Nejd in 1845.

Wallin was apparently deeply affected by the Eastern ethos and the spiritual allure of Islam. He told his associates in Finland that he found assimilation back into European society to be an insufferable form of oppression. It was postulated that Wallin converted to Islam during his travels. He was buried at the Hietaniemi Cemetery in Helsinki, and his name is engraved on his tombstone in Arabic characters.

Figure 64 - *Sir Richard Burton (1821-1890).*

The noted British explorer and travel writer Lieutenant Richard Burton of the Indian Army journeyed into Arabia in 1853, but his travels were limited to visiting only Mecca and Medina. He was an expert in linguistics and ethnology. As a social anthropologist by inclination, he had no interest in horse breeding.

However he did make the following notes on Arabian horses:

- The purest breed of horse was called *Attechi*; a very superior breed; it was also called *Kochlani*, the *Kohl-eyed*, with dark skin around the eyes. This is the true blue blood. The most pure of all the horses is called *Kohlani al-Ajuz* (the old woman).The term *Rabite* is the classical term for the *purebred Arabian horse*.

- The original home of the *purebred* horse is the vast plateau of Nejd.

- There are no birth certificates for the highest class horse. A simple comment from the owner was sufficient proof of purity.

- Stallions could be bought easily, but the Bedu would not part with a mare. Mares were usually owned by several men jointly.

- The general passion that the Arabs have for the horse explains the many wars and *ghazus* that have been fought over them.

- It was considered a high honor for a Bedu to steal a highly bred Arabian stallion from a foreigner or an infidel. This was not considered theft but rather was considered to be liberation of the noble horse from one who did not deserve to own it. However, the penalty for stealing a nobly bred horse from a fellow Bedu was severe. The recreant was chained by all four extremities with chains, firmly attached to four iron posts, chained in an isolated location. He was left to die.

- Arab horses are never taught to leap; this eliminated the risk of the horse making a sudden lunge at the edge of a stream.

In describing the natives he made this observation:

> "As a general rule, the expression of the Badawi face is rather dignity that cunning for which the Semitic race is celebrated...the ears are like those of Arab horses, small, well-cut, "castey", and elaborate."

Burton's gift for beautiful prose proved useful when he described the Arab's *kayf*, or sense of pleasure and delight.

> "The savoring of animal existence; the passive enjoyment of mere sense; the pleasant languor, the dreamy tranquility, the airy castle-building, which in Asia stand in lieu of the vigorous, intense, passionate life of Europe. It is the result of a lively, irrepressible, excitable nature, and exquisite sensibility of nerve. It argues for voluptuousness unknown to northern regions..."

And finally, he echoed the impression of Lady Anne Blunt when she first came to know the intoxicating atmosphere of the East:

> "Thus in England where laws leave men comparatively free, they are slaves to a grinding despotism of conventionalities, unknown in the land of tyrannical rule. This explains why many men…feel fettered and enslaved in the so-called free countries."

His book of adventures Pilgrimage to Al-Medinah and Meccah was published in England in 1855.

Burton was knighted in 1886.

The English adventurer William Gifford Palgrave (1826-1888) traveled extensively in the Nejd, visiting both Hail and Riad. Palgrave was in the British military as a young man, where his facility at languages was recognized. Always restless, he became interested in missionary work while stationed in India, and he was received into the Society of Jesus in Madras as a priest.

Figure 65 - *William Gifford Palgrave, used with the permission of the National Portrait Gallery, London.*

But Palgrave was a man whose spirit was multifaceted. Being intrigued with the fragrance of the East after reading the Arabic poem "*Antar*", he planned to see Arabia for himself. His purpose in traveling to Arabia was to gauge the capacity of the Arabs to be converted to Catholicism.

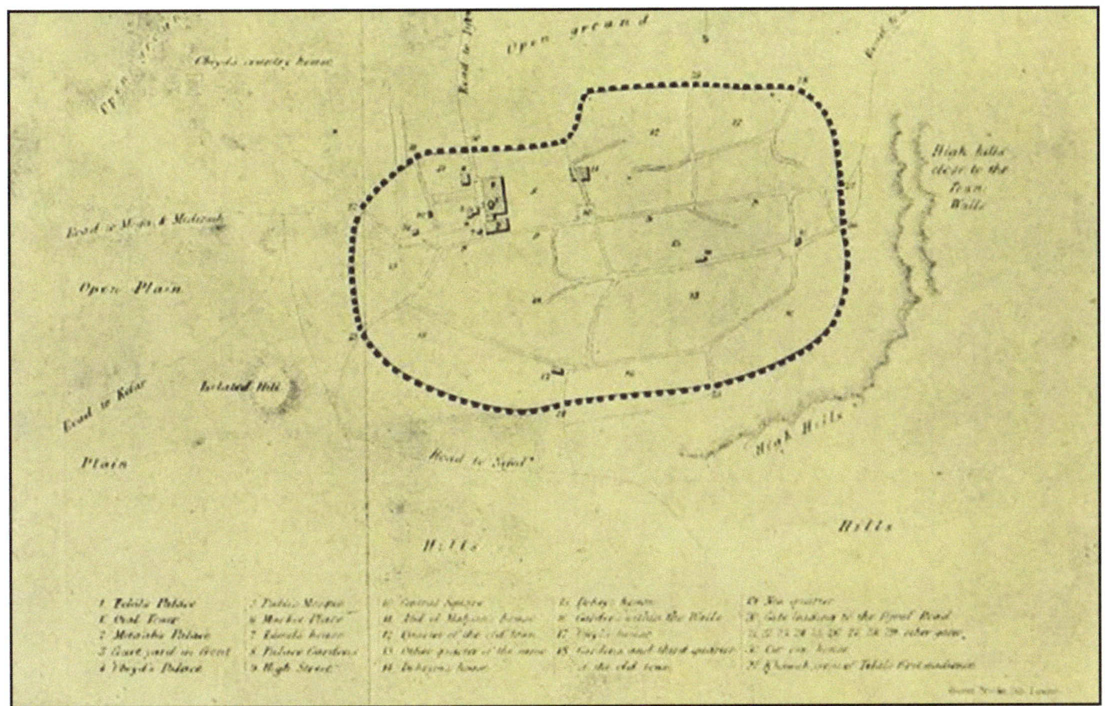

Figure 66 - *William Palgrave's city plan of Hail, 1862.*

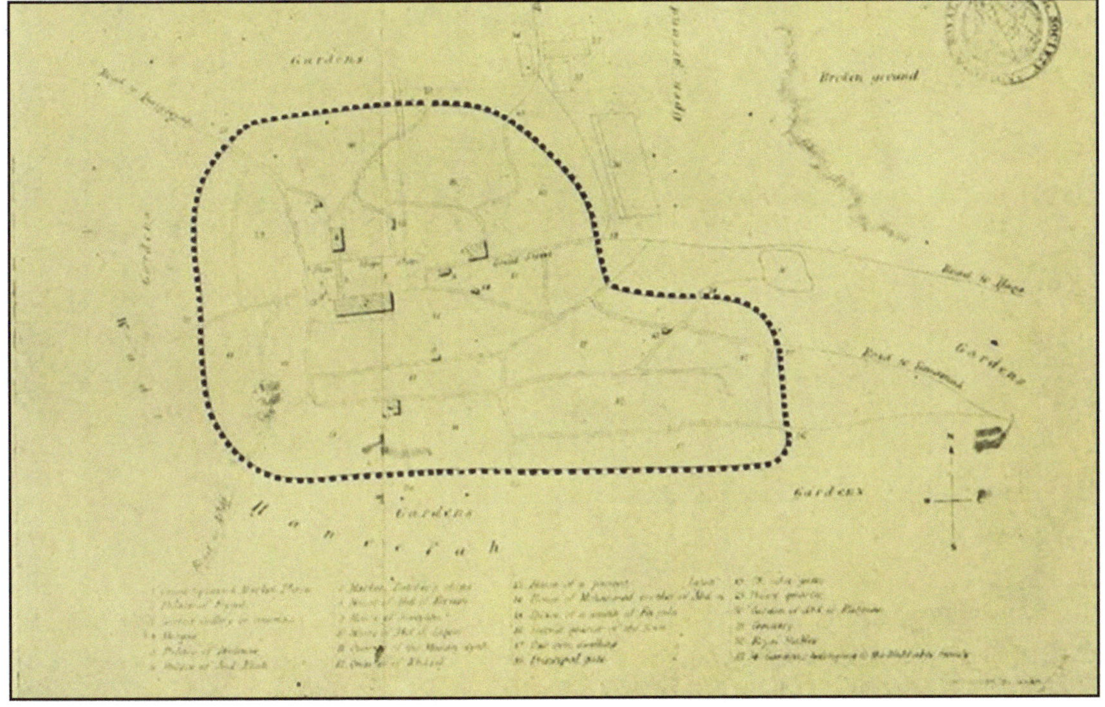

Figure 67 - *William Palgrave's city plan of Riad, 1862.*

Figure 68 - *William Palgrave, National Portrait Gallery, London.*

Palgrave planned to adopt the guise of a Syrian doctor for his journey into the desert. He called himself Saleem Abou Mahmood al Eys. Initially, he did not attempt to pass himself off as a Muslim but planned to simply make no claims of origin or religion. He thought this would provide sufficient safety.

Having reached Hail and taken in the surroundings, he became determined to go to Riad, to see the power of the Wahabbis for himself. But reaching Riad meant crossing the Nejd. He asked for advice about this passage and was told: "It is the Nejed; he who enters does not come out again." Its hills and valleys sheltered thieves and murdering bands of ruffians, who now were fired by the zeal of Wahabbism. To them, it was a religious duty to slaughter an infidel, and besides, it was a personal honor. Palgrave, undeterred, was able to obtain a letter of passport to Riad from the Emir of Hail, and this provided some measure of confidence for his companions.

Figure 69 - *William Palgrave.*

In 1862 he journeyed to the center of Arabia. Posing as a medical man, he was allowed to see the horses of the Saud ruler Faisal in Riyadh under the pretense of offering to treat a sick horse. Palgrave was not a horseman, nor did he have any great knowledge on the subject of *purebred Arabian horses*.

But the description that he left to posterity of the *purebred Arabian horse* resonates even today.

> "The horses of the Emir," he said, "were never tied by the neck or haltered, but were restrained from wandering by the use of leg irons. These iron rings were attached by a length of iron chain to a peg driven into the sand. There were only a few shelters and pens in use at the stud."
>
> "Faisal's entire muster is reckoned at 600 head. Never have I seen or imagined so lovely a collection. Their stature was indeed somewhat low; I do not think that any came up to fully 15 hands . . . but they were so exquisitely shaped that want of greater size seemed hardly, if at all, a defect. A little, a very little saddlebacked, just the curve which indicates springiness without weakness; a head broad above and tapering down to a nose fine enough to verify the phrase of 'drinking from a pint-pot'. A most intelligent and yet singular gentle look, full eye, sharp thorn like little ears, legs for and hind that seemed as if made of hammered iron, so clean and yet so well twisted with sinew; a neat round hoof, just the requisite for hard ground; a tail set on or rather thrown out at a perfect arch; coats smooth shining and light; the mane long, but not overgrown or heavy The prevailing color was chestnut or grey; a light bay, an iron color, white or black, were less common; full bay, flea-bitten or piebald, none…the slope of the shoulder, the cleanness of the shank, and the full round haunch...was a perfection and a harmony unwitnessed (at least by my eyes) anywhere else. Their

appearance justified all reputation, all vale, all poetry."

Very few foreigners were allowed to see the horses of the Saud Emir. A description of the Royal Stud written by Harry St. John Philby 50 years after Plagrave's visit provides more details. Philby was a key British adviser to Ibn Saud during the First World War and was central to the post-war conquests of the Saud family in Arabia.

During the 1920s he described the earliest days of the *purebred Arabian horse* farm at the wells of Kharj, built by Ibn Saud and supplied with horses which were obtained from the nomads.

> "At the northern end is a well four fathoms deep with a large flat earthern drinking trough attached to it. Each paddock is furnished with rows of circular mud mangers in which lucerne brought from the plantation is piled high twice a day, and to which the horses are loosely tethered. While round the walls runs a flimsy shed in which the animals are placed for protection against the sun during the hours of greatest heat."
>
> "At the time of my visit there were some 50 animals in the building - five or six fine stallions, a dozen promising young stock, the rest mares of varying quality with a couple of camels, and a donkey to keep them company. So far as I could see, the animals never received any sort of grooming or attention; they were never exercised and never on any account allowed to leave the building; the paddocks are cleared of refuse only at long intervals, and all day long the wretched beasts the pick of the stock of Nedj stand at their mangers in ever growing heaps of litter and manure, browsing in their rich fodder. Nevertheless, they suffer less than one might expect from this extraordinary treatment, and when they go forth to war their skins are glossy and their capacity to endure fatigue and privation seems unimpaired."

In 1863, Palgrave also had occasion to meet the Al Thani family of "Guttar" during his time in the eastern desert. He described Qatar as being a collection of small fishing villages inhabited by subsistence workers who lived in poverty. He saw no horses. Palgrave had an audience with Mohammad Bin Thani who was at the time governor of Bida, a small mud brick village that was to become Doha. He described the elderly diplomat as "generally acknowledged for the head of the entire province. He is a shrewd wary old man, slightly corpulent, and renowned for prudence and good humored easiness of demeanor, but closed fisted and a hard customer at a bargain, more the air of a pearl merchant than a tribal chieftain. He wore a Bengali turban of the date of Siraj al-Dawla (an Indian king who died in 1757) to judge by its dingy appearance".

Figure 70 - *Doha, Qatar, in 1908, from the Qatar Digital Library.*

One of the most memorable explorers and horse traders of the interior of Arabia was the Italian Carlo Claudio Camillo Guarmani. He was born in Livorno Italy in 1828, died in Genoa Italy 1884. The achievements of Carlo Guarmani are remarkable for their daring, in that he is the first European to bring horses directly out of the heart of the Arab heartlands. Guarmani was a Levantine Italian who was enthralled by the desert, its inhabitants, and its horses. For 14 years he made frequent forays into the desert; he perfected his ability of passing undetected among the Bedouin, dressed as an Arab, and capable of speaking Arabic.

In 1864 he received commissions from both Napoleon of France and King Victor Emanuel II of Italy to buy stallions from the tribes of inner Arabia. The horses were to be used to improve the quality of cavalry mounts. Interest in the value of pure Arabian blood had been excited among Europeans following the highly publicized sale of *purebred Arabian horses* from the stud of Abbas Pasha of Egypt in 1860.

Guarmani's journey took him in a circular route, beginning at Taima. He then went southeast to Khaibar (populated entirely by Blacks, their master having died during a smallpox epidemic), and then to the twin oases deep in the heart of the Nejd; Buraida and Anaiza. From there he proceeded to Hail.

Figure 71 - *Arabian Stallions, by Carle Vernet.*

Guarmani did not go to Riyadh, after hearing rumors that Faisal Ibn Saud was at war with the Ataiba Bedouin.

He bought horses from the Hutain and the Ataiba tribes but he did not buy from the Anazeh as he intended. He states that the Anazeh had just completed a large sale to the Indian markets and had few good horses left.

Many of the desert tribes that he encountered refused to show him their horses, fearing the dire effects of the dreaded “evil eye”. Guarmani found that the conventional Bedouin method of turning away inquiries from foreigners was to claim that the horses were scattered across the desert and could not be collected easily.

Guarmani also visited the Beni Sakhr tribe, who had about 50 mares, but none that interested the Italian.

After meeting the Rashidi Emir Talal Ibn Rashid in Hail, Guarmani left for Jauf, crossing the Nafud Desert with his string of horses, selected from the various tribes which he encountered.

After surviving a deadly Bedouin attack in Wadi Sirhan, Guarmani emerged from the desert with several horses and the material for an entire book. His publication Journey in Northern Nejd appeared in 1866. He also wrote a book on Arab horses entitled The Al Khamsa

His valor was commendable. His editor wrote of him ”Inured to fatigue and hardship, thoroughly conversant with local usages, dressed as a Bedouin and mounted on horseback, he penetrated far into the desert. There he spent long days in a tent, studying Arabian horses.”

Other less adventurous Europeans wrote books concerning their experiences with the Arab horse in the Middle East. Ermete Pierotti, (1820 to 1880) an academic mathematician and an engineer to the Pasha in Jerusalem, published a brief description in 1864 of Arabian horses that he encountered during his work in the Levant. He noted the high esteem that the Bedu accorded the noble Arab horse and wrote that each tribe guarded its mares jealously, suspicious of any stranger who happened by since anyone unknown to him might intend to steal a horse. The horses of pure nobility, he wrote, were becoming quite rare among the nomads.

When a noble blooded horse was sold, several tribal authorities attested to its pedigree and provided a written document to that effect. Pierotti was impressed by the seriousness and honesty with which the tribal members treated a sale. He describes the solemn assurances and rituals accompanying the sale of a *purebred* horse, stylized ceremonies attested to by most other European writers who have witnessed such events.

Figure 72 - *Ermete Pierotti*

Regarding care of the animals, he wrote that the young are always secured with leg hobbles to prevent them from wandering. The horses are ridden by age two. A mature horse stood about five feet tall at most.

They are fed hay, barley, and are given water only once a day. He noted that the Arab horses suffered frequently from foreleg disorders, inflammation, and weakness owing to the uneven rocky terrain on which they are ridden, and the harsh bit handling (the Turkish bit) that he witnessed when their riders stopped them short from a gallop. The stallions are rarely aggressive or vicious. Pierotti admired the extraordinary flexibility of the horse in motion; they are graceful light and picturesque. Their nimbleness and agility were useful attributes in battle. When left peacefully at home, the mares could safely be ridden by women and children.

Figure 73 - *Colonel Lewis Pelly, in 1882, Qatar Digital Library.*

There was no shortage of 19th century British military advisers who toured the interior of Arabia seeking military intelligence. Lieutenant Colonel Lewis Pelly of the Bombay Army was sent by the British government to appreciate the strength and expanding power of the Wahhabi movement. He was instructed to determine the intentions of the Al Sauds in relation to British commercial interests in Kuwait and Irak. He began his journey in 1865 in Kuwait, having been deposited there by military streamer.

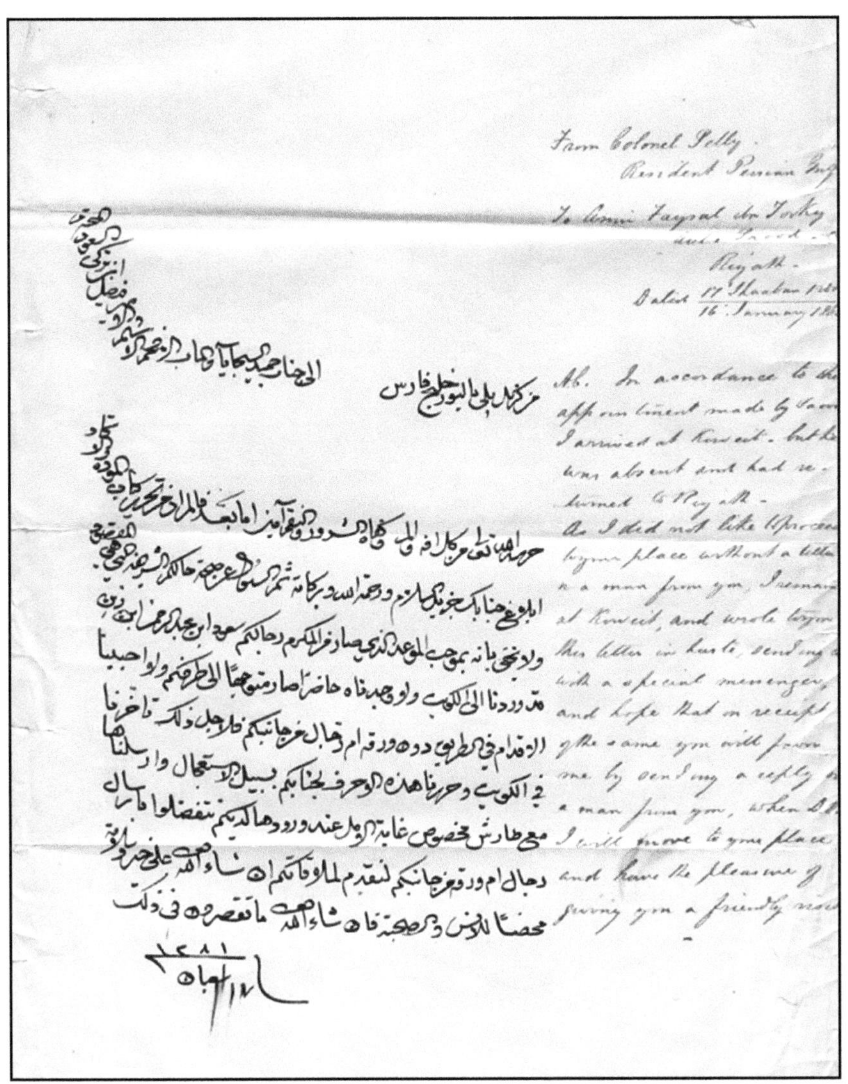

From Colonel Pelly.
Resident Persian Gulf

To Amir Faysal ibn Torky
... Saood ...
Reyath.

Dated 17 Shaaban 128[illegible]
16 January 18[illegible]

AC. In accordance to the
appointment made by [illegible]
I arrived at Koweit. but he
was absent and had re-
turned to Reyath.
As I did not like to proceed
to your place without a letter
or a man from you, I remain
at Koweit, and wrote to you
this letter in haste, sending it
with a special messenger,
and hope that on receipt
of the same you will favor
me by sending a reply or
a man from you, when I
I will move to your place
and have the pleasure of
giving you a friendly visit

Figure 74 - *On arriving in Kuwait, Pelly wrote to the Emir Faisul Ibn Turki al Sa'ud in the Nejd capital of Riyath. Pelly's letter begins by stating that he was disappointed that the Emir had not met him personally in Kuwait. Be that as it may, Pelly then asked the Emir for a guide so that he might have a "friendly visit" with the Emir in Riyath. The letter was then translated into Arabic and sent to the Emir. Images from the Qatar Digital Library.*

Figure 75 - *A street scene in Kuwait, 1918, image from the Qatar Digital Library.*

Colonel Pelly, despite the apparent disinterest of the Emir, traveled with a guide as far as Riyadh. Being a military man he was interested in the native horses of the region and recorded in his diary a few observations about the horses that he saw along the way.

*Colts were raised on camel's milk, said to be the perfect diet for the young horse. The horses were to be kept continually in the open air. He was ridden at 1.5 years of age.

*The Bedouin were partial to the very best horses that they owned, and would sooner part with one of their children that to sell one of their best mares.

*Bedouin horse riders were frequently encountered. When they dismounted they passed the lead rope through the animal's forelegs and tied it to the hocks. The horses stood perfectly still.

*The Emir of Riyadh, Saud, obtained his horses, said to be superb, by demanding that the local tribes pay their annual taxes owed to him in horses.

*According to the Bedouin, the blood of the asil horse, once tainted, cannot be purified.

*Pelly lists 5 breeds of Arabian horses in Nejd:

- Saglawiyah
- Kohaylat
- Obayyat
- Dahmat
- Wadhnat

He lists dozens of subsidiary "breeds".

Concerning the Emir himself, Pelly found that he was over seventy years of age, blind and infirmed.

> "The Imam rose, but with difficulty, took my hand and felt slowly over it. His face as remarkable, with regular features, placid, stern, self-possessed, resigned."

In response to Pelly's comment that the conditions of Riyath were isolated and spartan, the Imam responded:

> "...we are content. We are princes according to our decree. We feel ourselves a king every inch."

He was, as Pelly wrote, not well liked by his subjects, but he was uniformly respected. His judgments and punishments were strict; he was regarded by his subjects with a mixture of awe, fear, and distaste.

Pelly's report was published in 1865 as A Journey to Riyadh.

Figure 76 - *Charles Montagu Doughty (1843-1926), image used with the permission of the Royal Geographic Society.*

Charles Doughty was born in England in 1843 and was Cambridge educated, graduating from Caius College. He was a traveler, journalist, and author, best known for his Travels in Arabia Deserta, published in1888. He wandered in the desert for two years beginning in 1875. He presented himself as a medical doctor and was warmly received among the Bedu along with his bag of patent medicines. He took no special pains to hide the fact that he was a European, and was commonly referred to by the natives as the "Ingleysy". He did not affect Islam and was known by the natives as the "Nasrany".

Travelogues and anthropologic studies were very popular at the time in Europe; the sense of magic and remote mysteriousness of the Middle East was a constant source of intrigue and interest to Europeans of the Victorian age. Doughty's book was well received.

Charles Doughty traveled for several years among the central desert tribes which were concentrated around Jebel Shammar. Doughty found that the Al Rashidis, like the Sauds, had continued to obtain and breed high-quality Arabian horse stock. Since the ruling princes had no *purebred Arabian horse* herds of their own, they acquired them from the few small nomadic tribes of the plateau who continued to breed the true blood. Their stock was acquired, the English explorer Doughty writes, from other desert sources "gotten in two generations of the spoil of the poor Beduw".

He reported that Rashid's asil horses numbered in the hundreds:

> *"The Emir's horses are grazed in nomad wise; the forefeet hop-shackled, they are dismissed to range from the morning. Barley or other grains they taste not. They are led home to the booths, and tethered at evening, and drink the night's milk of the she-camels, their foster mothers."*

The Emir Rashid was a tyrant, ruling the people of the city of Hayil with an iron will. He also ruled dozens of small Bedouin tribes around Jebel Shammar. Doughty estimated that the entire population of Rashid's kingdom was about 20,000. His kingdom contained an estimated 30 oases. The city dwellers approved of the Emir's xenophobic behavior because he brought stability and safety to the inhabitants. For the Bedu, however, he was an onerous despot, taking from them taxes, tributes, and horses. They were his victims, not his beneficiaries.

He was the unquestioned ruler of Jebel Shammar, and he was recognized to be the greatest prince of the Nejd. Even the formerly great House of Saud was said to be khurban, ruined. The Wahabbi forces at er Riath (Riyadh) were also losing their power as well as their force of intimidation.

Doughty's book was the first to give the Western reader a sense of the nature of the Arabian horse as it existed in the desert. He wrote that the desert mare was gentle to handle and did not display an ill temper even when provoked. He said that the mares would enter the Bedouin's tents in the heat of the day, seeking shelter from the blistering sun, walking carefully between sleeping men until a place could be found to rest. Infants played on them. Children led them. To the Romantic Era English horseman accustomed to the ill-tempered, irritable, and irascible English Thoroughbred horse, these descriptions were astonishing.

Doughty was a scholastic and anthropologist at heart and showed little interest in horses or their breeding. He mentions having been shown horses of noble breeding but provides no details. The Bedu, he states, all agree that Allah created the Arabian horse from the soil of the Nejd. Doughty was skeptical about this claim. He observed that life on the Nejd plateau would be impossible for man or beast without the use of the dromedary, the only animal that could survive in the arid plateau without the management of Bedouin.

Among the smaller tribes, he saw high caste horses, which were mostly bay or chestnut in color. The color grey was rare, a black was never seen.

The *purebred* horses, he was told, were well suited for desert life; they could withstand heat, lack of water, and lack of sufficient forage. They were extremely tough, hardy, and resilient animals. This, said the Bedu, was in contrast to the Gulf Arabian horses (Northern) horses, which were impurely bred and subsequently unable to bear up under the tribulations that existed in the desert.

Doughty notes the Arabic word root *asl* in his glossary and states that the natives used the word asil to denote the wellspring or lineage of a horse.

All horse breeding in the central deserts took place in the small individual tribes that were scattered about the sandy wastes of Jebel Shammar. No breeding was done in Hayil by Ibn Rashid. He simply took the horses that he wanted from the smaller and defenseless serf-like tribes in *ghrazzus* or *razzias*, in which he and his salaried warriors appeared without warning at an isolated encampment, mounted on horses and camels. They took anything that they wanted and dealt viciously with those Bedu that resisted.

The Emir confiscated horses as he needed them. He was required by religious duty to send an annual tribute of 40 stallions to the Wahhabi leader in er Riath (Riyadh)

He also commandeered stallions from the Bedu for sale to the Indian market. India being a British colony, the demand for high-quality horse flesh was significant. Ibn Rashid collected the horses annually and sent them with a staff of servants to Kuweyt (Kuwait) for transport by sea to Bombay. He used the money from the annual sale of these horses to finance his government, which consisted mainly of his salaried raiders. He lent out 400 of his horses and 800 camels to them: they could be called on in an instant for a *ghrazzu*.

There were horse dealers that operated out of the Nejd who bought thin underweight horses at very low prices from the nomadic tribes, and then fed the horses until they were presentable for sale. These horses were then sold on the open market, usually at Kuwait, and usually sent on to British military customers in Bombay.

The Bedouin may have been isolated from the outer world but the isolation was not hermetic. Doughty frequently saw goods sold in the *suks* wrapped in old discolored issues of the Bombay Gazette, written in English.

Figure 77 - *Zarif, by Emile Volkers, bred by the Roala Bedouin, used with the permission of Archiv Olms.*

Figure 78 - *Wilfred Blunt. Image used with the permission of the National Portrait Gallery, London.*

The Bedu loved firearms, he wrote:

> Muskets that fired round balls were constantly in evidence and in constant use, either for target practice or simply for sport. The Bedu loved to compete with each other. The best gun powder, they said, was En-

glish, and they were most fond of the brand of black powder made at Hall in Dartford; some of them had powder cases embossed with this company's address.

Figure 79 - *The Blunts at a Midday Halt in the Arabian Desert, watercolor by Lady Anne Blunt.*

Wilfred and Lady Anne Blunt were both members of the British aristocracy, and they had developed an affinity for the culture of the Middle East and the horses that were to be found there. The Blunts visited the Rashid capital at Hail in 1879. They were warmly received by Mohammad Ibn Rashid and given full access to his collection of horses. Lady Anne described the Emir as having a "thin sallow and careworn face", but he greeted them graciously. He was dressed in "the regularly barbaric love of finery and he wears a gold and jeweled sword". He was very proud of his horses.

To their surprise, the Blunts saw a distressingly disorganized collection of horses in very poor physical condition. "As they stand in the yard, slovenly and unkempt, they have very little of that air of high breeding one would expect . . . Ibn Rashid's yard contained thirty or forty foals and yearlings, beautiful little creatures but terribly starved and miserable."

The mares were more impressive. Lady Anne commented on the fact that they were all about 14 hands, most were grey, but all had the most strikingly beautiful heads that she had ever seen. Overall, however, she concluded that the horses of Hail did not possess the signs of asil breeding that one would expect from a group of horses of Nejdi origin.

Ibn Rashid offered the comment that Ibn Saud once had an excellent herd of *purebred* horses, but most of these were now dispersed. The death of the Saud Emir Feysul proved to be devastating to the clan. Ibn Rashid himself had been able to acquire some of the finest Saud horses. The Blunts were shown a Hamdani Simri that had been obtained from the Saud herd, but they were unimpressed with her.

Figure 80 - *The Plain of Melakh and the Euphrates River, sketch by Lady Anne Blunt.*

The Emir invited the Blunts to a khawah one evening, and he offered his guests the opportunity to buy some of his horses. Wilfred Blunt deflected the offer with a short diplomatic speech and offered the counter possibility of bringing the Emir a cannon as a gift. At this suggestion, the Emir beamed with delight.

That same evening the Blunts met an elderly Bedu who related to the Blunts his memories of making several journeys to Cairo to deliver horses to Abbas Pasha. The Pasha preferred Saklawi, remembered the elder, and he preferred the Saklawis of the Shammar above all others. He paid as much as 12,000 rials for the best horses.

The Blunts left Hail without acquiring a single horse. At the time, they did not regret this fact. Lady Anne wrote that the horses could not gallop properly and that even the best of the Rashidi horses were "ponies". They would later come to regret their decision.

At this time the Blunts were beginners in the enterprise of breeding Arabian horses. They remained untutored for a long time, and they bought many desert horses that they later regretted purchasing. They were, and always had been, English Thoroughbred people.

Lady Ann explained:

> "We were still under the impression that the true Arab style should be avoided, as European prejudice at the time held that the racing type animal was ideal for use in European studs."

The Blunts came to Arabia with the notion that pure Arabian blood could be used to rejuvenate the English Thoroughbred in England. They were looking for Arabian racing stock to use in England for the specific purpose of breeding to English stock, with the goal of producing a better hunter and a faster racer. The Thoroughbred horse in England had begun to deteriorate, showing the signs of intense inbreeding. Many turf enthusiasts called for an infusion of new blood into the historically closed General Stud Book.

It was much later that the Blunts came to an awareness of the special qualities of the *purebred Arabian horse* and the need to preserve the Arab blood in its pure form. The Blunts could not easily overcome their innate dismay at the small stature and light bone of the *purebred Arabian horse*. They hesitated to buy these diminutive horses initially because they feared the ridicule of horse breeders in England when they returned home.

Figure 81 - *Wilfred Blunt and Pharaoh, by Lady Anne Blunt. Pharoah was born in 1876, bred by Barki Ibn el Derri of the Resallin Gomussa branch of the Sebaa Anazeh. Pharoah was from the strain of Seglawi Jedran of Ibn ed Derri.*

Figure 82 - *A Council of War in the Desert, sketch by Lady Anne Blunt.*

"An ounce of blood is worth a pound of bone." *Arab proverb.*

Despite the fact that the desert tribes were frequently at war with each other, there was a surprising amount of interchange between horse breeders on the peninsula during this period.

The diary of Lady Anne Blunt, which she began keeping in 1878, contains the following entries:

- Ibn Saud gave the mare Sherifa, who he apparently bred, to the Turkish governor of Mecca as a diplomatic gift. The governor subsequently gave the mare to Sheikh Takha of Aleppo. He soon died and his heirs quickly liquidated his assets, selling Sherifa to Mr. James Skene, British consul at Allepo. Skene eventually sold the mare to the Blunts, who bred her for many years and considered her head to be the finest that the desert had ever produced.

- Ibn Saud had the reputation among the desert tribes as a ruthless and bloodthirsty man. He was an avid collector of fine high caste desert horses and often bought them from the smaller tribes. However, if a sale could not be agreed upon, Ibn Saud would commonly send a *ghazu* consisting of his own warriors to steal the mares that he wanted.

- *Ghazus* were common, and Ibn Rashid kept a group of mares confined in a special ward of his palace at Hail for use in *ghazus*, which by their very nature were sudden and unplanned,

- Ibn Saud at one time had an excellent collection of *purebred* horses, and was widely known for his Hamdani Simri horses, said to be of unsurpassed quality.

Figure 83 - *The Start of a Ghazu, sketch by Lady Anne Blunt.*

Figure 84 - *The Shammar Tribe Moving Camp at Night, sketch by Lady Anne Blunt.*

- Abdallah Ibn Khalifeh of Bahrain had an excellent collection of well bred horses, and he was protected from raids and theft by the insular nature of his kingdom.

- Feysul Ibn Turki Ibn Saud bought shares in an El Nejib mare from Ibn Khalifeh. Of his share of the produce, he gave the oldest to Sherif Adballah Ibn Luwa, and he kept a younger grey mare. Eventually the El Najib line died out, except for some specimens in the stud of Ibn Khalifeh.

Lady Wentworth summed up the true station of the horse in the hands of the Bedouin.

> "The Nomad rider loves his horse, but it is a fierce, brutal love, a stern admiration which knows no mercy for failure, no sympathy for weakness, no kindness for age or wounds. 'For blood shall pursue thee!' And the end of each and all is the merciless tearing to pieces by the wild beasts and birds of prey, the carrion birds waiting for the last defenceless weakness to tear out the eyes."

Figure 85 - *Lady Anne Blunt, Anne Isabella Noel Blunt, 15th Baroness Wentworth.*

Blunt wrote that the discovery of the *falj* solved for him the puzzle of horse breeding in central Arabia. He wrote:

> "In the hard desert (Arabia Petra) there is nothing for a horse to eat, but here there is plenty. The Nafud accounts for everything. Instead of being the terrible place it has been described by the few travelers who have seen it. It is in reality the home of the Bedouin during a great part of the year."

Figure 86 - *Wilfred Blunt thought the falj (Arabic for water source) to be highly significant as a source of vegetation for horse grazing. During his time in the desert, he was puzzled over how horses could live there given the apparent lack of grass or berseem. He could not harmonize the legends that the asil Arabian horse originally came from Bedouin tribes living in the Nejd when the land appeared to be so inhospitable to life of any kind. He learned, however, that the enormous horseshoe-shaped depressions regularly seen in the Nafud contained rich vegetation growing on their sides, vegetation sufficient to sustain horses. The natives told Blunt that these depressions were called falj.*

In fact, the Arabic word intended may have been related to the word *falaj.* The plural form is *Aflaj*, a common place name in Arabia. This term refers to an ancient form of man-made desert irrigation in which long underground channels were dug by hand over great distances to allow water to flow by gravity from a subterranean spring to a lower elevation. This method of irrigation eliminated evaporation and was first introduced in Persia in the first millennium BCE. From there the technique spread throughout the Middle East. The Bedouin were probably implying to Blunt that these formations were the residua of abandoned and collapsed man-made tunnels. The pits are in fact common naturally occurring topographic features of all deserts, and they are formed by the action of wind.

The Nafud Desert is technically known as an erg, which is a geologic landform consisting of a broad flat wind-swept flat desert. The term is derived from the Arabic word '*arq*' which means dune field. The Nafud is unusual in that it is composed of a very ancient and very deep layer of red sand, the product of wind and rain erosion of the underlying sandstone sedimentary bedrock that had occurred over eons of time.

In the Nafud, which is a red sand desert, landforms consisting of the substratum of large diameter red sand grains are minimally affected by normal wind patterns, and these landforms may persist for hundreds of years. Once a red sandpit is formed, it persists over time, collects rainwater, and provides vegetation.

Figure 87 - *A large gathering of Bedouin. Raswan states that the natives rarely gathered in such great numbers in the desert, except in the event of a substantial downpour of rain. On these occasions, the bare desert bursts into a florid profusion of grasses and wild flowers. From the Raswan collection.*

From these two examples, it would seem that the Bedouin could breed and raise horses in the central parts of Arabia because they possessed the cultural and tribal knowledge of how to find vegetation for grazing. This is, in point of fact, what nomadism means. It is an incorrect view to suppose that the nomads roamed aimlessly about the barren landscape hoping to find grass. This was simply not the case. The elders knew the annual fluctuating patterns of the cycles of rain and drought throughout their customary grazing zones. There was never one place that always had rain and grass, but there was always some site somewhere that would provide grazing. The desert is a dry arid land, but not always, and not everywhere.

Figure 88 - *Jebel Shammar, watercolor by Lady Anne Blunt. The band across the middle of the page is intended to portray a mirage partially obscuring a view of a caravan of pilgrims bound for Mecca.*

In the case of the Arabian explorer Roger D. Upton (1827-1881) his interest was very specific. Upton wanted to go to the Nejd, the fabled heartland of Arabia and the source, many believed, of the *purebred Arabian horse*. Upton wanted to obtain first quality Arabian stallions to bring back to England. He was confident that the infusion of pure Arab blood into the English Thoroughbred stock would be beneficial in restoring the lost vitality that he so deplored in the Thoroughbred.

Figure 89 - *The ruins of a castle of Feysul Ibn Rashid at Jauf in the Emirate of the Shammar. Photo by S.S. Butler and used with the permission of the Royal Geographic Society.*

Figure 90 - *The south gate of Ha'il, photographed by Gertrude Bell in 1914. Image used with the permission of the Royal Geographic Society.*

Figure 91 - *Ha'il, March 1914, by Gertrude Bell. Used with the permission of the Bell Archives at the Newcastle University Library. The Blunts and Doughty would have seen Ha'il just as shown here.*

In his youth, Upton was in the Royal Lancers, a group of men known for excellence in the saddle. Known colloquially as the Delhi Spearmen, the Lancers were formed in 1715 as a division of the Queen's Dragoons. Their motto, *vestiga nulla retrosum*, typified Upton. Retiring from the military with the rank of Major, he spent his mature years deeply involved in the matters of the turf in England. He became an expert on Thoroughbred bloodlines and became an outspoken critic of the deterioration of the English Thoroughbred. In his view, the racehorses were losing their sprint speed. He decried the loss of proper structure and conformation. They were losing their "spirit". As a former Lancer, his opinions carried considerable weight among experts of the day.

Upton catalogued the foundation horses of the English Thoroughbred horse:

- 101 Arabian stallions
- 7 Arabian mares
- 42 barb stallions
- 24 barb mares
- 1 Egyptian stallion
- 5 Persian stallions
- 28 Turkish stallions
- 2 foreign horses

He proceeded to publish his findings in Newmarket and Arabia (London, 1873) in which he "pointed out the errors in the breeding of our horses". He singled out the undefeated Thoroughbred racehorse Flying Childers as the best horse to be produced by the breed. He attributed this fact to the quality of the sire of the horse, the Darley Arabian. Upton also singled out the racehorses Herod, Eclipse, and Trumpator as superior stock, again due to the closeness of Arabian blood in their pedigrees. He considered the *purebred Arabian horse* to be the "ideal of excellence".

The following year he set out to procure the Arabian stallions with which he would regenerate the English Thoroughbred. With the backing from two highly placed government ministers and $62,000 dollars worth of gold, he set off for the Levant and Syria. His journey, greatly facilitated by the expert aid of the British Consul General in Allepo J.H Skene, resulted in the work Gleanings from the Desert.

Upton's book contains a vigorous and highly readable account of the pureblooded desert horse as he came to know it in the heart of the vast peninsula. Much of his text regards strains.

Upton wrote:

> "It is only consistent and reasonable to believe that horses among the Arabs of the interior deserts have a better claim to genuine than such as have come to hand through foreign sources…"

"In the whole of Arabia, the Anazeh, a great race of Bedaween, dating back to remote antiquity... have the best horses."

"At a certain period in the history of the Arabs, in very remote times, an authentic breed was established by a selection from the general or universal race of Kuhl in Arabia."

Figure 92 - *Illustration from Newmarket and Arabia, by Upton, a reproduction of the famous de Dreux painting.*

Upton brought back several horses; a chestnut colt, the stallion Yataghan, and two mares, Zuleika and Haidee. He was also involved in the purchase of the mare Kesia for his patron. He bought these horses from the Gomussa, a branch of the Anazeh tribe. Haidee was a chestnut mare foaled in 1869, a Managhieh Hedruj. Kesia was a Heheilet Nowagieh born in 1865. Yatagan was a chestnut stallion born in 1870, of the Kehilan Halawi. He later acquired the stallion Alif for the Australian breeder A.A. Dangar. Zuleika was lost from view in the historical record, but the others made significant contributions to Arab horse breeding.

Figure 93 - *Mounted Arab warriors, from The Raswan Archives*

The Upton acquisitions did not revolutionize the English Thoroughbred horse.

William Tweedie was a Scotsman, (1836-1914) who spent his youth in the British Army in India. There he rose to the rank of Major General. During this time he developed an intense admiration for the Arabian horse, an affinity borne of the animal's utility to a military man. Later in life, he was assigned to serve as Consul General at the British Embassy in Baghdad, where he was posted from 1885 to 1891. From Baghdad, Tweedie made a number of journeys into the heart of the desert. He wanted to see for himself if the legends that the wellspring of pure Arab type was to be found deep in the desert were true. His book regarding his findings, The Arab Horse; His Country and His People was published by Blackwood Press in London and Edinburgh in 1894.

Tweedie saw much of the Bedouin way of life in the desert but was most struck by the fact that an understanding of the true nature of the Arabian horse could not be separated from its environment and its people.

Europeans, he said, had a distorted view of the Arabian horse:

> "The Arab horse is apt to be more a creation of the imagination than of observation and experience."

Tweedie was a logically minded military man, not given to romantic idealizations.

Figure 94 - *A Bedouin Reshma, from Tweedie.*

But he was also struck by the closed xenophobic nature of the Bedouin society itself. He came to believe that a great deal was hidden from the European explorers. The natives were very coy about allowing their best horses to be seen. "Is there not some reason to believe that the purer of the breed is inaccessible to foreigners?"

The Bedouin cultural hierarchy also played a role in the types of horses that Europeans would buy in the desert. Foreigners entering any encampment were quickly herded to the Sheikh's tent, where the Sheikh's horses (but not his best horses) were offered for sale. Much praise was made of the horses by the attendants. Tweedie found however that "a simple fellow who owns but one mare, and rides her, is more likely to have a genuine colt or filly beside his tent than the Sheikh."

Figure 95 - *Charles Huber, image from the Societe de Geographie de Paris.*

Regarding the sale of horses, Tweedie found that the Arabs were honest when dealing among themselves, but they considered the foreigners to be legitimate targets for fraud. They considered the English to be fools, whose knowledge of horses was miserably inadequate. The Arab horse seller would never condescend to name a price for a horse, insisting that the bargaining start with the customer. The foreigner could expect to be treated by the Arab with cunning and rapacity.

Tweedie was free to move about the desert without hindrance. In the cities, though, foreigners were viewed with alarm and suspicion. In Hail for example, Tweedie found that he was given no liberty within the walls, and was watched constantly by members of the Emir's bodyguard.

Charles Huber's journey into the Nejd was notable for more sober reasons. He was a scholar from Strassburg Austria, an academic epigraphist. He ventured into the Nejd in 1883 in search of the fabled Tieme stone, a stele that contained Aramaic writing from the 6th century BCE. After locating the stone near the ancient trading city of Tayma in northwest Arabia, conditions forced him to take the tablet to Hayil where it was confiscated by Ibn Rashid.

Figure 96 - *Sir James Boucaut of Australia, mounted on the Crabbet stallion, Messaoud.*

While leaving the Nejd to return home, Huber was murdered by his native guide. The murder was investigated by the office of the Vice-Consul in Jedddah and the assassination was announced at the Geographical Society of Paris. The Consul was able to ultimately obtain the precious artifact from the Emir. The Tieme stone is today located in the Louvre Museum in Paris. Huber was buried in the Christian Cemetery in Jedda, but his heart was removed and sent by the French consul for burial in the Pantheon. His death placed a chill on Arabian exploration for some years. Huber had no interest in horses.

As the 19th century came to an end, the sun had begun to set on asil horse breeding in the Nejd.

Early in the 20th century, Sir James Penn Boucaut, an Australian *purebred Arabian horse* advocate, viewed the deteriorating horse breeding conditions of the Arabian Desert with distress and wrote:

> *"An impure breed could never have maintained its essential sameness and characteristics so uniformly for so many thousands of years as the Arab has done, nor would men of all nations have so uniformly and so universally praised an animal which was not of surpassing excellence. If he passes away by human folly, you will never see his like again."*

Figure 97 - *Bronze Sculpture of the Egyptian Arabian Stallion Kaisoon, by Karen Kasper, image used with the permission of the artist.*

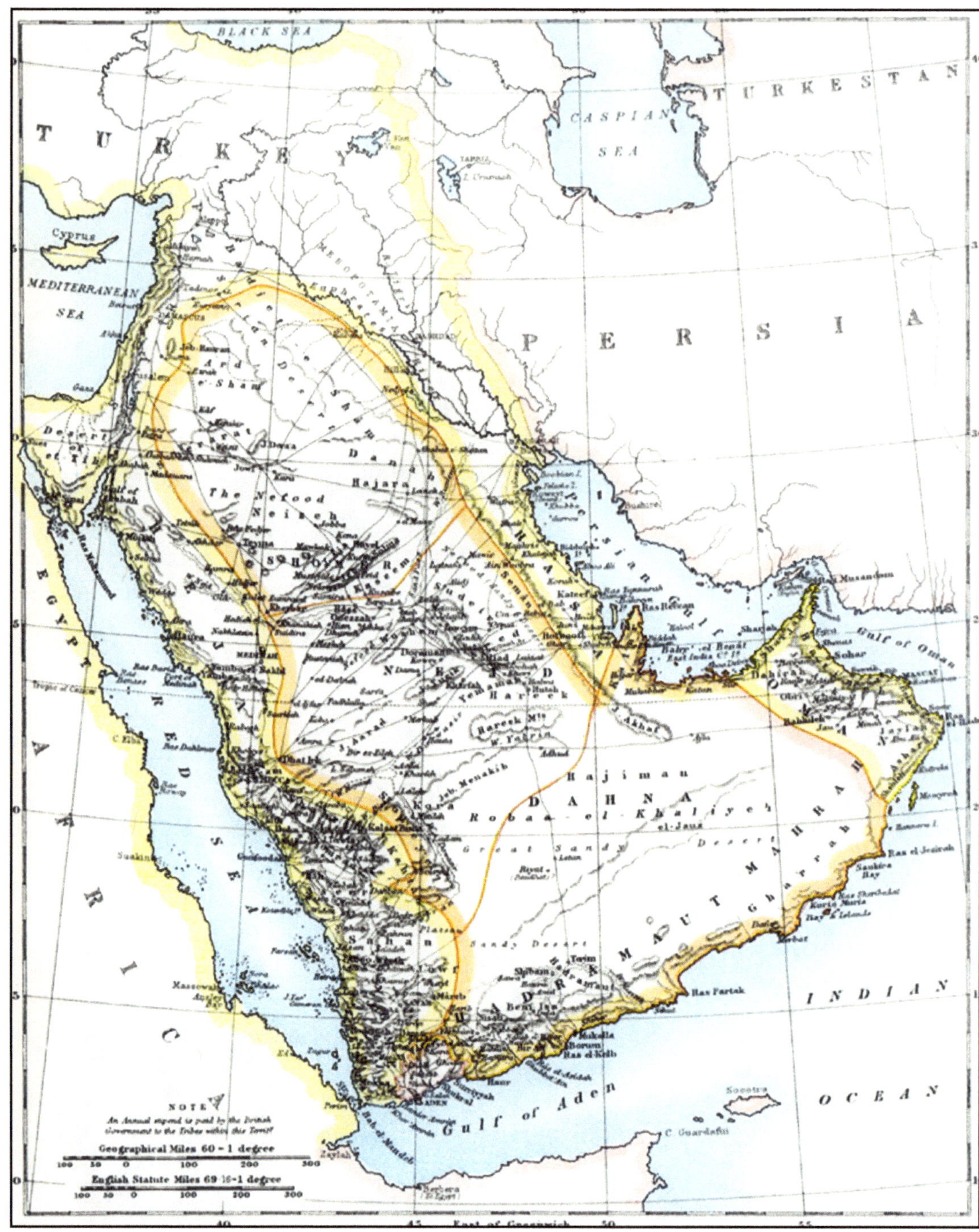

Figure 98 - *The Middle East, 19th century. As the Arabian Peninsula interior became more accessible to Western explorers and horse buyers, the demonstrably purebred desert horse had become a rarity among the desert tribes. Inter-tribal warfare made travel in the inner desert hazardous. The Bedouin tribes of inner Arabia was constantly fighting and quarreling. Some fought over grazing rights, some fought for political domination, and some fought merely for survival. Ghazzus were often carried out by a weaker tribe to surprise a stronger and potentially more vicious enemy. These were pre-emptive strikes and often were successful in preventing future strife.*

Figure 99 - *An Arabian Horse, by Alfred de Dreux, image in the public domain.*

Figure 100 - *Combat of the Giaour and Hassan, by Eugene Delacroix, Chicago Institute of Art, image in the public domain.*

**Beware of a silent dog and still water.*
or
Still waters run deep.

CHAPTER FIVE

CHAPTER FIVE

ARABIAN PENINSULAR WARS AND THE PUREBRED ARABIAN HORSE

This book deals with the genetic identity of the foundation mares of the Royal Agricultural Society (RAS) and Inshass stables, and with the subsequent matrilines of the Egyptian Agricultural Organization (EAO). It is natural that pedigree researchers and Arab horse breeders alike would need to know the sources of these mares or their progenitresses. Unfortunately, the precise sources of most of the root bloodstock will remain forever unknown. There is little documentary evidence of provenance beyond the names of the horses as they are recorded in the Egyptian records.

The origins of the tribute mares of the Inshass stud will be reviewed. Many gift horses were given to the Kings of Egypt, and King Ibn Saoud of the Kingdom of Saudi Arabia was the principle donor. He maintained a large herd of *purebred Arabian horses*, gathered from the desert, at his royal stud at Al Kharj, near Riyadh. Gift mares such as Hind, El Obeya Om Grees, Mabrouka, Kahila, and Nafaa were kept by the Egyptian King and included in the Inshass herd.Until 1952 this was the personal and private stud of Kings Fouad and Farouk. Most gift horses were considered by the stud management to be of doubtful blood purity and were discretely given away. In this chapter, the military and political events that led to the kingship of Ibn Saoud and the formation of his private stud at Kharj shed some light on the nature of the horses collected bred, and raised there.

The purpose of this chapter is also to survey the Arabian Peninsula political and military events that contributed in large part to the loss of equine pedigree documentation and led to the near extinction of the *purebred Arabian horse*. A review of the pertinent historical background of the 19th and early 20th centuries is presented, illustrating how such large numbers of high quality desert horses were translocated from Arabia to Egypt. The course of events that led to the rise of Ibn Saud and the beginnings of the 20th century royal Arabian horse stud at Al Kharj in the Nejd will also be reviewed. This historical background is particularly useful in view of the significant matrilines of the RAS/EAO that originated as gift horses to the Kings of Egypt from Ibn Saud.

The reasons for the deficiency in provenance are historical. There is an enormous lacune in the textual record regarding asil horse breeding on the Arabian Peninsula during the 19th century. Historical evidence suggests that all of the matrilines destined to found the RAS/EAO were brought out of the central Arabian Deserts during the 19th century and early 20th century. However, written incontrovertible proof is simply not available. Egyptian Arabian pedigree researchers can only assume that they were asil. The Arabian Horse Families of Egypt by Colin Pearson is an excellent example of Egyptian Arabian horse pedigree research, based on careful first hand examination of the existing records in the possession of the RAS/EAO. But Pearson found many gaps and inconsistencies in the records.

Mitochondrial DNA research is beginning to fill in many of the gaps.

The endless wars on the Arabian Peninsula led to the loss of pedigrees for the horses that were eventually used to form the RAS broodmare band. This limitation was compounded by the fact that the numerous smaller tribes of Arabia that bred horses did not keep written pedigrees of their horses. Of the large number of mares that were acquired by the RAS and Inshass during their formative years, most of the matrilines had little impact as foundation mares. Most are no longer extant. *Many were called but few were chosen.*

Fortunately, the Abbas Pasha Manuscript has preserved lengthy matrilineal pedigrees for several mares that could have been the El Dahma of Abbas Pasha listed in the RAS records. Pedigree connections to Ghazieh I of Abbas Pasha are suggested in the Manuscript, but the connections are neither direct nor definitive.

Figure 1 - *Photo of Ali Pasha Sherif. Image from the Forbis archives.*

However, one looks in vain for other documentary connections to living horses. From the time of the death of Abbas Pasha to the time of the auction of his horses by his son Elhami Pasha, six years had passed. Only scattered records of sales deaths or births for this six-year period exist. This deficiency is noteworthy. 90 stallions and 210 mares were eventually sold during the 1860 auction. Most were bought by European governments for use in their studs. A young man named Ali Bey, known to history as Ali Pasha Sherif, bought 40 of the horses. His continuation of the pure breeding of Abbas Pasha at his palace in Cairo guaranteed that at least some asil horses would reach the RAS early in the 20th century.

The paucity of pedigree records from Ali Pasha Sherif himself suggests that the Pasha was not particularly interested in documentation. He no doubt was unaware of the critical historical role he was playing in the perpetuation of the asil horse. Besides, Ali Pasha Sherif had no reason to mistrust the purity of the Abbas Pasha horses: their purity went without saying. Ali Pasha Sherif was a nobly born Egyptian with an innate dislike of the British. He considered it beneath his dignity to provide proof of purity to "Christian infidels". His adherence to the tenets of Islam was deeply ingrained. It was blasphemous to sell authentic asil horses to foreigners, although economic conditions often made the practice necessary.

There is additional historical data. At the request of Lady Anne Blunt, Ali Pasha Sherif provided several memoranda, which were brief *hujjaj*, of particular horses. These memoranda are recorded in Lady Anne's diary. For the stallion Aziz, for example, he gave the sire as Harkan, whose dam was one of the two Harkah mares sent to Abbas Pasha by Mohammad Ibn Qarmalah. Ghazieh is mentioned in a memorandum for Mousib as the "mare of Mufdi Ibn Rushud".

Even the celebrated RAS mares Bint el Bahrain, Roga el Beida, Venus (Yunis, Arabic for Jonah), and El Obeya om Grees have no known pedigrees. The mare Rodania stands out as an important and perplexing problem in the discussion of provenance, having been purchased in Syria and exported directly to England. She was never in Egypt. The less prolific matrilines, those founded by El Kahila, Mabrouka, Hind, and Nafaa were gifted to the Egyptian King by Saudi Arabian King Ibn Saud. Some of the gift mares were eventually incorporated in the equine breeding program of the Egyptian Agricultural Organization. They have no known pedigrees.

And then there are the matrilines founded by El Samraa and Bint Karima. These two lines still exist, but their tribal sources and pedigrees are unknown. Their matrilineal descendants are now very few in number.

Mitochondrial DNA research is beginning to provide solid evidence regarding the relationships between strains, historical families, and genetic matrilines. These three designations are not identical. The results have altered the way in which these three designations are to be understood. This data forms the basis for a much needed preservational program for endangered matrilines.

The RAS/EAO foundation mares Nafaa el Saghira, Gamila (a Hadba Enzahi), Gamila Manial (a Koheila Mimrieh), Badria, Samha, Dameh Nejiba, Nura, Freiha, Selma and even the fabled Jellabiet Feysul have been lost in pure matrilineal EAO form.

The world continues to lose matrilines of desert-bred Arabian horses at an alarming rate. Gone are the Saklawi Ejrifi, the Ubbayan Sharrak, the Jilfa Sattam al Bulad, the Kuhaylan Ajuz, and the Wadnan Khursan.

The study of EAO matrilines is especially urgent because of the continuing loss of matrilineal diversity through extinction. Since genetic diversity is essential for productive breeding, conservation of the surviving matrilines occupies a position of supreme importance. Today, even the most apparently heterogeneous horses of RAS/EAO descent share a large number of ancestors that are repeatedly found in their pedigrees. This fact is demonstrated by a study of the pedigrees of present-day pure EAO horses tracing back to their earliest recorded ancestors.

The horse breeding tribes of the desert left few durable records. There are no primary sources. There are, however, a number of authoritative firsthand accounts of *purebred Arabian horse* breeding activity on the peninsula during the 19th century which the student of the Egyptian Arabian horse will find interesting, informative and above all, convincing evidence that *purebred* horse breeding continued during these troubled times. The many nomadic tribes that were not subjected to Ottoman control continued to breed the classic antique type Arabian horse. The Ottoman government, despite its best efforts, simply could not destroy the way of life of a people who would not submit to its vicious and burdensome yoke. It was not in the Bedouins' nature to be controlled or governed.

Today, the study of the Bedouin *hujje* is, for the most part, an academic exercise. No RAS/EAO foundation mare has a hujja. There are very few horses living today, in Egypt or elsewhere in the world, that can be connected to any hujja preserved from Middle Eastern sources. A few are mentioned in memoranda supplied by Ali Pasha Sherif. The pedigree information that is available on these RAS/EAO horses has been reconstructed by researchers using fragmentary evidence. The Abbas Pasha Manuscript, published by Forbis and Sherif add much to the facts regarding the lineages of the Egyptian horses.

Figure 2 - *Bedouin warriors, from* The Raswan Index.

It seems remarkable that this would be so, but it is so. Political and military chaos engulfed the peninsula from 1811 onward. The reason for this complete disjunction between the Arabian Desert horses of the past and the living Egyptian Arabian horses of the modern day was the Ottoman wars against the Arabian tribes in the early part of the 19th century. After Ottoman rule collapsed, ensuing inter-tribal warfare and the violence attending the consolidation of Saudi Arabia under King Ibn Saud led to further losses in both horses and written records. The trail in the search for continuity in the verifiably accurate pedigree of the Nejd horse has gone cold. The desert horse had come to the verge of extinction.

The early 19th century was in many ways the beginning of a "slow emergency" for the *purebred Arabian horse*. The Arab horse of the highest caste breeding was once a valued feature of life among the Bedouin. Rich or poor, noble or common, each Bedouin had or desired to have an asil horse. All of this changed in the year 1811. The desert horse began to die out, the victim of a century of unremitting war in the desert. Strain names that had once numbered in the hundreds in Bedouin lore began to disappear. Many matrilines died out. What remains today of this once great race of horse is now dangerously close to complete annihilation. While this observation may seem exaggerated, alarm over the continued close breeding of an already closely bred collection of animals is not entirely unfounded.

Before 1811, the tribes of the interior of Arabia lived an ancient way of life, existing in a state of isolation, constantly warring and skirmishing amongst themselves but unaffected by intrusion and interference from the outside world. Nominally a part of the Ottoman Empire, Deserta Arabia was for all practical purposes free from the influences of any outside forces. The horses that they bred were safe from outside influence, and, so far as can be told, they were purely and conscientiously bred by their pastoral caretakers. The tribes maintained their isolation by fiercely defending their land against any attempt by outside powers to impose control over them. After 1811, however, the Bedouin horse became the victim of its own success.

The actual root cause of the conflict can be traced to events that occurred a century earlier. During the late 18th century Wahabbism emerged; this was an Islamic fundamentalist and revisionist movement led by preacher and scholar Muhammad Ibn Abd al-Wahhab (1703-1792). Al-Wahhab was a religious zealot living in Diriyah, the home of his protector and advocate, the powerful desert Emir, Ibn Saud.

The Saud family itself had arisen from obscure origins, being descended from Bedouin natives of the peninsula. Few facts are known about the man named Saud who founded the dynasty. More is known of his son, Muhammad Ibn Saud Ibn Muhammad Ibn Migrin, who died in 1765. The family lore states that the Saud line is descended from the Anazeh Bedouin and was originally known as the Al Migrin.

Figure 3 - *The Tuwaiq Escarpment, shown here from the west, cuts through the Nejd plateau. It is known locally as Jebel Tuwaiq. While the western edge is precipitous, the eastern slope forms a slow gradual decline into the region of Al Diriyah, located just beyond this escarpment.*

In 1744, Muhammad Ibn Saud was Emir of Diriyah, a small village in the Nejd. He gave sanctuary to a young, obscure, and zealous Islamic scholar named Muhammad Ibn Abdul Wahhab whose mission was to purify Islam. His fellow Bedouin, in his view, had fallen from the straight and narrow path into heathenism. He taught a form of Islam that was strict and rigid in its interpretation of the Qur'an. His mission was to establish a fundamentalist form of Islam and force his views upon the wayward. Ibn Saud was deeply influenced by the reformational sincerity and revolutionary spirit of the young man. Ibn Saud was fired with religious zeal of fanatic proportions.

Being an ambitious Emir, Ibn Saud was also cognizant of the usefulness that a religious radical might have for a young prince bent on conquest and expansion. The Emir and the reformer proved to be a potent combination, and the First Saudi state was born of their collaboration, based on a radical fundamentalist interpretation of the doctrine of Sunni Islam.

Al-Wahhab was a strict constructionist Islamicist who interpreted the Qur'an in a rigid and literal sense. He interpreted the words of the Prophet in a very conservative manner and decried the deterioration in piety that he saw in the world of his day. He was a Sunni, and he rejected the ideals and practices of Shi'a Islam as heretical. He suppressed the Sufi sect of Islam as corrupt, due to its veneration of saints and holy places of pilgrimage. He was outraged by the perceived moral decline among the desert Bedouin. He railed against the liberal religious modernization and innovations introduced by the clerics of the day. He was a firebrand and a revolutionary.

He was opposed to blind religious obedience to established religious authorities and pious muftis. He sought to sever all ties with traditional Islam by denouncing established religious scholars, and he destroyed shrines and holy grave sites of the saints. Like the Prophet, he was an iconoclast. Wahhab was so extreme in his opinions and strict enforcement concerning proper Islam that his own father was eventually compelled to write a book called <u>The Divine Thunderbolts Concerning the Wahhabi School</u>, condemning his son's religious zeal and fundamentalist behavior as excessive.

But the words of the teacher galvanized the Al Saud family. As history had demonstrated, no force can motivate the Arab people to great deeds and actions more effectively than religious zeal and enthusiasm.

The theology of Al-Wahhab was simple and sect-like, advocating ultraconservative principles based on a highly selective interpretation of the text of the Qu'ran and a puritanical frame of mind. He wished to destroy all personal human identity in favor of emphasizing the oneness of God. He based his beliefs on the concept of the doctrine of tawhid, the unity of God. This was meant to place primary significance on the central Islamic concept of monotheism, *Al-Allah Al-Wahid. Al-Wahhab* aimed at the abolition of personal glorification and abhorred the practice of shirk or idolatry. He considered this a major sin. Man's egotistical and self-centered way of life was to be annihilated in order to attain a true understanding of the oneness of Allah.

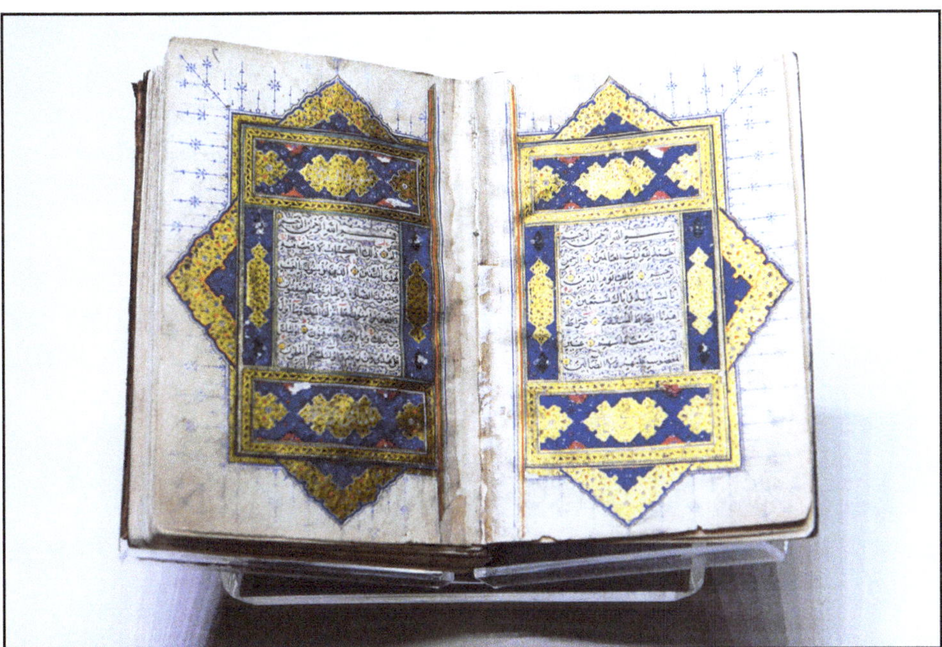

Figure 4 - *Holy Qu'ran 1598.*

Al-Wahhab was exceedingly strict with regard to shirk and its shades of meaning. The greater shirk is open and unapologetic worship of any god other than Allah, particularly the worship of polytheistical deities. Lesser forms of shirk involve the humanization of Allah. In pure Islam, Allah and the Prophet could not be depicted or portrayed in any way, and the name of Allah may not be used in association with any human, animal, or deity. According to the Qu'ran committing the greater shirk is the only sin that is unpardonable.

The lesser *shirk* refers to the sin of idolatry that is hidden or concealed from public view. This, Al-Wahhab insisted, referred to the sins of personal pride, arrogance, and insincere assertions of piety. A person who openly professed a serious commitment to Islam but conducted himself in ways that violated the tenets of the faith was guilty of the lesser *shirk.*

The term *Wahhabism* was first applied to the movement by the Ottomans and was later adopted by the British. Wahhab himself regarded the term as disparaging. He did not use it and he did not like others to use it. He preferred the term *muwahhidun* or Unitarians. Later adherents to the cause of Wahhab preferred to use the generic term of *salafists* translated as the "ancestors", referring to the "old religion" or pure Islam. This same concept of spiritual unity had sprung up independently throughout the world of Islam during this period. Wahhab was not alone in his insistence on a return to the uncorrupted faith of the Prophet. Other Muslim clerics called for a revivification of the true spirit of the Prophet's teachings which emphasized humility and a spiritual longing for unity with Allah. The true spirit adamantly condemned any non-Arab as beneath contempt, and lest they contaminate the masses, were summarily beheaded when found within a Muslim country. This behavior had significant repercussions for the future of Western exploration, and especially for the acquisitions of horses from Arabia.

The princes of the Al Saud family and this holy man Wahhab set out to reform their wayward Muslim Bedouin brothers by force of arms. It was no accident that this expansion of power had political ramifications for the Al Saud clan. Born out of both religious zeal for Islam and virulent hatred for the Ottoman occupational forces, this uprising was, almost coincidentally, the first attempt by native desert Arabians to form a unified state; a state that was destined to be a Saudi state.

By 1803, the warring Wahabbi rebels and Saud military forces had gained control of eastern Arabia, and the holy shrine in Karbala. By attacking Ottoman interests they attracted the attention of authorities in Constantinople. In 1807, the Ottoman functionary in Egypt, Muhammad Ali, was ordered by Sultan Mustafa IV to suppress the reformational rebellion. The Egyptian ruler was initially reluctant to invade the land of his fellow Muslims, and for a time he ignored the order.

By 1811, the Sultan in Constantinople had become increasingly alarmed, and he emphatically ordered the Egyptian Viceroy, the Wali, to initiate an offensive campaign against the rebels and extinguish the Wahabbi spirit. The Wali was, after all, an Ottoman subject. But he was also a Muslim, and he feared war against another Muslim people. However, the progressive expansion of Wahhabi influence west into Mecca and Medina, all the way to the coast of the Red Sea, was alarming enough to the Wali of Egypt that he was forced to respond.

The Wali (governor) of Egypt, Mohammed Ali, was a Turkish speaking Albanian. Born as Mehmet in Kavala, Macedonia in 1769, he rose from obscure origins. He came to the Sultan's attention during years of skillful service in the Ottoman army in Egypt.

He was effective as a leader, and he was intelligent. He rose through the ranks rapidly and achieved the rank of commander. His popularity with the Egyptian civilians made him the natural champion of a movement to oust the hated Ottoman-appointed head of the Egyptian government. An uprising led by the Ulema of Cairo succeeded in deposing the reigning governor of Egypt Ahmad Khurshid Pasha. Mohammed Ali soon rose to the position of Wali. He consolidated his power in Egypt by the infamous murder of his political and military adversaries the Mamelukes at the Cairo Citadel on March 1, 1811.

Figure 5 - Sublime Porte, Topkapi Palace, Constantinople.

Figure 6 - *Mohammad Ali the Great of Egypt. Image in the public domain. Alternately charming, jovial and tyrannical, he was the first modern Egyptian monarch.*

The Ottoman-Saudi War of 1811-1818 was a protracted conflict between the Saudi Wahabbis and the forces of Mohammed Ali, under orders from the Sultan in Constantinople. Prior to this period, the Arabs living in the heart of Arabia had been fairly moderate in their interpretation of the meaning of the Qur'an. The desert Bedouin interpreted Islam in a rather liberal manner. At heart, the Bedouin were animists, not religious in temperament. And this type of pedestrian Islam had been the rule in central Arabia for over 1000 years. But revolution was in the air, and the desert Arabs had a Zealot.

Figure 7 - *Mohammad Ali the Great of Egypt.*

After attaining this high position of power in Cairo, Mohammad Ali's relationship with the degenerate Ottoman government in Constantinople became progressively more tense and antagonistic. He was viewed as a threat. As Mohammed Ali himself once said "I am well aware that the Ottoman Empire is heading by the day toward destruction . . . on her ruins, I will build a vast kingdom . . . up to the Euphrates and the Tigris."

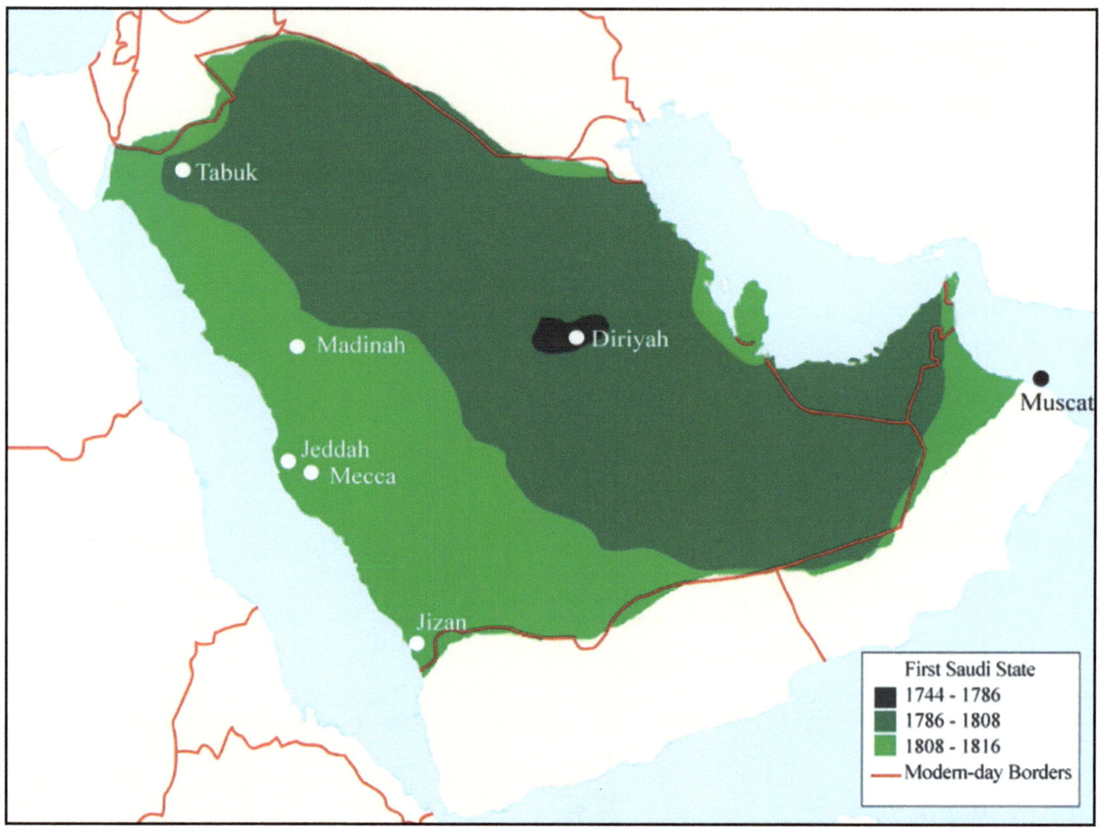

Figure 8 - *The First Saudi State.*

Mehmet was ambitious. He was clever. He was modern. He was barbaric. He wanted Arabia and he wanted Mesopotamia, but he was not to have either. It suited his grand design that the tribes of Arabia be controlled. All the same, he was Muslim and was reluctant to attack other Islamic peoples for fear of committing blasphemy.

For the time being, however, the Sublime Porte's edicts could not be ignored. Pursuant to the orders from the Sultan, Ali organized an Egyptian military force led by his son Tousson to invaded Arabia. His mission was to bring an end to the insurrection and extinguish the Wahabbi influence over the Saud family. The conflict began in 1811 and lasted until 1818. The Egyptian forces of Mohammed Ali were aided by an alliance with the Arabian Desert Emir, Prince Feysul Ibn Dauwish, a man who bore hereditary hatred toward the Saud clan.

Prior to the war, the native inhabitants and nomadic tribes of the interior of Arabia were controlled by two principle competing Bedouin chieftains: Prince Ibn Saud and

Prince Feysul Ibn Dauwish of the Mutayr. For decades, a state of tense dynamic equilibrium had kept the peace, but these two men and their families were bitter enemies, the result of a long-standing blood feud. They also owned (or controlled through family ties and tribal alliances) many of the asil horses of the central Arabian Desert. Saud had the last few remnants of the Hamdaniyat strain, and Feysul Ibn Dauwish owned the few remaining Jellabiyat, which he had obtained from Ibn Hithlayn of the Ajman Bedouin.

Figure 9 - *A lone Arab lancer, from The Raswan Archives.*

The invasion of Arabia by the Egyptian/Ottoman coalition was not altogether unexpected. Relations between the desert princes and the Egyptian government had been strained for many years since Mohammed Ali had regularly sent troops into the peninsula to extract many of the high caste nobly bred desert Arabian horses from the Bedouin tribes. Ali did this under the pretext that the horses were owed to the Egyptian government as a form of taxation or tribute. In truth, Ali needed these mounts to supply his own army and cavalry, which, along with the navy, he had modernized. But beyond this need for cavalry mounts, he had developed a passion for the beauty of the desert asil horse. He wanted the most highly bred horses for himself. They were to form the basis for his own personal Arabian horse collection.

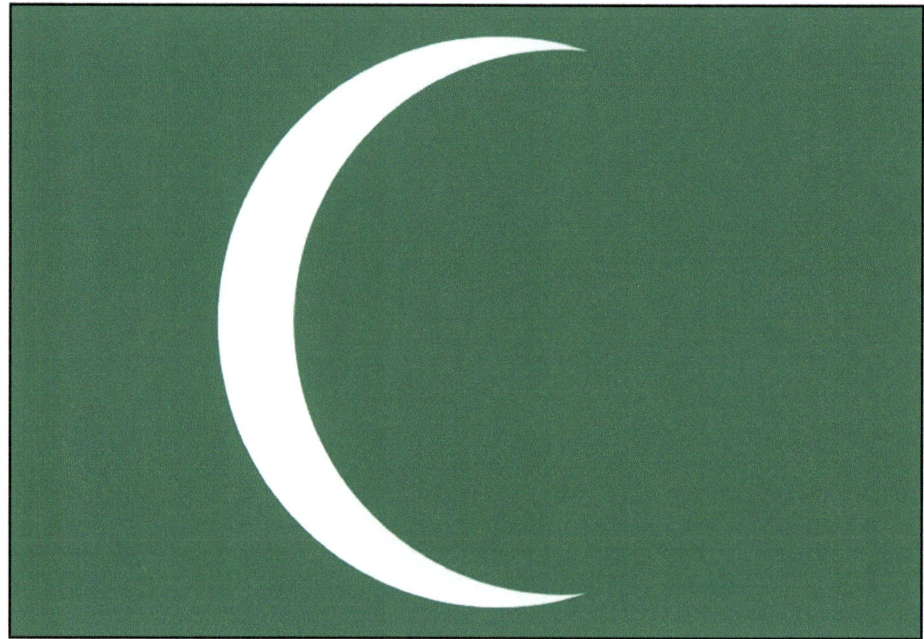

Figure 10 - *The flag of the early Saud era.*

A long and stuttering war followed. It was a war not of large scale battles. It was a war of widely scattered cavalry skirmishes in which the Egyptian commander Tousson was materially aided by the Arabian Prince Feysul Ibn Dauwish. Intermittent battles left the desert scattered with the stench and putrefaction of dead and dying mares and the lifeless bodies of the men killed. From 1811 until 1814, there was no decisive victory for either side, although the tide of events tended to favor Saud. However, Saud died suddenly in 1814, and his son Abdullah came to power.

Abdullah had no wish to continue to resist the Egyptians and sought to end the carnage by appealing to Cairo for peace. A peace treaty was formulated was signed. According to the terms of the treaty, Abdullah was required by Mohammad Ali to surrender 222 of his best quality Arabian horses. These horses were taken to Egypt and became a part of the horse breeding studs of Mohammed Ali in Cairo.

Despite the appearance of complete Egyptian/Ottoman supremacy, the war was never really decided on the battlefield. A history of these events by the European biographer Edouard Gouin (L'Egypte au XIXe Siecle, 1847) states that the outcome of the Egyptian campaign against the Wahhabis was always in question, right up to the end. The assistance of "El Duish", the Saud's blood enemy, was an important factor in the victory of Ibrahim and the Turks. According to Gouin, the outcome was uncertain. The sudden death of Saud was determinative.

However, the treaty of 1814 was not effective in controlling the radicalized Saud family and the Wahabbis. After a brief period of peace, hostilities resumed, now under the leadership of a reluctant and diffident Saud Prince Abdullah. In response, Mohammed Ali sent his son Ibrahim Pasha, not Tousson, into the desert at the head of the Egyptian forces. Once again Prince Feysul Ibn Dauwish, keen on bringing about the destruction of the house of Saud, was the key ally aiding the Egyptians against the Saud family.

Warriors fighting on horseback played an important role in the battles that took place during the subsequent skirmishes, but warriors mounted on camels were a more common sight. A particularly deadly struggle between camel-mounted forces took place at the village of Byssel, a small town about 4 days march southeast of Mecca. The Sauds were supplemented by all of the chiefs of the mountains of Yemen. Burckhardt wrote an account in his journal of an 1815 battle which he witnessed at Byssel: "the unified armies of all the southern Wahabbi chiefs who attacked Mohammed Aly Pasha consisted of 25,000 men, but...had with them only 500 horsemen, mostly belonging to Nejd." Faisal Ibn Dauwish, fighting on the side of the Egyptian Ibrahim, had few horsemen.

In 1816, Tousson died suddenly. His death was attributed to the "plague". His son, young Abbas, later told Sir Charles Murray, British consul general to Egypt, that it was an "undisputed fact that my father had been poisoned by Mohammed Ali, who had grown jealous of Tousson's popularity with the army and considered his own safety thereby endangered."

For years, bloodshed, destruction, and carnage swept across the desert highlands as Ibrahim and the Turks pursued the elusive Saud/Wahhabi band of warriors. The Nejdean Desert Bedouin of central Arabia that were aligned neither with the Ottoman government nor with Saud and the Wahabbis experienced severe privation as the war engulfed their homeland. Many of the Bedouin of Nejd came under Wahabbi rule early in the course of the war, and any Bedouin who owned a horse was compelled to fight with the Wahhabis. This disincentive to horse ownership caused many Bedouin to sell their horses to foreigners through the Basra market. They parted with these treasured animals only under duress. However, the proud Bedouin were reluctant to sell their precious animals to infidels.

Figure 11 - *A Bedouin warrior on camel back, armed for battle with a traditional lance, a sword and a dagger. Image from the Library of Congress.*

Figure 12 - *Bedouin warriors, from* *The Raswan Index*.

The only alternative for the nomadic tribes was to keep their horses and fight in support of the Egyptian forces. But even then, they were fearful that if they fought in alliance with the Egyptians, the Wahabbi leaders under Saud would confiscate their horses under the pretense of charges of disloyalty or disobedience. This practice did at times occur, and the Islamic zealots sold the confiscated animals to buyers outside of the country, using the cash to finance the costly war.

Times were tumultuous and uncertain. The common desert dweller and Bedouin horse breeders and owners feared not only their Arab brethren, the Wahhabis, but they also feared the Egyptian invaders. The troops sent by Mohammad Ali had acquired a reputation for taking the best horses from the desert tribesmen. Ibrahim Pasha acquired, through conquest, force, or intimidation, many of their nobly bred horses in the course of his campaign, requisitioning them from the defeated tribes. In response the Bedouin began to sell even their best horses to foreigners, feeling more secure with money, which could be safely hidden. By the end of 1818, the supply of first quality Arabian horses from Nejd had begun to diminish.

In 1818, Prince Abdullah Ibn Saud Al Saud was defeated by the Egyptian army. For several months the Egyptians besieged the Saud forces, now confined within the walls of Diriyah. Abdullah eventually capitulated on September 9, 1818, and his

capital at Diriyah. Seige cannons pounded the walls of Diriyah into powder. He was taken to Constantinople and ultimately beheaded. He was charged with desecration of the holy cities and many mosques.

Figure 13 - *Ibrahim Pasha, in 1845.*

The Wahhabi cleric, Sulayman Ibn Abd Allah, grandson of the Wahhabi founder, was also executed. The Ottomans considered the clerical leaders to be more dangerous than the Al Saud rebels and dealt with them harshly. The Grande Porte feared radical Islam far more than it feared any desert prince. The Sultan was well aware that no force motivated the Arab Bedu to action more effectively that religious inspiration. From Cairo, Mohammed Ali instructed his son Ibrahim to confiscate all of the Saud horses from the family stud at Turayf, seizing them as the spoils of war, and return home.

Figure 14 - *Ibrahim Pasha in old age.*

The desert insurrection of the Saudi prince was suppressed. Hostilities ceased. In 1819, the Egyptians began the long journey out of the Nejd, with Ibrahim at the head of a substantial military force. His triumphal return to Egypt represented a massive transfer of pure blooded Nedjean horses out of the desert. The Egyptian army was accompanied by Prince Feysul Ibn Dauwish and his mounted warriors, who joined the expedition to Cairo to oversee the transfer of a large number of Feysul Ibn Dauwih's own horses, now also the possession of Ibrahim. The hubristic attempt proved to be the undoing of the over-confident Ibrahim, who lost most of his men and all of the captured Saud horses to disease and exposure on the ill-fated journey back to Egypt. Feysul's cavalry and the core of his own breeding stock also perished during the ill-fated journey.

Figure 15 - *Abdullah bin Saud Al Saud.*

Ibrahim Pasha confessed later to his friends in Cairo that his misfortune was, he felt in his mind, the result of Divine retribution. In retrospect, he believed that he had transgressed against holy laws when he robbed the Bedouin of their treasure of

Arabian steeds, the inheritance of their father, Ishmael, the "Angel horse of Jabrail (Gabriel)", and "the flower of the desert breed". His fate, he thought, was his punishment from God, a just punishment for his grievous sins.

The peninsular wars proved to be a turning point in the fortunes of the horses of the Nejdean tribes. The loss of the 222 Saud horses from the treaty of 1814 and the eventual loss of all of the remaining Saud horses in 1818 left the heart of Arabia devoid of most of the animals that had been the source of the Nejdean's power and survival. This was, of course, the goal of the Sultan in Constantinople, who wished not only to pacify the desert tribes but to deprive the troublesome Bedouin of the means to carry out future wars. Without horses, the warriors of Al Saud would be less of a threat.

Carl Raswan, writing in 1961 observed that "The disastrous blow dealt to horse breeding in Arabia was so great that to our day the various tribes of Arabia have struggled in vain to recover their former wealth of Asil animals... the Mutayr Bedouin have no more than thirty first-class broodmares, and the Ruala... do not possess more than seventy excellent broodmares among them."

With the execution of Abdullah Ibn Saud Al Saud in Constantinople in 1818, the First Saudi State came to an end. The Egyptian forces had suppressed the Wahhabi uprising and Ibrahim left a residual force of garrisoned troops in Diriyah to maintain the peace.

The loss of Diriyah would have been devastating to a clan less resilient than the Al Sauds. Diriyah had been the largest city in Arabia up to that time, and its thick adobe walls were only breached by the prolonged cannon fire of the troops of Ibrahim. After the assault the city lay in ruins, and the Al Sauds were forced to flee. But they would return.

The peace did not last long. The dominance of the Turks and the Egyptians was soon challenged by the eternally recalcitrant Al Saud clan. The Second Saudi State was initiated when the forces of a new and aggressive Saud leader Turki Ibn Abdullah captured Riyadh, seizing control from the Egyptian forces of Mohammad Ali that had been left to defend the region. Soon, the Saud clan was once again in control of most of the peninsula. The Second Saud state, known as the Emirate of Nejd, existed from 1818 to 1891. A new capital was established at Riyadh, a small village near the ruined and abandoned Diriyah.

Figure 16 - *Equestrian statue of Ibrahim Pasha by Charles Henri Cordier, in the Midan Opera Square in Cairo.*

Figure 17 - *The ruins of Diriyah following the 1818 conquest of the forces of Mohammad Ali of Egypt. Many of the ruins of the old city, including the Salwa Palace and the Saad bin Saud Palace, have been restored.*

Figure 18 - *The ruins of Diriyah. After weeks of intense bombardment, the city walls were breached. The inhabitants were left to the barbarous treatment of the Egyptians.*

Figure 19 - *The House of Saud.This building at Diriyah that has been reconstructed and is now a national museum.*

Figure 20 - *Saad Ibn Saud Palace, Diriyah, now restored as an historical site.The Al Saud war horses were stabled in a section of the main walled court.*

Figure 21 - *The restored mosque of Muhammad ibn Abd al-Wahhab in Diriyah. This was the center of Wahhabi teaching and proselytizing for decades. The mosque was destroyed by Ibrahim Pasha, and in 1818 the Al Saud clan fled, leaving the "shell of their old capital behind them, an enduring reminder of the frontiers of the possible."*

The influence and authority of the Al Saud clan continued to emanate from their new capital in Riyadh. But all was not peaceful among the Al Saud brethren. This period was characterized by feuding and quarreling among the Saud elders and power struggles for dominance among the young men. These conflicts often boiled over into armed combat. The ascendancy to power of the Saud Emir Faysul in 1843 marked the apogee of Saud dominance in the century. When he died in 1865, the clan devolved into a period of confusion and deterioration. No one could replace the strong hand of Feysul. The Al Sauds devolved into frank civil war, as rival clan leaders sought to force the issue of rulership of the Saud tribe. The protracted civil war proved to be the undoing of the Saud family. They eventually became prostrate from the endless bickering and factional warfare.

During this same time, a new Nedjean desert tribal force was gaining strength and prominence in central Arabia. The Al Rashidis, centered in the northern Nejd at Ha'il, were becoming more powerful, ambitious, and more aggressive. The Emirate of Jabal Shammar was the ancestral land of the Al Rashidis, the virulent enemy of the Sauds. Established in 1836, the Rashidis controlled Jabal Shammar and the region around Al Jauf in the north. The capital of the Rashidis was at Ha'il, and they were closely allied with the Ottomans. Throughout the existence of the Emirate, the Rashidis quarreled among themselves for power and rulership. Most of the Rashidi Emirs were assassinated by factional competitors within a few years of obtaining the supreme office. They were nominally Wahhabis but were liberal and peaceful in their adherence to the faith.

Figure 22 - *The Second Saudi state (1818-1891).*

The absence of a generally accepted line of succession eventually proved to be the undoing of the clan. The Law of Primogeniture was often challenged by the Rashidi brothers eager for power, but the attempts to expand the control of the clan was crippled by constant in-fighting.

The first Emir was Abdullah bin Rashid. He was followed by his son Talal bin Abdullah, who ruled for twenty years, from 1848 to 1868. Talal's Emirship was characterized by liberal and tolerant policies regarding foreigners and members of the Shia sect of Islam, a sect hated by the Sunni Rashidis. Gifford Palgrave observed that Emir Talal achieved an atmosphere of equanimity by adopting a demeanor of tolerance of all religious faiths.

From 1868 to 1921 there were ten Rashidi Emirs. The last Emir would surrender to Al Saud many years later.

Created in 1836, the Emirate expanded its scope and power over ensuing decades. The Emirs of Jabal Shammar ruling from Ha'il quarreled constantly and bitterly with the Al Saud clan over control of the Arabian heartland. They gained ascendency not through wars and battles but through popular sympathy toward the natives of the Nejd and a liberality which the Wahabbis and Saud found anathema. During the 1860s, the Rashidi clan, backed by the Ottomans, seized the province of Kaseem without bloodshed The English adventurer William Palgrave was a close observer of the Jabal Shammar inhabitants.

In 1865 he wrote:

> "The inhabitants of Kaseem, (the rich agricultural heartland of the Arabian Peninsula) weary of Wahhabee tyranny, turned their eyes toward Telal, (Emir of Ha'il Telal bin Abdullah, 1846-1868) who had already given a generous and inviolable asylum to the numerous political exiles of that district. Secret negotiations took place, and at a favorable moment the entire uplands of that province-after a fashion not indeed peculiar to Arabia-annexed themselves to the Kingdom of Shommer by universal and unanimous suffrage."

Despite the loss of Kaseem to the Rashidis, there was little change in conditions on the peninsula. The two tribes, Rashidi and Saud, had co-existed in the Nejd for decades in a state of tense but static equilibrium. The state of peaceful co-existence appeared to be stable. However, a minor conflict arose over the arrest by the Al Saud family of the Rashidi leader Ibn Sabhan, regarding a civil dispute over a trivial matter concerning Zakat. The Rashidis, long impatient to regain control of Nejd, and long wishing to renew the blood feud with the Al Sauds, used this insult as a pretense for war. They consequently mounted a sustained military offensive against the Al Saud clan. The matter was settled at the Battle of Mulayda, January 24, 1891, fought in the Qassim region. The Al Sauds, led by Abdul Rahman bin Faisal and his allies (the Mutair and the Otaiba), were vanquished by the Al Rashidis, who were led by Muhammad Ibn Rashid. With this defeat, the Second Saudi State came to an end. Riyadh was occupied by the most viscerally hated enemies of the Saud family.

Sources from the time state that the victorious Rashidis captured 300 of the blue-blooded desert bred war horses of the Al Saud clan.

Figure 23 - *Flag of the Emirate of Ha'il, based on the Turkish insignia. Only the colors were changed.*

Lady Anne Blunt's diary contains a description of this action, as it was related to Lady Anne by Colonel Ross, a British army officer. He told the Blunts that the Sauds rose up against the Rashidis, leading to a "sanguinary and decisive battle in which the Sauds were finally defeated and many princes were killed".

Following the Battle of Mulayda, the embittered and quarreling Sauds were dispersed among loyalist desert tribesmen for protection. Most of them sought asylum with their longtime friends and protectors the Al Murrah Bedouin. Later the Saud family moved to Kuwait. Abdul Aziz Ibn Saud, future King of the Kingdom of Saudi Arabia, was 15 years old at the time. He was an extremely intelligent young man, and while in exile he studied statecraft daily at the majlis of the Emir of Kuwait, Mubarak al Sabah.

Figure 24 - *The geographic extent of the Emirate of Ha'il at its zenith.*

Purebred horses continued to be bred by the widely dispersed and dwindling migratory desert tribes. In the north, the Anazeh and the Shammar perpetuated the time-honored tradition of breeding pure-blooded stock, but horses were kept more for prestige and raiding purposes than for commercial reasons. The Muteyr, Beni Khalid, and the Ajman tribes continued to breed horses in the north near the Persian Gulf. The Al Murrah occupied a region further south, near the edge of the Empty Quarter. Near the Red Sea and inland from Mecca the Harb, the Otaybah and the Qahtan continued their ancient nomadic way of life.

But the desert's *purebred Arabian horse* was inexorably in decline as it continued to face threats both from without and within the peninsula.

The events in the Middle East during the dawn of the 20th century accelerated the process of the disappearance of the *purebred Arabia horse* from Arabia. The "slow emergency" was becoming more serious.

As the 19th century drew to a close, life on the Arabian Peninsula seemed trapped in an ancient past. This fact was noted by Lady Anne Blunt when she wrote a political forecast for the peninsula in her book Bedouin Tribes of the Euphrates, published in 1879. She decried the iron grip of the Ottomans on the desert dwellers, and she expressed her sympathy for them, taxed into oblivion by the Sultan. Then she adds: "But will no other power appear in the desert?"

In the year 1900, there was no reason to suspect that Arabia was on the verge of momentous political change. The Rashidis were the unchallenged dominant central power, but the peninsula was also populated by numerous smaller tribes, all of whom were independent and all of whom were, as a rule, nomadic and ungovernable. The Ottoman government was its nominal ruler, but any observer could see that the day of Ottoman ruin was near.

There was no reason to expect that the peninsula would ever become a united kingdom. There was no reason to suspect that the Saud family would have any future role to play in peninsular politics. The Sauds were now powerless, living as expatriates. At the time, the leading candidates for control of the political future of the peninsula were the Hashemites, the Beni Hashim, of the Hejaz, a family descended in patriline from the prophet Mohammad. Their claim to religious authority was unquestionable, and by extension, their future role in governing the Arabs seemed secure. The Sultan was the region's supreme leader, ruling from behind the Grande Porte in Constantinople, and he remained the Custodian of the Two Mosques, at Medina and Mecca. The Sultan was a Muslim, and the faith cemented an uneasy relationship between the Sultan and his subjects. The Sherif of Mecca was appointed by the Sultan, as the position was passed down from one generation to the next within the male line of descent of the Hashemites.

Figure 25 - *Qasr al-Masmak, built around1865 by the Rashidis as, after the Battle of Mulayda, they consolidated their position of superiority on the peninsula. The fort was built during the reign of Mohammed ibn Abdullah ibn Rasheed. The fort, now in the Riyadh historic district, was a fortification intended to defend against the Al-Saud clan, unseen but not forgotten. The fort was garrisoned by Rashidis, who were supported by the Ottoman Sultan. It was this fort that was captured by the young Amir Abdulaziz bin Abdul Rahman bin Faisal Al Saud in 1902.*

Quite unexpectedly, a remarkable event took place.

In 1902, the defeated Saud ruler Abdul Rahman's son, Abdul Aziz (Ibn Saud), gathered his scattered family together, urging them to leave the safety of their protectors. Organizing the men for battle, he and his men stormed into their occupied homeland, forcing the Al Rashids out of Riyadh and into historical obscurity. On the night of January 15, 1902, Ibn Saud led a small group of 40 men over the city walls of Riyadh. They subdued the inhabitants and executed the governor of the city, Ajlan, in the public square. Ibn Saud then began a long and costly war of conquest with the ultimate intention of unifying the entire Arabian Peninsula under his rule. The conquest was to occupy his full energies for the next thirty years.

Figure 26 - *The extent of the Ottoman Empire around 1900. The individual vilayets were the administrative units of the Ottoman Sultan, governed by his appointed vila, who independently governed the inhabitants of the vilayet regarding law, taxes, and foreign matters. The great central waste of Arabia, where Ottoman rule was weak, is again noted. The unique designation of Egypt indicates its relative independence from interference from Constantinople. It was not a vilayet and was only nominally controlled by the Sultan.*

In 1903 Ibn Saud began regional attacks on the Rashidis. In 1906 the Al Saud clan was defeated Abdulaziz ar-Rasheed at Mehanna, dispersing the Rashidis and expelling the Turks as well. The final battle of this campaign took place near Al-Qassim in 1907 with the defeat of the remaining Al-Rashidis by Saudi forces.

This had important strategic consequences in that Ibn Saud now had control over the important oil fields around Basra. Equally important was the control that he gained over the Shatt El Arab, the water route to Basra. He sought to expand his control over the entire peninsula, with the ultimate goal being to conquer the Hejaz and expel its ruler Sayyid Hussein bin Ali, the Sherif of Mecca.

The principal impediment to Ibn Saud's ambition was not just his desert brethren but also the Sultan of Constantinople. The outbreak of World War I in 1914 eliminated the latter impediment.

Figure 27 - *An Ottoman Army Encampment, by Adolph Schreyer. Image in the public domain. During the late 19th century, the Ottoman military forces were in a parlous state. The Empire had been known as the "sick man of Europe" for decades, and this decline was nowhere more evident than in its military. Ruined by the losses of the Crimean War (1853-1856), the Army was further strained by losses in the Russo-Turkish War of (1877-1878). Money to finance military re-building was insufficient, and the low literacy level of the troops made control and command functions ineffective. Local education in Turkey was inadequate, and most Ottoman Army officers were trained in France. There was no effective central administration system and no organized system of logistical planning.*

The Ottoman control of crucial parts of the Balkans is noted. The religious conflicts between Muslims and Christians in this region date to this era.

The capture of Riyadh gave the charismatic young leader a large following, and he solidified his control on the population by constant armed pursuit of the Al Rashidis who remained in the Nejd. By 1912, he controlled the east coast of Arabia and all of the Nejd. To unify the land, however, he had to unify the people. In 1913 he founded the *Ikhwan*, a military-religious brotherhood of radical battle-hardened warriors. They were to be soldiers for the faith and soldiers for Ibn Saud, which they regarded as one and the same. They were religious fundamentalists, and savagely radical. This army of warriors gave Ibn Saud more military power and, as a result, more broadly based popular support among the tribal Bedouin. The principles on which Ibn Saud's subsequent successes were based were coercion and the threat of violence.

Ibn Saud's intention was to create an elite hardened fighting unit to support his plans of peninsular conquest. In order to create this corp, desert tribesmen were selected for their fierceness in battle, and these men were then trained together as units of military collectives composed of men from many different tribes. This technique was used to sever tribal loyalties and to inhibit the eruption of blood feuds among the men.

Figure 28 - *Flag of the Ikhwan. The inscription is the shahada, or the testimony of acceptance. "There is no god but God, and Muhammed is his messenger." The message of the banner was clear; those who would not adhere to the Wahabbi interpretation of the Qu'ran and profess the testimony of acceptance would die by the sword.*

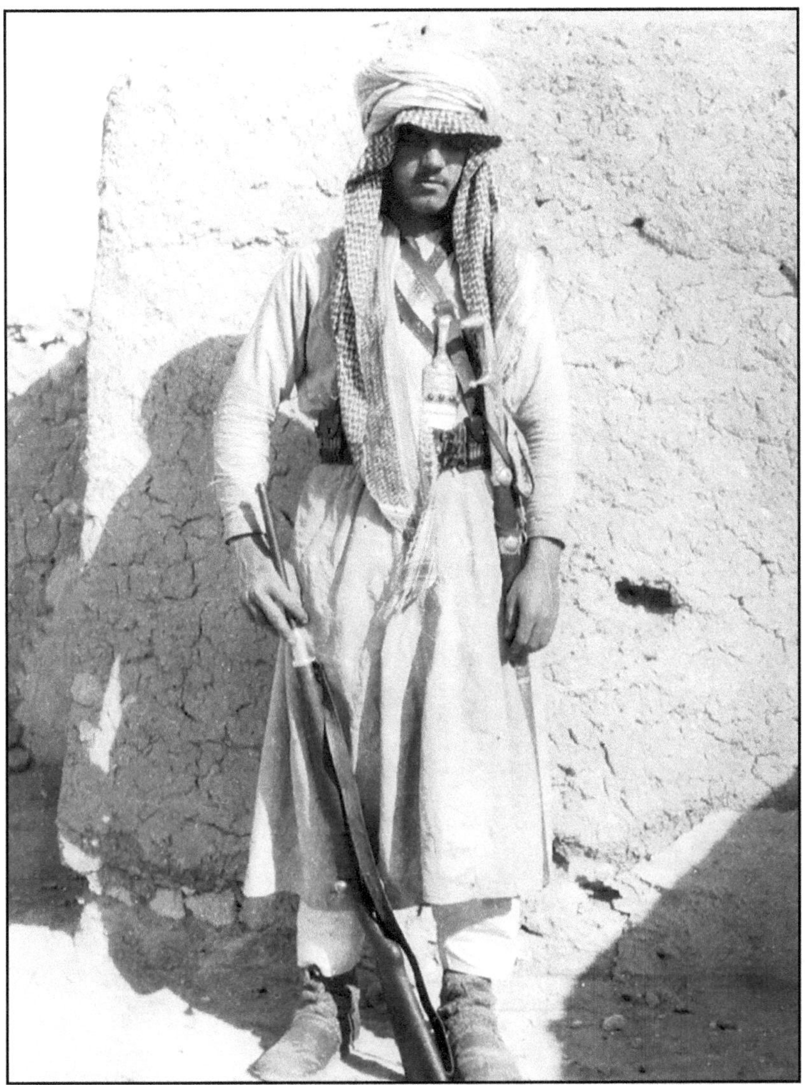

Figure 29 - *A desert tribesman dressed in the traditional Ikhwan manner. The white turban was a potent symbol; the average Bedouin wore a roped headdress. Photo used with the permission of the Royal Geographic Society, London.*

Using the terrorizing presence of his fighters, Ibn Saud sought to break down tribal allegiances among all the wandering Bedouin, forcing them to settle in oases and farms. The settlements were organized as hujar, the plural of hijra. The terminology was selected by Ibn Saud to suggest to the Bedouin that they were making a holy journey from ignorance to a condition of sanctity and true Islamic devotion. The largely illiterate Bedu were taught to be "proper Muslims". Strict Wahhabi theology was enforced, smoking and drinking were prohibited, and idolatry was to be eliminated from the peninsula. Settlement of the tribes was intended to make them more governable.

Figure 30 - *Wahabbis of the Arabian Desert, white turbaned. Photograph made by British agent Gerald de Gaury in the 1930s. Photograph used with the permission of the Royal Geographic Society.*

Religious leaders were sent out among the population to preach the arch-conservative Muslim beliefs of the Wahabbis. Sunni Islam was taught, and Shiites were forced to convert under threat of death. Shiite holy sites and places of pilgrimage were destroyed.

The *Ikhwan*, Arabic for brotherhood, was comprised of groups of Bedouin from various tribes who were converted to the Wahhabi doctrine by *Ikhwan* ulama, or clerics.

These men became dedicated to the purification of Islam, which in their view had become contaminated by their fellow Bedu who had failed to adhere to strict Muslim law. The *Ikhwan*'s conservative Islamic theology led them to regard all non-Muslims as traitorous infidels who must, under strict Islamic law, be executed. They regarded this as a religious duty. In this way, the deserts of Arabia would be purged of impurity and become sanctified. Fellow Bedouin Muslims who failed to adhere to *Ikhwan* practices were dealt with just as harshly.

The *Ikhwan* fought in the traditional manner that had been practiced in the desert for centuries, attacking the enemy while mounted on camels and horses. Their weapons were the lance, the sword, and the dagger. Their organization as a fighting force was irregular, and their methods in battle were notoriously savage. Captives of the *Ikhwan* were dispatched in uniform fashion, by having their throats cut.

Captain W.H.I. Shakespear a British political agent, was dispatched from Kuwait in December 1914 to offer Ibn Saud a treaty of alliance with the British. Shakespear was chosen for the mission because of his long-term association with Ibn Saud, as well as his familiarity with the political dimensions of the inter-tribal battles.

He traveled for 19 days to reach the encampment of the Al Sauds. After concluding an agreement with Ibn Saud, he chose to stay on with the Bedouin army and participate in an impending battle with the Bani Shammar tribe, allies of Ibn Rashid.

Figure 31 - *Ibn Saud.*

Figure 32 - *Ibn Saud's army on the march near Habl. The photo was taken by W.H.I. Shakespear on June 8th, 1911. This photograph is used with the permission of the Royal Geographic Society, London.*

Figure 33 - *Ibn Saud's army assembled near Habl. The photo was taken by W. H. I. Shakespear on March the 8th, 1911. This photograph is used with the permission of the Royal Geographic Society, London.*

Shakespear explored and mapped much of the region during the campaign, and photographed the *Ikhwan* forces on the march. He was killed on January 15, 1915, during the battle at Jarrab, a small village east of Buraydah, age 37, suffering three bullet wounds. The Saudi forces were defeated at Jarrab and driven from the field. In official reports, Ibn Saud expressed profound regret "for the loss of one whom he regarded as a brother".

WESTERN ARABIA; FOMENT IN THE HEJAZ

The Hejaz was the crown jewel of Arabia that Ibn Saud needed to obtain in order to complete his conquest of the peninsula. It was a place of supreme spiritual importance to all Arabs because of the holy cities Mecca and Medina. Since the 10th century, the ruler of Mecca and the supreme authority in the Hejaz, the Sharif, was a Hashemite, which was an Arab clan that took its name from Hashem, one of the Prophet's male ancestors. The line of descent derived from the Banu Hashim, a branch of the Quraysh tribe. All Hashemites trace their lineage to the Prophet, and this fact gave them supreme respect and authority throughout the Muslim world.

In 1908, the Hashemite Sharif Ali Abdullah Pasha died, and the Ottoman Sultan in Constantinople, the "Custodian of the Two Holy Mosques (Mecca and Medina)" appointed a new Hashemite ruler. Hussein bin Ali was appointed to the office of Emir of the Hejaz, making him the Sherif of Mecca. His role in the coming World War would be pivotal.

The outbreak of the Great War meant that Ibn Saud would have to wait for a time before advancing his agenda.

World War I began with the invasion of Serbia by Austria-Hungary on 28 July 1914. The War pitted Britain, France and Russia against an alliance of Germany, Austria-Hungary and the Ottoman Empire. The Ottomans, now under the leadership of the "Young Turks", were emboldened by the global war and eager for an opportunity to repel the British from their former territories.

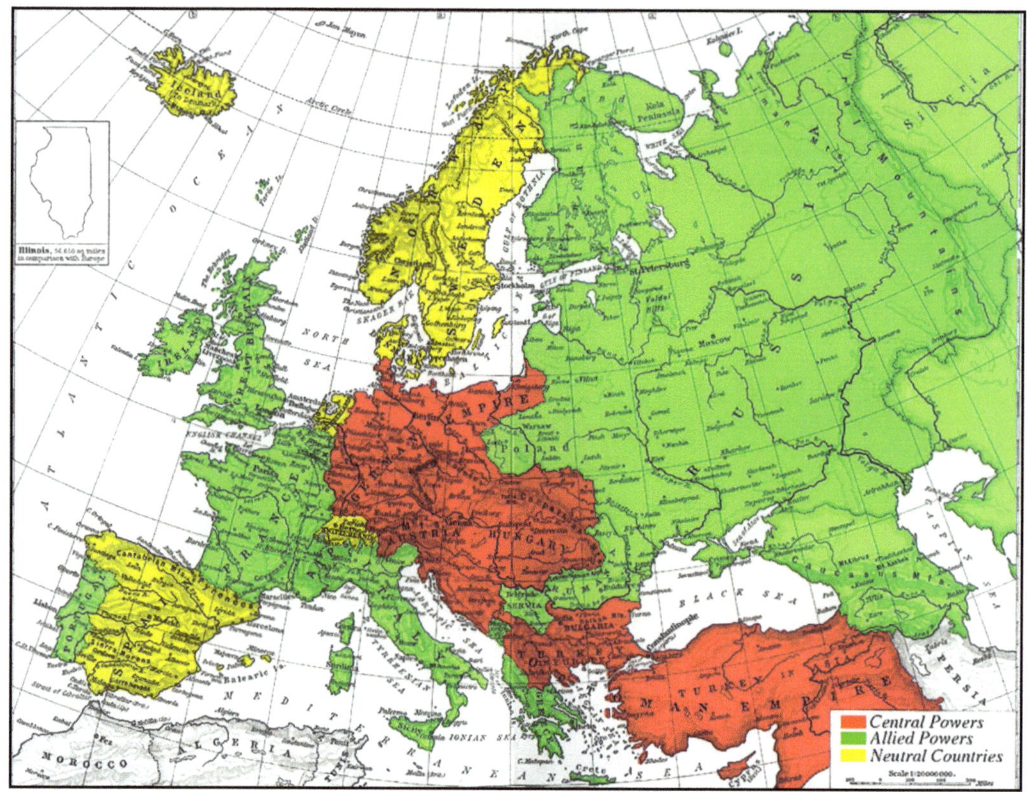

Figure 34 - *The Combatants of World War I. The Ottoman alignment with the Central Powers of Germany and Austria-Hungary is illustrated. This alliance extended German military power into Mesopotamia and the Levant, bringing the War to the borders of Deserta Arabia.*

The Young Turks, (*Jon Turkler*) having deposed Sultan Abdul Hamid II, the last Sultan of the Ottomans, used the war as an opportunity to motivate the scattered Muslims of the Middle East to rise up in a holy war, a jihad, against the hated British, a nation that they reviled only slightly less than Russia.

Constantinople radio broadcasted the following edict to the anxiously waiting Arab world.

> "O Muslims, know that our Empire is at war with the mortal enemies of Islam; the governments of Muscovey, Britain, and France. The Commander of the faithful summons you to the jihad."

The Turkish offensive was swift. In January 1915, the German military command led a Turkish force from Palestine across the Sinai, with the intention of capturing the Suez Canal and, eventually, reclaiming Egypt from the British. The attack failed, but the British were made aware of their vulnerability in the Middle East. The foreign office in London began to search for answers. Control of the Canal was essential to British interests.

Figure 35 - *Sayyid Hussein bin Ali, Sherif of Mecca, 1854-1931, was at one time King of the Hejaz and later leader of the Arab Revolt. He was a Hashemite, a blood descendant of the Prophet, and as such enjoyed great respect throughout the Arab world. He rose to the highest position of rule, even as the control of the Ottoman government was rapidly waning. His sons, Abdullah and Faisal, were key political figures in the new Arab world after the fall of the Ottoman Empire. His third son Ali was inclined to war rather than diplomacy and was essential in the capture of Aqaba.*

Figure 36 - *Mecca and the Ka'aba, late 19th century. Photograph from the Library of Congress Collection.*

Figure 37 - *Worship at the Ka'aba Mecca, late 19th century. Photograph from the Library of Congress Collection.*

The Hashemite claim to legitimacy in the Muslim world was a chronic thorn in the side of Ibn Saud. Throughout the scattered fragments of the Islamic world, many Islamic leaders hoped for a time of unification of the faithful, unification under one ruler, one Arab. The Arabs of the peninsula may have disliked being part of the Ottoman Empire, but they fervently hated the British. But the Al Sauds and the Ottomans were Muslim brothers, and this fact deeply troubled the Hashemites. The faithful inhabitants of Mecca feared the prospect of fighting on the side of the infidel British against their fellow Muslims.

In the broader world of Islam, the direct descendants of the Prophet were received with great respect and were considered to be the rightful possessors of the highest religious authority. This fact would be part of the dilemma as the seeds of Arab nationalism began to germinate. But with a war on, the British government determined that they could most effectively manipulate the Hashemites as pawns in the impending campaign.

First, however, the Sherif had to be convinced that his best future interest resided with the British.

The origins of the revolt involved a great deal of duplicitous diplomacy and bribery on the part of the British Crown during World War I. The British needed to find a way to lure the Arab leaders away from their Ottoman allegiances (and by extension their German/Austria-Hungary allegiance) and convince them to fight on the side of the Four Great Powers.

The British Secretary of War Lord Kitchener was the first to go on the record:

> "If the Arabs rise up against their Ottoman overlords, Britain will give the Arabs every assistance against foreign aggression."

The notion that Britain would embark on a war against the Ottomans from the south was the brainchild of the young Winston Churchill. The European War with Germany was initially disastrous for the French and their ally, Britain. Early in the War, the outcome was very much in doubt. The Allies were able to stop the German advance into France but conditions had deteriorated into costly trench warfare. To the allies, the unthinkable was becoming a distinct possibility. A German victory seemed likely.

The war in Europe was grinding on, destroying an entire generation of French and British soldiers. And the war was at a stalemate: Passchendaele, Verdun, Ypres, the Somme. The carnage was unprecedented, and there seemed to be no way to break the deadlock.

The First Lord of the Admiralty Winston Spencer Churchill wrote to the British Prime Minister "Are there not other alternatives to sending our armies to chew barbed wire in Flanders?" Churchill was determined to mount a naval assault on the Ottoman Turks, which in time became the disastrous and deeply regrettable Gallipoli campaign.

Desperate, Britain adopted the alternate strategy of creating a diversionary war against Germany's ally, the Ottomans, far from the horrors of Central Europe. The plan was to attack the Central Powers from the Mediterranean, through the Ottoman Empire, but by land, through the Levant.

The plan was simple. The British were already in control of Egypt, having annexed that country in 1882 in what was referred to as the "veiled protectorate". The Ottoman government was still recognized by the government of Egypt, but the Egyptian Khedive was allowed no role in the military affairs of the country which had become controlled entirely by the British.

The British intended to mass their forces in Egypt and move north across Sinai, into Palestine, and force the Ottomans out of the Levant. They would then seize Jerusalem, capture Damascus, and then advance into Anatolia itself. This maneuver would force the Germans to divert men and material from the Western front in order to meet the challenge in Anatolia.

But the British needed the support of the Arabs in order to do this, because these areas in the Levant all belonged, in one sense or another, to the Arabs. The British needed the sanction of the Arabs in order to legitimize what they intended to do. No Arab leader possessed more spiritual or secular authority than did the Hashemite Sherif, a direct descendant of the Prophet.

Arabs throughout the Middle East could see that the Ottoman Empire was near its demise, and the dream of a unified Arab Empire began to form in the minds of Arab intellectuals and the Arab people alike, a people who were now widely dispersed and disorganized politically. Most significantly, they were eager to throw off the oppressive yoke of both the Ottomans and the British.

But there was Hussein's hereditary enemy Ibn Saud to consider.

The British for their part were motivated to dominate of the region because of their resolute insistence on keeping the Suez Canal open to international traffic. The canal was the key to the maintenance of the British Empire. And for the British politicians of this period, the maintenance of the British Empire was paramount.

Figure 38 - *Group of female Arab Nationalists demonstrating in Cairo, 1919. Throughout the Islamic world, a groundswell of popular support grew from the people who had been under the oppressive domination of foreign powers for centuries.*

Sherif Hussein was ambitious. He did not just want to be King of the peninsula; he wanted to be King of the entire Arab peoples. He dreamed of being the ruler of a great unifying renaissance of Islam worldwide. He dreamed of a future caliphate. His goals were in direct conflict with the Ottoman intentions, and the collision of purpose was complicated by the fact that they were both an Islamic people. Arab aspirations to regain the lost power, prestige, and wealth of the ancient past fueled the expectations of progressive, nationalistic minded Arabs in the early 20th century.

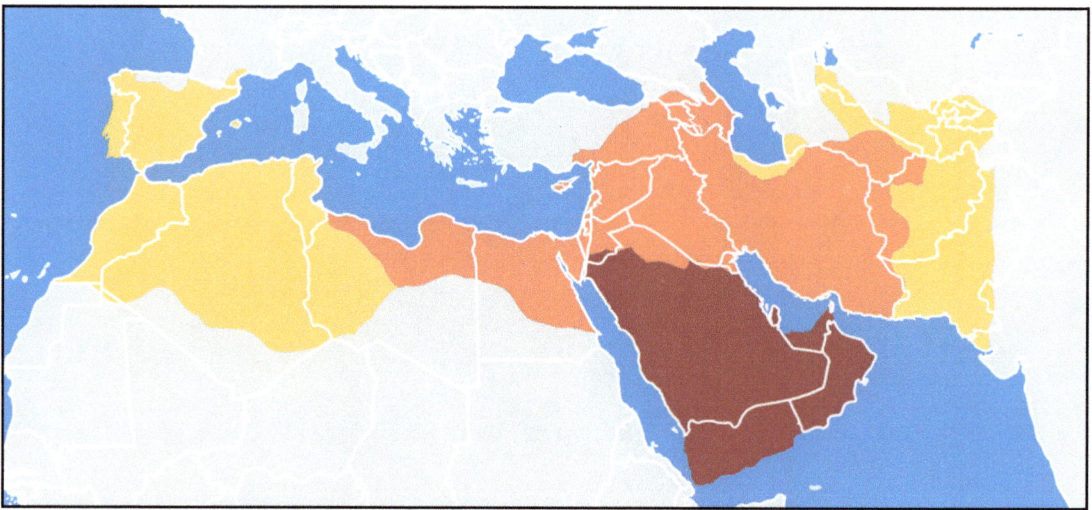

Figure 39 - *The extent of the Arab Caliphate in 750 CE. The conquests of the forces of the Prophet Mohammad are shown in brown. The expansion of the early caliphates is shown in orange. The displaced Ummayads are shown in yellow.*

The glory of Islam at the apex of its power had been a remarkable achievement. The Ummayads were displaced to Andalusia (modern-day Spain) by the conquest of the Abbasids, a rival Arab clan who established a new caliphate, ruling from the newly built city of Baghdad. The dual Arabic caliphate system existed for centuries, from 750 to 1258 CE. The educational, intellectual, artistic, architectural and public works advancements were the marvel of the age. At a time when the King of England was illiterate, the Caliph corresponded with his governors regularly.

The splendor and magnificence of the Muslim cities during the caliphates and the sophistication of the educational centers and libraries were superior in all respects to other cultures of the time. In Europe, by contrast, the populace led brutish, barbaric, and short lives. And Medieval Europeans were largely illiterate.

Figure 40 - *The Great Mosque in Cordoba, built by the Muslim Caliph Abd Al-Rahman. Construction began in 784 CE.*

During the 10th century, regular postal service existed from Damascus, providing mail service throughout the Caliphate. In Cordoba, there were paved streets and city street lights The sophistication and urbanity of Cordoba was the envy of all Europe.

The past splendor and accomplishments of the Arab people, now only a shadow of its former glory, were still remembered by the scholars and imams of the Middle East.

Hussein envisioned not just Arab independence but a full renaissance of the past glory of the Arab people, with he himself as its sole ruler. In effect, he wanted to re-establish the Caliphate, which had been eliminated at the time of the Ottoman conquest of Arabia. Based on his credentials, he was considered by the British to be the ideal person to manipulate towards their own ends. The Sherif himself thought that cooperation with the British represented his best chance at achieving Arab freedom.

Figure 41 - *Dome of the Eagle, (Qubbat an-Nisr) the central dome of the Ummayad Great Mosque in Damascus.*

Figure 42 - *The Ummayad Mosque in Damascus. This site is the fourth holiest for the Muslim. Construction on the Mosque was begun soon after the establishment of the Ummayad Caliphate around 634 CE.*

Figure 43 - *Al-Mustansiriya University, Baghdad, Iraq. Established in 1227 by the Abbasid Caliph Al-Muustansir, it was the first Islamic University and one of the oldest universities in the world.*

Figure 44 - *Dome of the Rock, Jerusalem. The original mosque was completed in 691 CE by the Umayyids. The exposed central stone in the mosque is believed to be the site from which the Prophet ascended to the seven heavens during the Night Journey.*

Figure 45 - *Al Aqsa Mosque, Jerusalem. The site has been a sacred location since the first mosque was built here by the Umayyads in 705 CE. It has been destroyed by earthquakes and re-built several times. A holy shrine to Islam, the Prophet was transported to this site during the Night Journey.*

Convincing the Muslim Sherif to fight against his fellow Turkish Muslims was complicated. The Sherif's first duty was ostensibly to Islam, and his people had to be convinced in the righteousness of killing their brethren Muslims, the Ottomans. The Sherif justified his decision to pursue the Al Thawa Al-Arabiyya by making the claim that the Young Turks who now ruled the Ottoman Empire were apostates, recreants who had deviated from the path of true Islam. This made their defeat a religious duty.

The seduction of the Sherif of Mecca was brought about by secret negotiations between the Sherif and the British High Commissioner in Cairo Sir Henry McMahon. In a letter to Hussein dated 24 October 1915, McMahon stated: "Great Britain is prepared to recognize and support the independence of the Arabs." Hussein had personal aspirations to be the sole ruler of the Arab people and the land of Arabia, and the support of Britain seemed to be the support that he would need to achieve this goal. He was lured by the promise of finally achieving his life-long goal of unification of the Arab peoples and the restoration of their dignity and prosperity. This promise had originated with Lord Grey in the Foreign Office in London.

The British were desperate. The failure of the Allied Gallipoli campaign, begun in April of that same year, was proving to be a disaster for the British and their allies Australia and New Zealand. At Gallipoli, the Turks were defending their homeland and they fought with a ferocity and tenacity that would soon lead the British to withdraw from the peninsula in utter defeat. The British foreign office needed a new plan of attack against the Turks.

The Sherif was initially opposed to accepting the British promise at face value. He did not wish to see Muslim fighting Muslim. His son Abdullah was in favor of the venture. Faisal was ambivalent, aware of the British reputation for deception and political intrigue.

An inducement of gold was offered to Hussein by the British. Military advisers, among them T.E. Lawrence, were also sent to Arabia.

After much debate and with deep reservations on the part of Hussein, the Arab Revolt began in June 1916, consisting of a force of about 70,000 men. The Hashemite forces with allied tribesmen began a guerrilla campaign aimed at repeated attacks against the Turkish rail connection between Medina and Damascus. First Jiddah fell, and then Mecca. Medina was stubbornly held by the Turks.

The Sherif's sons soon took Taif, Yenbo, and Rabagh.

Figure 46 - *The 1917 Guerilla Campaign: The Arab Revolt. This photo was taken by Colonel Lawrence as the Ageyl bodyguard began to move north in the first leg of the campaign. Emir Faisal bin Hussein al-Hashimi and Sherif Sharraf are shown. Photograph used with the permission of the Imperial War Museum London.*

Figure 47 - *The camp of Ali bin Hussein, 1918, during the early days of the Arab Revolt. Photograph used with the permission of the Imperial War Museum, London.*

Figure 48 - *Colonel Lawrence's Ageyl bodyguard. Photo taken by Lawrence, used with the permission of the Imperial War Museum, London.*

Figure 49 - *Airplanes of the Royal Air Force tethered during a wind storm in the Hejaz. The planes were used to support the Arab Revolt. Image used with the permission of the Imperial War Museum, London.*

The British did more than to send gold and Colonel Lawrence to Hussein. They also sent the English No.14 Squadron of the Royal Flying Corps under the command of Captain Henderson, along with two aeroplanes and several armored cars. At first Sherif Hussein refused to allow them into the country, fearful that so many Christians fighting with the nomadic and devoutly Islamic Arabs would lead to dissolution of the inter-tribal alliances. Rumors circulated that the Turks intended to crucify any Christians that were captured in the war. It was a gamble and the stakes were high.

The planes were eventually allowed to join the Sherif's forces. They were used to scout the forward positions for enemy troop movements, and dropped handmade gelignite bombs on the Turkish encampments. The armored cars proved to be too heavy for movement on the sand, so the armor was removed. The Rolls Royce cars were soon found to have inferior traction and the Crossley tender was substituted. The Arab coalition under the command of Ali eventually captured the port city of Aqaba, expelling the Turks.

Figure 50 - *The Hejaz Railway, at Al 'Ula, about 380 km north of Medina. Built by Ottoman Sultan Abdul Hamid with German engineering assistance, the track connected Damascus with Medina. It was initially envisioned to facilitate the travel of pilgrims journeying to Medina and Mecca. The Sherif of Mecca, conscious of his precarious hold on power, regarded the construction of the railroad as a threat, since it provided Constantinople with a way to quickly supply the Ottoman/German garrisons in the Hejaz. Image from the Library of Congress.*

Figure 51 - *An Ottoman train station along the Hejaz railway*

Figure 52 - *As he lay on the tracks, Lawrence took this photograph to document one of his many explosive detonations along the Ottoman Railway tracks to Medina. Lawrence was never reluctant to glamorize the wild and frenetic nature of his raids with the Arabs; he wrote to his mother that the raids took about 10 minutes and were very similar in nature to the Buffalo Bill Cody's Wild West extravaganzas that he had seen in England. This particular detonations took place near Deraa. Image used with the permission of the Imperial War Museum, London.*

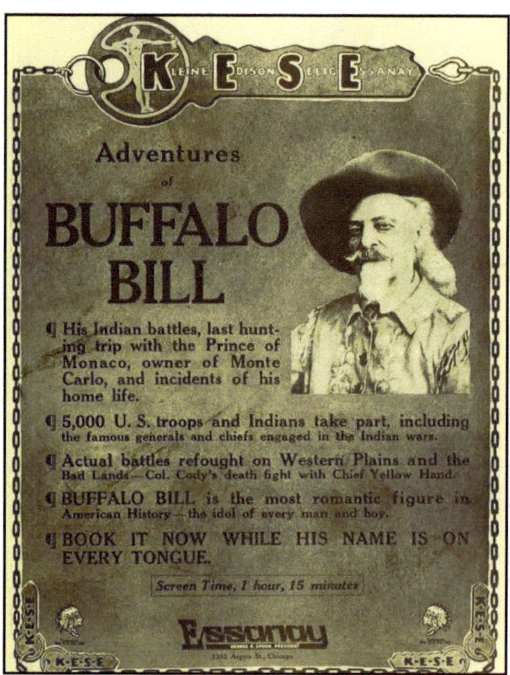

Figure 53 - *A poster for a cinema presentation of the Adventures of Buffalo Bill.*

Lawrence described Auda as noble powerful and proud. "Only by means of Auda Abu Tai could we swing the tribes from Ma'an to Aqaba so violently in favour that they would help us take Aqaba and its hills from their Turkish garrisons."

The fighters under him were acclaimed for their "tradition of desperate courage and a sense of superiority that never left them". The Hashemite fighters led by Hussein's son Ali were also critical to the success of the attack on Aqaba.

Following the capture of Aqaba, the Bedouin forces turned their attention north. However, the political landscape began to change as the Arab Revolt reached the Levant. On November 2, 1917, the British Government issued the Balfour Declaration, stating that His Majesty's government favored the formation of a Jewish state in Palestine following the successful issue of the war. On November 26, 1917, the Manchester Guardian published a news story in which the details of the secret treaty between Britain, France, and Russia were disclosed. In the story, the plans to carve up the Middle East into sectors that would be controlled by the Great Powers was revealed. During late night sessions at 10 Downing Street during 1915 and 1916, the arbitrary national borders of the present-day Middle Eastern States were drawn up. This news was not warmly received among the leaders of the Arab Revolt, and they nearly abandoned the campaign. To placate them, General Allenby gave them more rifles, armored cars, and air support, formally designating them as the right flank of his military advance into Palestine.

Figure 54 - *Emir Saud Ibn Rashid, the fifth man from the left, with Turkish troops at Bir-Ali in 1917. Photograph from the Fikri Pasha collection, used with the permission of the Imperial War Museum, London*

In 1918, the joint Arab forces turned north and supported General Allenby's march into Palestine. The British decimated the Turks at the Battle at Megiddo in Northern Palestine in September 1918, and Allenby's Army entered Damascus on 1 October 1918. The Ottoman Turks capitulated on October 31, 1918. The War in Europe ended on November 11, 1918.

Figure 55 - *Auda Abu Tai, the famous desert sheikh of the Howietat. Photograph made in 1921 and used with the permission of the Library of Congress.*

The War was over. Faisal entered Damascus at the head of a triumphal parade. The next day he met with General Allenby at the Victoria Hotel and was warned by Allenby that the Arab role in Syria would be very limited. Despite this warning, Faisal set up an Arab government in Damascus, appointing himself as governor. His position would last only 16 months. Syria was given to the French, and the Arabs were ejected from the city, under threat of military force.

Figure 56 - *T.E. Lawrence, 1918. Photo used with the permission of the Imperial War Museum, London.*

"A man may achieve what he wills, but he cannot choose what he wills."

Arnold Schopenhauer

The loss of Damascus was a bitter defeat for the Hashemite Sherif, still in Mecca and still dreaming of Arab unification and the resurrection of the glory of the past Caliphates.

The Versailles Peace Conference began on 18 January, 1919. The Hashemites were to be disappointed. The outcome of the conference was not altogether unexpected, the treaty between the Powers having been revealed two years earlier. Sherif Hussein refused to sign or ratify the Treaty of Vesailles.

Figure 57 - *Sir Arthur Balfour (1848-1930) British Prime Minister, and later Foreign Secretary. His academic training was in philosophy, and he was fond of remarking that "nothing matters very much, and few things matter at all."*

The successes of the Arab Revolt failed to create an independent Arab state for the simple reason that the British had no intention that it should ever succeed in becoming independent. Unknown to the Arabs, the post-war plans for the partitioning of the Middle East had already been secretly drawn up by the British and the French. The Sykes-Picot agreement split the region into two arbitrary halves, with France obtaining the northern half (A) and Britain possessing the southern half (B). Palestine, colored yellow on the map, was envisioned as an "international" city. Irak, colored pink on the map (Figure 58), was to be British. Irak was important to the British because Irak had oil.

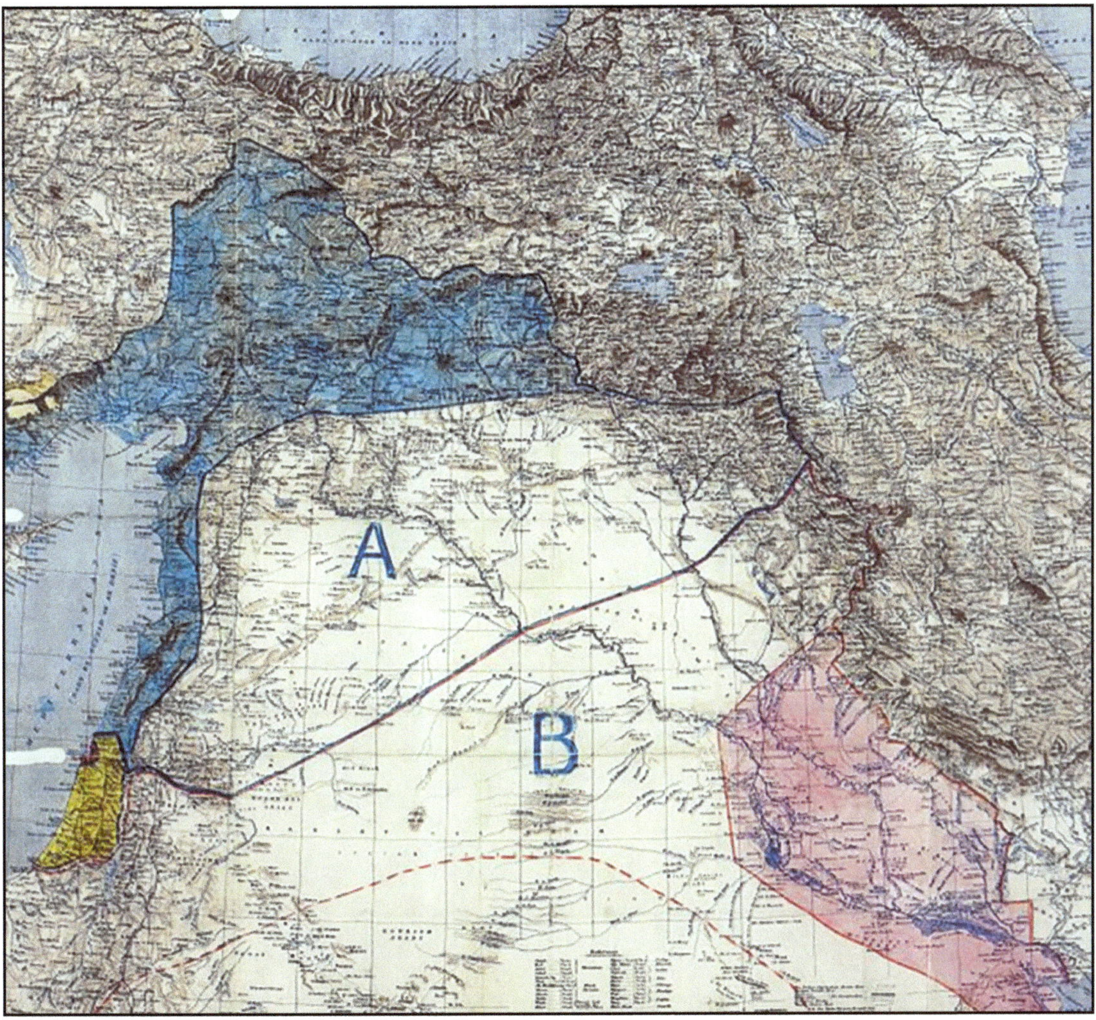

Figure 58 - *The map of the Sykes-Picot Agreement.*

There was to be no nation for the Arabs, not even a homeland. All of this was made public during the Versailles Peace Conference. At the conference, the United States President Woodrow Wilson was distressed to learn the details of the brutal accord but was powerless to intervene. Wilson's vision of a world in which all peoples were free to be autonomous and self-governed would not be realized. Wilson was forced to be content with the formation of the ill-fated League of Nations.

The British were fully aware of the consequences of the conflicting promises that it had made to various entities during the war. These secret promises were made with full awareness of their duplicity. British policymakers remained unperturbed. Regarding the region of Palestine, the British never planned for the Arabs to play any political role, despite the promises implied in the McMahon correspondence.

Figure 59 - *Faisal at the Versailles Peace Conference. Photograph from the Library of Congress.*

The cynicism and brutal dishonesty of the British policymakers in regard to this matter was summarized by Lord Balfour at the post-war Versailles Peace Conference.

> "As far as Palestine is concerned, the Powers have made no declaration of policy, which, at least in letter, they did not intend to violate... the Zionists are more profoundly important than the desires and prejudices of the 700,000 Arabs who now inhabit that ancient land."

Colonel Lawrence himself held the view that the Arabs were more unstable than the Turks.

> "If handled properly they would remain in a state of political mosaic, a tissue of small jealous principalities incapable of cohesion."

In this he was proved wrong.

Figure 60 - *Prince Faisal at theVersailles Peace Conference in Paris, 1919. Lawrence stands behind him. Photo used with the permission of the Library of Congress*

Following the end of the First World War Faisal was placed on the throne as King of the newly created Iraq. British interest in Iraq was due to the oil fields of Mosul and Basra, with the essential ports along the Shatt el Arab needed to transport the oil for use by the British Navy around the world. The British Navy had converted their entire naval fleet from coal steam to oil-fired coal steam early in the 20th century. The use of oil as a source of fuel for internal combustion engines was still in its infancy. The oil fields of eastern Arabia were unknown at the time. In the final division of land, the British also retained ownership of the Palestinian port of Haifa, as this was the ideal terminus for a land route to haul Iraqi oil by truck to the Mediterranean.

Figure 61 - *April 1921, Amman, Jordan. Col. T.E. Lawrence with Emir Abdullah, son of the Sherif Of Mecca. Abdullah was now King of the newly created country of Transjordan. The Englishman Sir Herbert Samuel (in the white topee) was appointed British High Commissioner of Palestine by the British Parliament, making him the first Jew to govern the region for over 2000 years. He was a committed Zionist and is credited with making the first substantive efforts towards the creation of the state of Israel. Photo used with the permission of the Library of Congress.*

Figure 62 - *T.E. Lawrence wrote The Seven Pillars of Wisdom, in which he recorded his observations of the Arab character, noted during his involvement in the 1916 Arab Revolt. Lawrence possessed a sensitive and artistic nature. The account has a psychological flavor and is essential to a Western reader who wishes to obtain a glimpse behind the generally impenetrable curtain of the eastern psyche. "Wisdom hath builded her house, she hath hewn out her seven pillars," Proverbs 9:1.*

Figure 63 - *Arab mounted warriors accompanying the new King Abdullah, Transjordan, 1921. Photo used with the permission of the Library of Congress.*

Lawrence served in the British Foreign Office following the war, advising Winston Churchill on Middle East affairs until 1921. He traveled to the region by air on several occasions, surviving a nearly fatal crash landing in 1919 in Rome at the aerodrome at Roma-Centocelle.

Lawrence later admitted that he was fully aware of the duplicitous intentions of the British Foreign Office toward the Arabs. He had given his word to Hussein that the British would honor their commitments, and when this did not occur, he lost the confidence and the trust of his erstwhile Arab friends. The sense of guilt and betrayal at his complicity in the deception proved to be an unbearable burden.

Figure 64 - *King Faisal of Iraq.*

In 1922 he enlisted in the Royal Air Force under the assumed name of John Hume Ross, working as an aircraftsman, in maintenance. In 1925 he left the RAF and joined the Royal Tank Corp under the assumed name of T.E. Shaw. In 1925 he again joined the RAF and stayed on in minor capacities for ten years. He retired from the RAF in 1935. He died in a motorcycle accident 2 months later.

The ambitions of Ibn Saud were interrupted but not altered by the First World War. He had pledged his support to the Ottoman Turks in 1914, but he became the center of intrigue as the British government attempted to obtain his cooperation against the Turks with promises of expanded power after the war had ended. His close affiliation with the Wahhabi brotherhood was a point of contention among the other desert Emirs. In exchange for British money, supplies, armaments, and protection, Ibn Saud agreed to cease aiding the Turks and to instead raid and harass the remaining Al Rashids still present in the Nejd. This decision led to a strong alliance between the Al Rashidis and the Ottoman Turks.

Figure 65 - *Following the end of the war, Faisal's brother Abdullah was given Kingship status by the British in the newly minted country carved out by the British called Transjordan. T.E. Lawrence served as diplomatic liaison.*

Figure 66 - *King Abdullah in Jordan.*

The British government, notoriously and persistently playing one faction against another, sent Hillary St. John Bridger Philby to serve as British adviser to Ibn Saud during the war. He was given the task of organizing an Arab Revolt against the Turks in the west of Arabia, protecting the oil fields near Basra and the Shatt al Arab. This waterway, the confluence of the Tigris and Euphrates, flowed into the Persian Gulf and was the only seaport suitable for transporting oil for the British navy. This was prior to the introduction of the internal combustion engine, and British warships were still steam powered. Oil was used to drench the coal used to operate the boilers, thus increasing the efficiency of the system. Philby was given a warm acceptance by Ibn Saud and over time Philby began to believe that a united Arabia would only succeed if Ibn Saud ruled as supreme leader of the country. Philby's landmark journey from the Persian Gulf across the Rub Al Khali established the feasibility of Ibn Saud's plans to destroy the Hashemites by attacking the Hejaz from the desert.

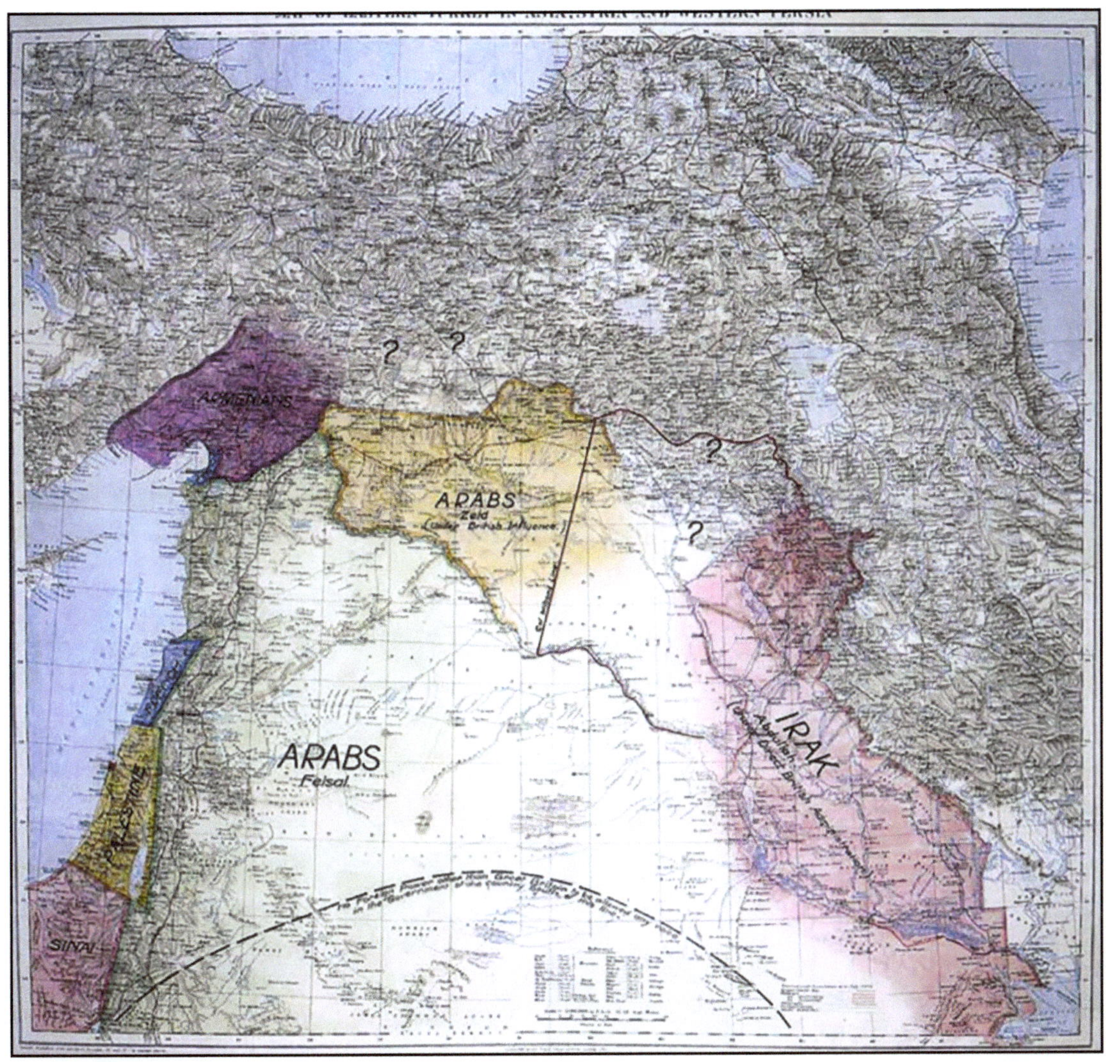

Figure 67 - *Map presented by T.E. Lawrence to the Eastern Committee of the British War Cabinet, 1918, showing his proposed lines of division for the Middle East. The final official division of states was very different.*

POST WORLD WAR I UNIFICATION OF THE ARABIAN PENINSULA

Figure 68 - *Harry Philby in Riyadh.*

Figure 69 - *A map of Arabia made by Harry St. John Bridger Philby of his exploratory journey from the Persian Gulf across the Rub Al Khali and to Jidda. This journey established the feasibility of the approach planned by Ibn Saud to cross the desert and to overwhelm the Hejaz in a surprise attack from the desert. This image is used with the permission of the Royal Geographic Society.*

Figure 70 - *Harry S.J. Philby. Photograph used with the permission of the Royal Geographic Society.*

Figure 71 - *Harry S.J. Philby in the desert. Photograph used with the permission of the Royal Geographic Society, London.*

World War I was over, and the Ottoman Empire was destroyed. Sherif Hussein had failed in his attempt to become the king of all Arabs. The time had come for Ibn Saud to pursue his ambitions of peninsular unification. All of his major impediments had been eliminated.

- Al Hasa is an oasis region in eastern Arabia located 60 km from the Persian Gulf. It was an important region because of the abundant freshwater springs, and extensive agriculture, which had existed there for millennia. In 1913, Ibn Saud conquered this region, expelled the Turks, and annexed it to his growing kingdom. He also appropriated portions of the land of Emir Sheikh al Sabah to Nejd. This led to a lengthy border dispute that was, after much disquiet, settled in favor of Ibn Saud in 1922 by British negotiators in an agreement known as the Uqair Protocol.

Figure 72 - *Political and military events which occurred on the Arabian Peninsula, 1914 to 1926. Many of these events are described by Dr. George Kheirallah, a personal observer close to King Ibn Saud, in his book Arabia Reborn. The light green lines show the movement of the troops of the Arab Revolt under the leadership of the Hashemite ruler of the Hejaz, Sherif Hussein of Mecca during World War One. The blue arrows indicate the campaigns and battles fought by the Saud army against the Hashemites after the war.*

- Al-Karj is located 77 km south of Riyadh. Natural springs made the region an important site of agriculture. Its inhabitants had been refractory opponents of the predatory aspirations and nationalistic activities of the Saud family since the 18th century and were among the last on the peninsula to fall under their rule. The freshwater wells there were especially important to Ibn Saud, and he eventually conquered the region. He established his Arabian horse breeding program at the wells.

- The Battle of Turabah was fought in 1919. The *Ikhwan* had been conducting border raids against the Hashemites in Transjordan, and at Turabah they attacked and defeated the forces of King Hussein and his son Ali. Following Turabah they carried out raids against the two sons of the Sherif, Abdullah in Transjordan and Faisal in Iraq. The *Ikhwan*, indoctrinated by Ibn Saud, had a venomous hatred of the Hashemites, who they considered to be corrupted Muslims. Turabah was notable in that the British government, alarmed at the rapacity and rapid progress of the *Ikhwan*, formally instructed Ibn Saud to advance no further toward the east.

- The year 1919 brought with it a severe blow to Ibn Saud. His eldest son Prince Turki and trusted field commander died from the 1919 influenza pandemic. Ibn Saud's wife Juahara died also.

- The Asir region was difficult to control and difficult to conquer, being composed of a multiplicity of minor sheikhdoms. Sheikh Seyyid Mohammed of the Idrisi was a powerful leader of the scattered nomadic tribes and warriors in Asir. He initially opposed British involvement in peninsular affairs during World War I, siding with the Turks. However, in 1915 he reversed course and signed a protection treaty with the British, whose interest in the area consisted of the critical need to maintain control of the port of Aden. The re-coaling depot at Aden was essential to the maintenance of the British shipping activities between India and the Mediterranean. The Idrisids fought on the British side against Turkish forces until the end of the First World War.

- The Tuhamat Asir, wishing to avoid bloodshed, voluntarily yielded power to Ibn Saud in 1920.

- The conquest of the Asir region occupied the interest of the *Ikhwan* under Ibn Saud for two years, from 1921 to 1922. Six thousand Nedji *Ikhwan* and four thousand Bedouin troops were needed to defeat Ibn Aidh, the chief of several competing rulers in the Asir. His was finally surrounded at his fortress at Harmalah where he surrendered.

- Sheikh Seyyid's capital was Sabia, which was little more than a village of huts. Sabia was located in the southern Asir, north of the coastal city of Jizan. His authority covered a small inland area, but the Turks were never able to successfully subdue the Sheikh, and he remained independent until his death in 1920. Ibn Saud eventually absorbed the area under the Treaty of Taif in 1934.

Figure 73 - *Abdullah bin Mut'ib, Hammud Emir of Jabal Shammar from 1910 to 1920. His successor, Abdullah bin Mut'ib, surrendered to Ibn Saud.*

- The conquest of Ha'il took place from 1920 to 1921. Being a walled city with defensive towers, it could only be taken by a prolonged siege. Ibn Saud delegated this task to his brother and his son, and with little bloodshed, the town eventually capitulated. The 10,000 Ikhwan warriors involved in the siege had to be restrained by Ibn Saud, lest they massacre the inhabitants, an action that could have had future foreign diplomatic consequences.

- After the fall of Ha'il, Ibn Saud contracted a case of erysipelas, which affected his face and left eye. The disease was left untreated, resulting in the loss of vision in his left eye and disfigurement of his noble appearance.

- The Sultanate of Nejd was created in 1921 as a result of the fall of Ha'il. Ibn Saud consolidated his power and transformed the scope and influence of the Emirate of Riyadh. He declared himself Sultan over Nejd and its dependencies. In 1924 Ibn Saud held a meeting with the *Ikhwan* in which he declared a jihad against Sharif Hussein and his clan. The *Ikhwan* accepted the assignment with alacrity.

- The Ta'if incident of 1924 consisted of a battle between the Hashemite forces of Hussein and the *Ikhwan*. The Hashemites, led by Hussein's son Ali, had captured Ta'if in 1916 as part of the Arab Revolt. The Hashemite controlled city was subsequently attacked by the Saud-backed *Ikhwan* led by Sultan bin Bajad and Khaled bin Luwai. Ali, defending the city in the company of Sabri Pasha, fled when faced with the *Ikhwan* army, leaving the city, abandoning it to its fate. The *Ikhwan* forces entered the city unopposed and, in a revenge massacre, killed several hundred of the unarmed city residents. After a second bloody battle at Al-Hada pitted the Hashemite Ali against the *Ikhwan*, the fate of the Hejaz was sealed.

- In 1925 the *Ikhwan* had approached close to Mecca, and Sherif Hussein fled, leaving the city in one of the few motor cars in the Hejaz, carrying with him much of the English gold that he had accumulated. He first went to Jeddah, and then to Aqabah. Although he intended to make a stand in Aqabah, the British government insisted on his complete exile from Arabia. He was informed by official channels that "HMG insists that you depart (Aqabah) in three weeks." HMG did not favor Al Saud in the process but could see that any attempt by the Hashemites to retain power would inevitably result in further bloodshed. This, HMG decided, was not in the best interest of Great Britain's economic plans for the region. In 1925 Hussein was removed from Aqabah against his will on board the British light cruiser Delhi and sent to forced exile in Cyprus. He told the Arab public that his only mistake was in trusting the British. Hussein declared his son Ali King, and Ali withdrew to the port of Jiddah, which he defended for a year, with his garrison of two thousand men. They were surrounded and besieged by forty thousand *Ikhwan* fighters. Ali, the last King of the Hejaz, surrendered to the *Ikhwan* on 12-22-25 and was evacuated, aided by British diplomatic intervention, on the HMS Cornflower. He was taken to Aden and then to Iraq He had no subsequent role in politics, and died there in 1935.

- The final assault on the heart of the Arab world was a delicate and sensitive matter for Ibn Saud. In an attempt to mollify the pro-Hashemite loyalists, the *Ikhwan* entered Mecca dressed as civilian pilgrims, as if on the hajj. This tempered the hostility of the Meccans and moderated the response of the British, who retained an interest in the final outcome of the campaign.

Figure 74 - *Ali Bin Hussein (1879-1935). Ali was the eldest son of Hussein, the Sherif of Mecca.*

- Ibn Saud delayed entering the city himself for two months. He did so as an unarmed civilian on pilgrimage. His timing was critical to his favorable acceptance by the international community as the legitimate ruler of the Hejaz, and then of Arabia itself. He was becoming a diplomat, leaving the bloody peninsular battles far behind him.

Figure 75 - *The Mosque of the Prophet in Medina, Al Masjid al Nabawi. The green dome houses the tombs of the Prophet, Abu Bakr, and Umar. The Mosque remains a holy site of prayer and pilgrimage for the world's Muslims*

Figure 76 - *Medina in the late 19th century. The Mosque of the Prophet in Medina, Al Masjid al Nabawi. The green dome houses the tombs of the Prophet, Abu Bakr, and Umar. The Mosque remains a holy site of prayer and pilgrimage for the world's Muslims.*

- The last hurdle for the *Ikhwan* was the siege of Medina, held by a small group of men still loyal to the Sherif. After resisting for ten months, the city surrendered to Ibn Saud's son, the fourteen-year-old Muhammad. The Saud victory consolidated his grip in the peninsula and the process of unification was complete. The *Ikhwan* were intent on taking the Holy City by storm, but Ibn Saud, mindful of his international standing and reputation, forbid it. The *Ikhwan* entered Medina facing little resistance. True to Wahabbi theology, they viciously destroyed all holy site and relics, forbidding the Muslim inhabitants to kneel and pray at the tomb of the Prophet. Ibn Saud would not allow the *Ikhwan* to destroy the tomb itself, aware of the delicate sensitivity of the Hashemite people and the world of Islam at large to the barbarous activities of the *Ikhwan*. Saud was treading a fine line, and he was now in full view of Western diplomats and journalists. He had thus far avoided attacking cities with foreign embassies, and he did not want photographs of the *Ikhwan* warrior's activities to be spread around the world.

- Having gained control of Mecca with little bloodshed, Ibn Saud issued the following edict to the people: "My greatest concern will be the purification of the holy area of the enemies of the faith who hate the Islamic world, namely of Husain, his children, and his followers." Ibn Saud was neither vague about his justifications nor modest about his reasons for conquering the peninsula and vanquishing the descendants of the Prophet from Arabia. He claimed that the Hashemites were corrupt Muslims. His authority in making this pronouncement was he, himself, alone. He was convinced that Allah was his benefactor, favoring the Al Saud family in their endeavors to purge the peninsula of blasphemy.

The *Ikhwan* eventually become a problem for the emerging statesman Ibn Saud. After 1926, Saud controlled all of the land that he wished to control, and he was content to consolidate his power and to create a stable government for his new country. He was becoming conscious of the need to appear to the outside world as a stable and reasonable leader. He wished for no more bloodshed, and he urgently needed to quell the ferocious rapacity of the *Ikhwan*. His growing reputation in the Arab world as an intolerant sectarian required his attention. However, the *Ikhwan*, the key to his earlier success, were intent on pursuing the infidels into surrounding regions. Their goal was not just to destroy non-Arabs, but to annihilate every Arab who did not accept their strict Islamic interpretations of the Qur'an and conduct their lives accordingly. Ibn Saud forbade them from doing this, having achieved his goal of uniting the peninsula as the country that was now called Saudi Arabia. He did not wish for the *Ikhwan* to expand their reign of terror and create international resistance to his kingship. The conflict over this issue could not be resolved peacefully. Two *Ikhwan* revolts against Ibn Saud took place.

On March 29, 1929, the matter was settled at the Battle of Sabilla. The northern *Ikhwan* Army, led by Faisal al-Dawish, now in rebellion against Ibn Saud, fought the last camel-mounted battle to be seen on the peninsula. They rode into battle against their King armed with lances and swords. The *Ikhwan* shunned modern technology and mechanization which they considered to be decadent, degenerate, and anti-Islamic.

Ibn Saud, however, had machine guns. He did not consider the modern machines of warfare to be anti-Islamic. And he used them. A lesser known fact was that the Royal Air Force actively took part in aerial attacks on the *Ikhwan* troops on numerous occasions. They did so not in support of Ibn Saud but rather to protect British interests in Transjordan and Kuwait. The British government did not formally support Ibn Saud but it did recognize his authority as King. Al-Dawish was severely wounded during the battle and Sultan bin Bajad escaped. The few surviving rebels attempted to harass the Saud army, but Ibn Saud surrounded the last pocket of resistance in January 1930, forcing their surrender. The leaders, ad-Dawish and Nayif, were imprisoned in Riyadh. They died the following year. Sultan bin Bajad was killed in 1931.

Figure 77 - *King Ibn Saud's bodyguard, early 20th century, Library of Congress collection.*

Figure 78 - *Abdullah bin Saud, Abdulaziz bin Abdul Rahman bin Faisal bin Turki bin Abdullah bin Muhammad bin Saud, 1876-1953. Photograph used with the permission of the Royal Geographic Society.*

Even after the *Ikhwan* had been disbanded, the rule of Ibn Saud was not entirely unchallenged. Scattered pockets of resistance continued among the smaller Arab tribes for several years. In 1933 a dispute arose over complaints from Imam Yahya regarding the border between Arabia and Yemen. On this occasion, Ibn Saud sent his son Faisal to settle the matter. He approached Yemen following the Red Sea coastal road, armed with machine guns, artillery, and armored cars. Imam Yahya promptly ceded the regions of Najran and Yam without further fighting.

Ibn Saud declared himself King of a united Arabia in 1932. His coronation was supervised by the British adviser Harry St. John Bridger Philby. Ibn Saud chose to continue the use of the *Ikhwan* flag to represent his new country; he changed the color of the background of the flag to green, the color most sacred to Islam. The reinvention of Ibn Saud had begun.

He set up his court in the Murabba palace in Riyadh and ruled his new nation from there until his death in 1953. He remained an adherent of the strict and conservative form of Islam; he considered this to be the only force with sufficient strength to bind the Arabian people together as a nation.

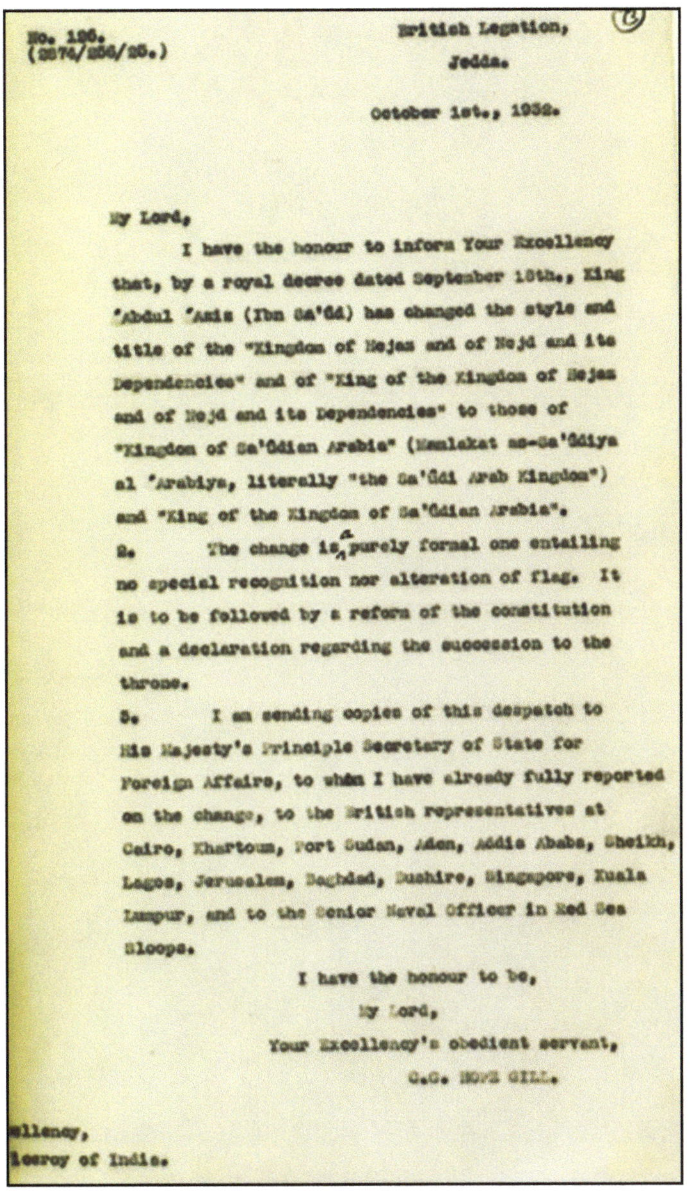

No. 196.
(2874/256/26.)

British Legation,
Jedda.

October 1st., 1932.

My Lord,

I have the honour to inform Your Excellency that, by a royal decree dated September 18th., King 'Abdul 'Aziz (Ibn Sa'ûd) has changed the style and title of the "Kingdom of Hejaz and of Nejd and its Dependencies" and of "King of the Kingdom of Hejaz and of Nejd and its Dependencies" to those of "Kingdom of Sa'ûdian Arabia" (Mamlakat as-Sa'ûdiya al 'Arabiya, literally "the Sa'ûdi Arab Kingdom") and "King of the Kingdom of Sa'ûdian Arabia".

2. The change is a purely formal one entailing no special recognition nor alteration of flag. It is to be followed by a reform of the constitution and a declaration regarding the succession to the throne.

3. I am sending copies of this despatch to His Majesty's Principle Secretary of State for Foreign Affairs, to whom I have already fully reported on the change, to the British representatives at Cairo, Khartoum, Port Sudan, Aden, Addis Ababa, Sheikh, Lagos, Jerusalem, Baghdad, Bushire, Singapore, Kuala Lumpur, and to the Senior Naval Officer in Red Sea Sloops.

I have the honour to be,
My Lord,
Your Excellency's obedient servant,
G.G. HOPE GILL.

...llency,
...iceroy of India.

Figure 79 - *October 1, 1932. Official dispatch to the British Foreign Office notifying them that Ibn Saud had merged Nejd and the Hejaz into one political unit, denominated himself King, and changed the name of the country to Saudi Arabia. This dispatch was made by the British Legation in Jeddah and sent to Sir John Simon, Foreign Secretary in London. From the Qatar Digital Archives. It is an irony of history that the change should be regarded as "purely formal".*

Figure 80 - *Flag of Saudi Arabia based on the iconography of the Ikhwan banner. The message of the Shahada, along with the sword, summarizes the high Saudi Arabian ideals. The green color was a reference to the color traditional to Islam. The banner of the first Fatimid Caliphate, used until 1171, was green.*

Figure 81 - *The Murrabba palace outside Riyadh, the first palace of Ibn Saud, Abdulaziz bin Abdul Rahman bin Faisal bin Turki bin Abdullah bin Muhammad bin Saud, 1876-1953. Photograph from the 1930s by Gerald de Gaury. Photo used with the permission of the Royal Geographic Society.*

Figure 82 - *The Murraba Palace, Riyadh. Image used with the permission of the Royal Geographic Society, London.*

The desert warrior was faced with a new dilemma. He was now the leader of a nation among the nations. He felt the responsibility of guiding his people into a new age. The Western world was particularly critical of the new King in view of the harsh methods that were used in the unification process. Arabic laws and punishments were judged by Western standards to be cruel and barbaric.

Ibn Saud used the press effectively to counter the criticisms of the Western world. He knew that to obtain legitimacy in the eyes of the world's governments, he must explain and justify his actions:

> "Bedouin have to be treated in a very hard way for only then do they learn their lesson, and ... once they have learnt it they will never forget it. We teach them the hard way, not to be cruel, but out of mercy. And once we have punished them we shall not in the mercy of Allah have to do it again as long as we live."

Figure 83- *Bedouin horsemen, from The Raswan Archives.*

"We have heard that your government favors light punishment. Do not you build large houses where you keep your evil-doers? Have you many of these houses? And are they not full? Well, that, to our mind, is what is cruel. You punish year in and year out and yet it does not seem to be effective. I punish in such a way that it has not to be repeated."

"You have also doubtless heard many stories about the fanaticism of the Wahhabis. It is good that you should know the truth about our creed and about our brothers. We believe that Allah The Exalted One uses us as His instrument. As long as we serve him we will succeed,

> no power can check us and no enemy will be able to kill us. Should we become a useless weapon in His hands then He will throw us aside, *wa sanahmiduhu*, and we shall praise Him."

Ibn Saud's power and success was the result of his belief in divine guidance. He was sustained by a profound belief in the divine guiding force of Allah that never left him.

The asil desert horses that survived the military and civil upheavals of the era were now in the hands of scattered and generally unlettered Bedouin. After 1932, with the peninsula unified, the numbers of independent nomadic tribes began to diminish. Most Bedouin had become settled city dwellers. Among the surviving traditional tribes, however, desert horse breeding continued.

Figure 84 - *The natural wells at Al Kharj in the 1930s, prior to the modernization. Image used with the permission of the Royal Geographic Society, London.*

The Muteyr of Nejd were known for their excellent quality Krush and Dahman horses.

The Utaybah, originally from the Hejaz, migrated to the central deserts. Some of their horses had been acquired by the Blunts (Hadban) and Ali Pasha Sherif (Shueyman).The Dhafeer of southern Arabia continued to breed pure desert horses and contributed the mare Donia to Egyptian breeding

The Harb was a large tribe in the Hejaz, who bred very good quality desert horse stock. Ali Pasha Sherif obtained the stallion Harkan, a son of Zobeyni, from this tribe.

The Sba'ah tribe was a prime source of pure blood for buyers in the 19th century. They provided Kesia for Upton, and Queen of Sheba for the Blunts. Unfortunately, no durable records of pedigree of the horses obtained from the Bedouin remain from this period.

Figure 85 - *Saudi mares. Horses at the stud of Ibn Saud at Al Kharj. Photograph taken in the 1930s, by Gerald de Gaury, used with the permission of the Royal Geographic Society.*

As the peace of Saud became the law of the land, the King's interests turned to the past. He began to collect *purebred* desert horses, maintaining and breeding them at his newly established stud at the wells of Al Kharg. George Khierallah, who knew King Ibn Saud well, wrote in 1952 that the King kept large stables at Al Kharj, carefully tending to the welfare of his pedigreed Arabian horses. Kheirallah found the horses to be entrancing. He wrote that the Arab horses were the most graceful and intelligent persons-not horses-in the world. They were part of the Arab family. Ibn Saud, he wrote, was a true Nedji, and he delighted in his horses.

Figure 86 - *The last horses of the Managia strain, central Arabian Desert, 1930. Photograph by Gerald de Gaury, used with the permission of the Royal Geographical Society, London. The Darley Arabian, a foundation of the English Thoroughbred, was from the Managia strain.*

The origins of the horses that were used to create the horse breeding stud of Ibn Saud at the Al Kharj wells in Nejd are unclear. This fact is noteworthy since any assessment of the foundational matrilines of the RAS/EAO involves this question. Several important RAS/EAO/Inshass mare lines came from Ibn Saud, gifted to the King of Egypt. These mares were El Kahila, Mabrouka, Hind, and Nafaa.

There are reports that Farhan al-Ulayyan, a slave of the Fad 'aan, was tasked with the job of securing *purebred* desert horses of the first quality from the remains of the impoverished Anazah, mainly from Syria. It is reported that he obtained these horses for the stud of the Saudi King.

Figure 87 - *King Ibn Saud's breeding farm at Al Kharj, 1930s. Photo by Gerald de Gaury, used with the permission of the Royal Geographic Society, London.*

Ibn Saud often used these horses and their progeny as diplomatic gifts to regional governments and nobility. Several of these horses were gifted to the Egyptian King during the first half of the 20th century and were added to the Inshass Stud in Cairo. Some of these horses were later incorporated into the breeding program at the EAO at the time of its creation in 1952.

The Saudi mares that left durable matrilines at the EAO in Cairo were:

1. El Kahila, a dark bay mare, born in 1921, a Kuhailan Krush

2. Mabrouka, a bay mare, born in 1930, a Saklawiyah

3. Hind, a grey mare born in 1942, sired by an Obeyan el Saifi stallion out of a Saklawi mare

4. Nafaa, a grey mare born in 1941, sired by an Obeyan el Saifi stallion, out of a Koheilan mare

Saudi gift mares that were given to the Egyptian King that left no matrilines or patrilines at the EAO include the mares Durra, Rezkia, El Galabi, and the stallions Mabrouk, Saadaa, and El Zarka.

Figure 88 - *A group of horses at the stud of Ibn Saud. Image by Gerald de Gaury, a British agent in Arabia in the 1930s. Photo used with the permission of the Royal Geographic Society, London.*

Figure 89 - *A group of fillies at King Ibn Saud's horse breeding farm. Photo by British agent Gerald de Gaury in the 1930s, used with the permission of the Royal Geographic Society.*

Figure 90 - *A group of fillies at the stud of Ibn Saud at Al Kharj. Photo by Gerald de Gaury, British agent, from the 1930s, used with the permission of the Royal Geographic Society.*

**I say its a bull, he says I should milk it.*

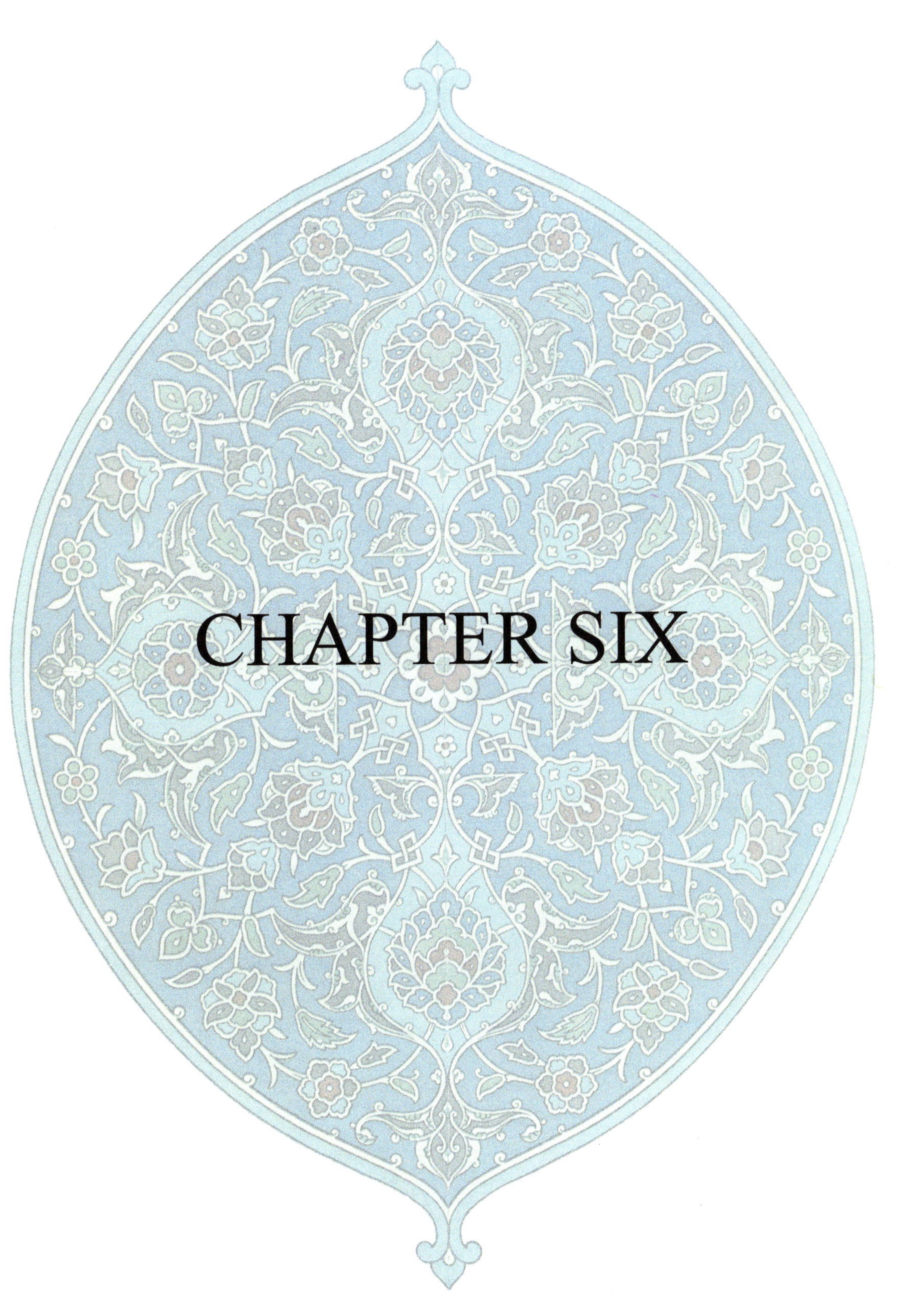

CHAPTER SIX

CHAPTER SIX

THE ASIL ARABIAN HORSE IN EGYPT

The purpose of reading history is, as Herodotus found, to draw lessons from human actions, to examine the motivations of these actions, and to learn the consequences of these actions. History instructs the reader about the responsibilities that personal actions entail. History teaches the reader to determine what is permanent and enduring, and what is fleeting and transitory; what is meaningful and what is trivial. For Herodotus, the central theme of history was the human flaw of *hybris*, "outrageous arrogance." He also focused on *ate*, "moral blindness." These factors, he wrote, were the cause of much human suffering. This was so in Arabia.

From 1818 forward, it was difficult to reliably trace a living Arabian horse pedigree to its origin in the Nejd or elsewhere in Arabia. The devastation and chaos of the Saud-Ottoman War brought down a curtain on reliable documentary connections with the past centuries of horse breeding in the Nejd. The political instability and warfare among the desert princes of the 19th century became an obstacle to organized *purebred* horse breeding. The vast majority of incontestably *purebred Arabian horses* were now to be found in Egypt, where political and domestic life was more settled. Nejdean horse breeding never recovered from the constant clan strife and warfare that splintered Arabia.

Figure 1 - *Mounted Arab Charging, by Horace Vernet, used with the permission of the Princeton University Art Museum.*

The greatest number of *purebred Arabian horses* was now in the hands of the Egyptian royal family. The Dynasty of Mohammad Ali the Great of Egypt became the conservators of the desert breed in exile, and many of the horses that they collected from Arabia and bred in Cairo were used to form the nucleus of the RAS/EAO. The history of the EAO and its horses is intimately connected with the activities of the members of the Egyptian Royals.

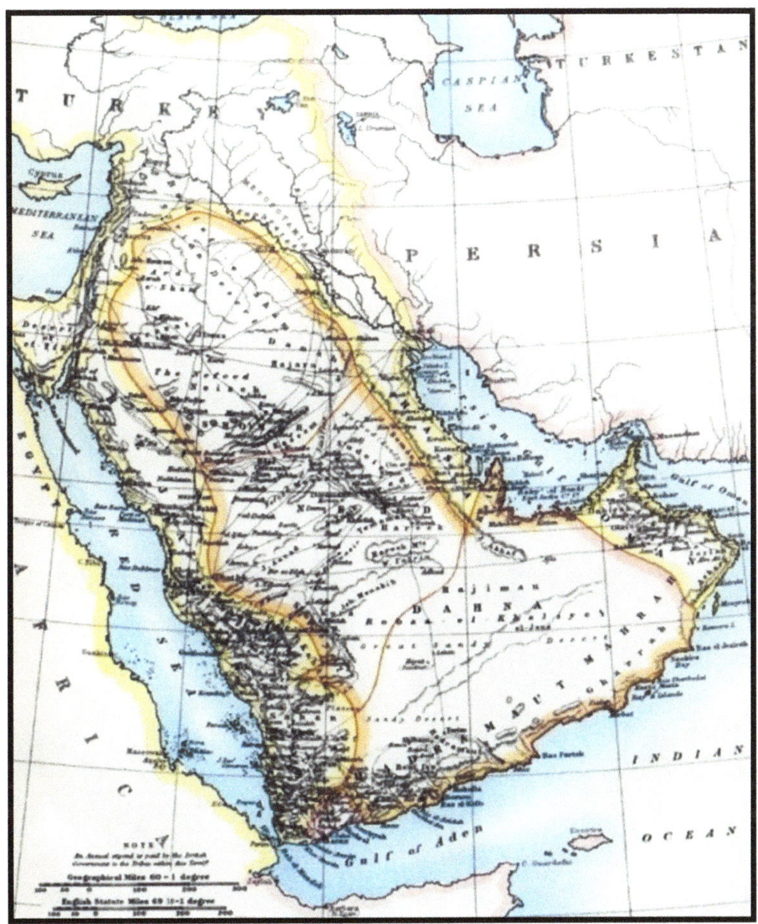

Figure 2 - *By 1840, Arabia had been thoroughly explored by Europeans, and much was known about the inhabitants and their horses. The details of the interior of the peninsula, so long unexplored and uncharted, were gradually becoming known to mapmakers.*

In 1848, the Wahabbi nemesis Ibrahim Pasha died, but by then the peninsular incursions of the Egyptian army had long since ended; Nejd and the central Arabian Deserts had been the homeland of the authentic desert Arabian horse for centuries, but the Nedjean horse diaspora was now nearly complete. The last half of the 19th century was to be occupied by the endless struggle between the Rashidis in Hail and the Sauds in Riyadh for absolute dominance on the peninsula. Horse breeding suffered.

The practice of obtaining high caste Arabian horses from the desert tribes was becoming more difficult. Authentic *purebred* specimens were rare. Ibrahim Pasha's successor and the son of Tousson was Abbas Hilmi I, Wali of Egypt and Sudan, Hejaz, Morea, Thasos, Crete (reign 1848-1854). During his lifetime, Abbas Pasha systematically stripped the Arabian Desert tribes of many of the few remaining first-quality Arab horse stock. Abbas established two breeding farms near Cairo for his constantly expanding herd, Dar El Beyda, and Abbassieh.

Figure 3 - *Statue of Ibrahim Pasha in Cairo, in the Opera Square, early 20th century, from the collection of the Library of Congress.*

Figure 4 - *Aghil Aga, a Koheilan Adjuse from the Anaze Ruala, imported for the Austrian government in 1856 by Col. Von Brudermann, from a lithograph by Von Stolz, photo by Raswan. In the latter half of the 19th century, it was still possible to find and purchase horses of this quality directly from the desert tribes, but the supply was rapidly dwindling. Photo from* The Raswan Index.

Once the horse was no longer in the hands of the Bedouin, the economic value of the rapidly vanishing *purebred* horse and the inevitable issue of pedigree validation became a matter of considerable interest to those individuals involved in the trade in Arabian horses. Instances of deception and misrepresentation on the part of horse traders were frequent, and as a result, a means to legitimize and validate parentage in the commerce of Arab horses was sought.

By the middle of the 19th century, the need for formal written genealogic records became more pressing.

As a child, young Abbas was sent to live among the Bedouin, a practice common to Egyptian royalty; its intent was to harden the children for the rigors of their future occupation of rulership. Abbas Pasha had witnessed the harsh barbarity of the war with the desert princes, and later he himself became a collector of their finest quality Arabian horses. Abbas, son of Tousson and grandson of Muhammad Ali, spent many years in Arabia, in the company of his father. Abbas was in close contact with the desert Bedu during these years. He was well liked by the Bedouin, and he became intimately familiar with their customs, manners, and particularly their ideas about horse breeding. This personal contact early in life served as his guide in the horse collecting of his later years.

Maturing to the title of Viceroy of Egypt, Abbas began breeding *purebred Arabian horses* on a large scale. His stud was located at Heliopolis and was modeled on the modernized methodology of the Shoubra stud of Mohammad Ali. Disease was minimal and mare fertility was above average. Breeding records were kept. The practice of shackling horses was abolished.

Later in life Abbas built an even more palatial stud in the desert, about half way between Cairo and Suez. This was the Dar el Beyda stud, known for its marble walls and enclosures, an extravagant work of art, the likes of which the world would never see again.

Abbas often sold Arabian horse stock to Europeans. Major Tweedie wrote that the Viceroy had over 1,000 horses at his stud near Cairo. In 1840, a sale was held at his stud at Dar el Beida. 110 mares and 180 colts were sold. The governments of France and Australia were among the principle buyers. Abbas I recognized the importance, indeed the necessity, of providing his European buyers with documentary evidence of the Arab horse's pedigrees; buyers had begun to insist upon authenticity, and Abbas Pasha took great pride in the genetic purity and authenticity of his asil horses. But he needed written documents to prove this fact to increasingly skeptical European customers.

To demonstrate his commitment to the purity of his horses, as well as to enhance his standing in the eyes of the Europeans, he compiled his personal research of the pedigrees of the horses that he had obtained from the Bedouin into a monumental work known as the Abbas Pasha Manuscript. Abbas' Mameluke emissary Ali Jamal Ad-Din Ash-Shamas-hirji Bey compiled the text from his personal interviews with the tribal leaders from whom Abbas had wrested the finest specimens of the few asil horses remaining in the Arabian Deserts. This manuscript is a historical document of singular importance. It consists primarily of stories and familial lore told by the sheikhs to the Pasha's Mameluke about the ancient origins of the horses they knew so well.

The focus of the testimony in the manuscript is on familial recollections of the continuity of Arabian horse origins and chain of ownership, narratives that tied the horses of that day to their ancestors of the very remote past. The Saqlawi Jedran mare Ghazala Ash-Shakra, the dam of Ghazieh I, whose descendants were important elements in the formation of the RAS, is documented here.

Figure 5 - *Abbas Pasha. Reproduction from the author's collection.*

The book sheds light on the origins of many of the mares that were obtained by Abbas Pasha and bred for generations in Cairo. These mares were the forebears of many of the foundation stock that were used to form the RAS. They include Ghazza, Ghazieh, Hajla, Harka, Jathimah, Jellabiet Feysal, Miskah, Samha, Shalfa, Shueyma, Sueyd, and many others. More on this topic can be found in the Abbas Pasha Manuscript by Forbis and Sherif. It is a work essential to the serious student of the Arabian horse.

Lady Anne Blunt was allowed to see the original copy of the manuscript at her Sheykh Obeyd in Egypt. The book had been found in the effects of the deceased El Hami Pasha and came into the possession of his daughter. The Pasha's daughter knew nothing about the contents of the book, but knowing of Lady Blunt's interest in these horses, offered her the opportunity to see it. There were several volumes of the book, each bound separately, and they were each about 6 inches by 9 inches. Lady Anne perused a few of the volumes, noting the names of Feysul Ibn Turki, Ibn Sbeyni, and other noted Arabian breeders and Arabian horse strains. She surmised that since she was viewing the original, then the version that she had seen in the possession of Ali Pasha Sherif a few years earlier must be a copy of the original.

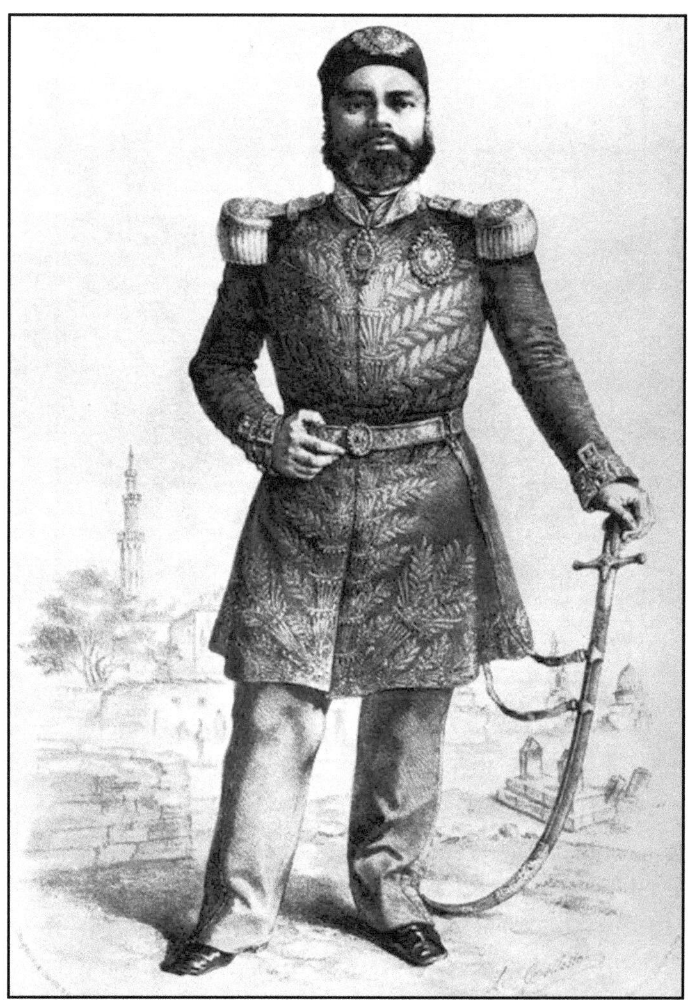

Figure 6 - *Abbas Hilmi I Pasha.*

The increased focus on written documentation was attributable to the growing awareness of authenticity on the part of European buyers and agents. Western buyers insisted on being provided with plausible and reliable ancestry of horses that they wished to buy. The Arab horse pedigree came to be required by the European horse trader (often acting as intermediary) as well as the European horse buyer. The barbarity of the Ottoman wars had annihilated the core of first quality *purebred Arabian horses* from the Nejd, leaving only small pockets of desert purists who had been able to retain a few horses of the first rank. In particular, the Khalifa family, rulers of the small island of Bahrein, were spared the destruction of the Ottoman wars and continued to preserve asil horses, most notably the Jellabiyat.

Figure 7 - *Ali Pasha Sherif.*

Despite the obsessional desire of Abbas to create a collection of Arabian horses unrivaled by anyone, he was, in the end, thwarted by geography. The horses of Abbas Pasha became caught up in the destructive forces of the disease-ridden ecosystem of the Nile River valley. By the time of Abbas' demise, few horses were left alive. Many asil horses had been taken from the deserts of Nejd to Egypt, but few had survived.

The Abbas Pasha herd was inherited by the Viceroy's eighteen-year-old son, El Hami Pasha, who had no interest in them. A total of 480 horses were disposed of after the death of Abbas. Most of the animals died from disease and neglect.

Seven years later, thirty of these horses were bought by a young nobleman by the name of Ali Pasha Sherif, who was able to breed large numbers of first quality horses using the stock descended from the Abbas Pasha collection.

Ali Pasha Sherif (Ali Bey) was the son of an Egyptian career diplomat who was at one time the governor of Syria. Early in his career Ali Bey was an artillery colonel in the Egyptian Army of Mohammed Ali. He was an ambitious, capable, and intelligent man who soon became an important administrator in the government of Mohammed Ali, serving as Prime Minister and later as Wali of El Sham and Arabistan (Syria and Arabia). His offices brought him into contact with the horse breeding tribes of the central Arabian Desert, and he developed a strong affinity for the Arab horse. Using the horses obtained from the estate of Abbas Pasha, he began breeding pure-blooded Arabian horses. At its peak, the herd was said to consist of over 400 horses. He added more horses to his herd from the dispersal sale of the horses of Khedive Ishmail in 1878.

In 1870, a pandemic of African Horse Sickness swept through Egypt, threatening this last small nucleus of *purebred* desert stock. Ali Pasha Sherif sent his horses to the safety of Upper Egypt and the majority of them survived. However, as time and age bore down on the Pasha, the core of his *purebred desert Arabian horses* was nearly lost, due to years of intensive inbreeding, disease, and neglect. The Pasha himself suffered from increasing mental instability and diminishing wealth. On the death of Ali Pasha Sherif in 1897, the residua of a once prolific race of horses were sold at auction. His efforts had not been in vain. The commitment of the Pasha to these horses ensured that the blood of the pure Nedjean horse would be preserved for posterity.

It is ironic that the country that preserved the desert horses of the 19th century for posterity should have been such a hazardous and unhealthy place for horses to live. The problem in Egypt was the abundance of groundwater and the inevitable consequence of standing groundwater: insects.

Central Arabia had been an ideal environment for the health of horses. The central Nedj was a rocky wilderness, located at high elevations, and was normally very dry, and with little standing groundwater. Egypt was ecologically different. The Nile Valley was near sea level, and due to the annual floods and crop cultivation, standing groundwater and damp soil was abundant. Insects were plentiful. The refuse of densely packed human habitation provided the waste material on which insects thrived.

African Horse Sickness, also known as *pestis equorum*, is an insect-borne disease of equines and is endemic to North Africa. It is caused by an *Arbovirus*, a large and varied group of viruses. The pathogen is from the genus *Orbovirus*, a member of the family *Reoviridae*. The virus contains double stranded RNA and is composed of 18,000 nucleotides. These viruses are spherical and are about 150 nm in diameter. There are 22 species of the *Orbovirus*.

There are 4 families of *Arbovirus*. The Togaviridae family causes VEE, WEE, and EEE. The Flaviviridae family causes Yellow Fever and West Nile disease.

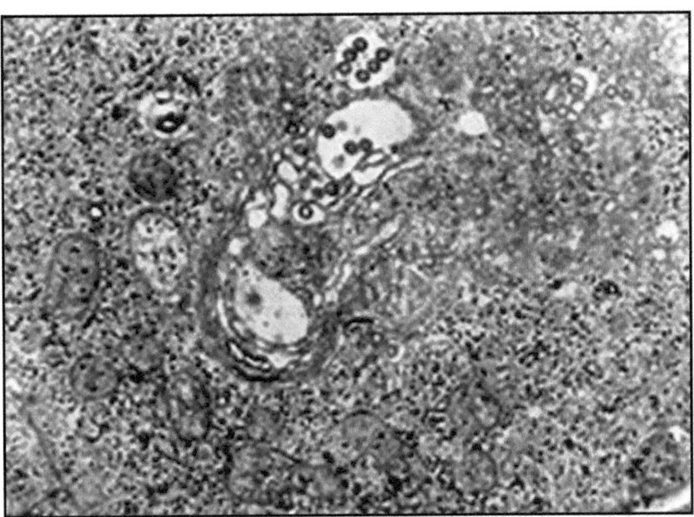

Figure 8 - *Photomicrograph of tissue taken from a victim of African Horse Sickness. The small spherical objects are viruses, and the clear areas indicated tissue necrosis, liquification and cell death.*

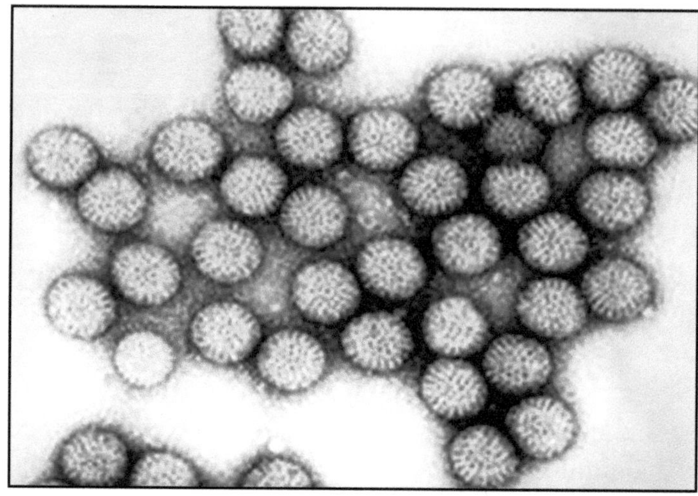

Figure 9 - *A photomicrograph of a group of Rotavirus, one of the members of the Reoviridae family.*

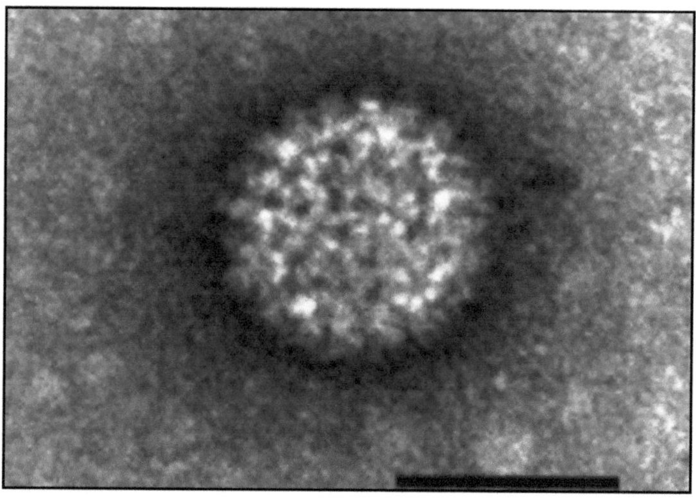

Figure 10 - *A single Orbivirus under high magnification. The scale bar represents 50 nm.*

Transmission of the virus is through the vector *Culicoides*, a genus of biting midges. The disease does not affect humans.

The main midge species involved in the transmission of African Horse Sickness is *Culicoides imicola*. Apart from serving as vectors of the *Arbovirus*, midges can also transmit the pathogens *Onchocerca* and *Plasmodium*.

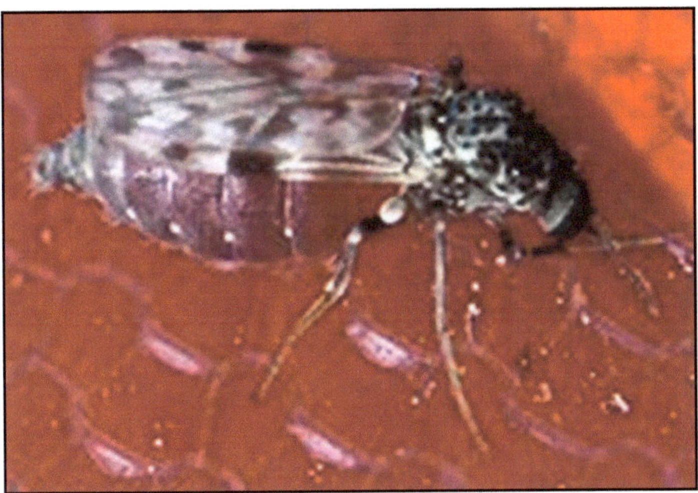

Figure 11 - *Culicoides, a biting midge.*

The first reports of African Horse Sickness are found in historical records from the 17th century when the disease was reported to be endemic in Equatorial Africa, Sub-Saharan Africa, and East Africa. Epidemics of African Horse Sickness in Egypt followed.

Culicoides feed on blood and are usually found in the vicinity of marshes, stagnant water, and garbage. Females lay their eggs in moist refuse or damp soil.

The virus is killed by cool temperatures, and over winters in dogs and camels, which serve as hosts for the virus. They seem to suffer few ill effects themselves. Sheep, goats, and buffalo are also hosts for the virus.

The disease in horses has both acute and subacute forms. In the acute phase, the animal experiences high fever and respiratory symptoms, shortness of breath and a refractory cough. Pneumonia, pulmonary edema, and inflammation of the heart and its surrounding tissue develop, and marked supraorbital swelling, pathognomonic for the disease, appear. Over 90% of animals succumb to the disease in about a week.

In the subacute form of the disease, fever and facial swelling are prominent but develop more slowly, over one to two weeks. Cardiac involvement is significant in the form of myocarditis, pericarditis, and heart failure. The mortality rate from this form of the disease is about 50%.

The differential diagnosis of African Horse Sickness includes *Equine infectious anemia*, *Equine viral arteritis*, *Anthrax*, *Trypanosomiasis*, *Piroplasmosis*, and *Purpura Hemorrhagica*.

Control of the disease consists of the immediate slaughter of affected animals and the isolation of healthy animals in *Culicoides*-free regions. Usually this means moving the animals away from human habitation, to a higher altitude where the climate is drier and cooler and the soil is less damp. Remote regions free of human habitation are considered to be an essential refuge for horses during an outbreak. This is because of the propensity of the midge to lay its eggs in human refuse.

THE EGYPTIAN ROYAL FAMILY AND THE HOUSE OF MOHAMMED ALI THE GREAT, PRESERVATIONISTS OF THE DESERT ARABIAN HORSE OF THE NEDJ

The numerous histories written about the preservation of the authentic Arabian horse in Egypt tend to place emphasis on the importance of the masterful and obsessive Egyptian Arabian horse breeders Abbas Pasha and Ali Pasha Sherif. In fact, other members of the Ali dynasty played equally significant roles in obtaining, and breeding the true asil desert horses.

The success of the Ali dynasty with its Arabian horses, important though it was, was limited, despite the grand examples of the few aristocratic horses that have been handed down to posterity. Thousands were bred, and only a handful persisted as bloodstock. The French orientalist Achille-Constant-Theodore Emile Prisse d'Avennes left this first-hand report of his observations of the studs of Mohammed Ali and Ibrahim Pasha:

> "So great is the ignorance of the Egyptians and the Turks of everything pertaining to the breeding and raising of horses that these establish-

ments failed utterly to accomplish anything of note."

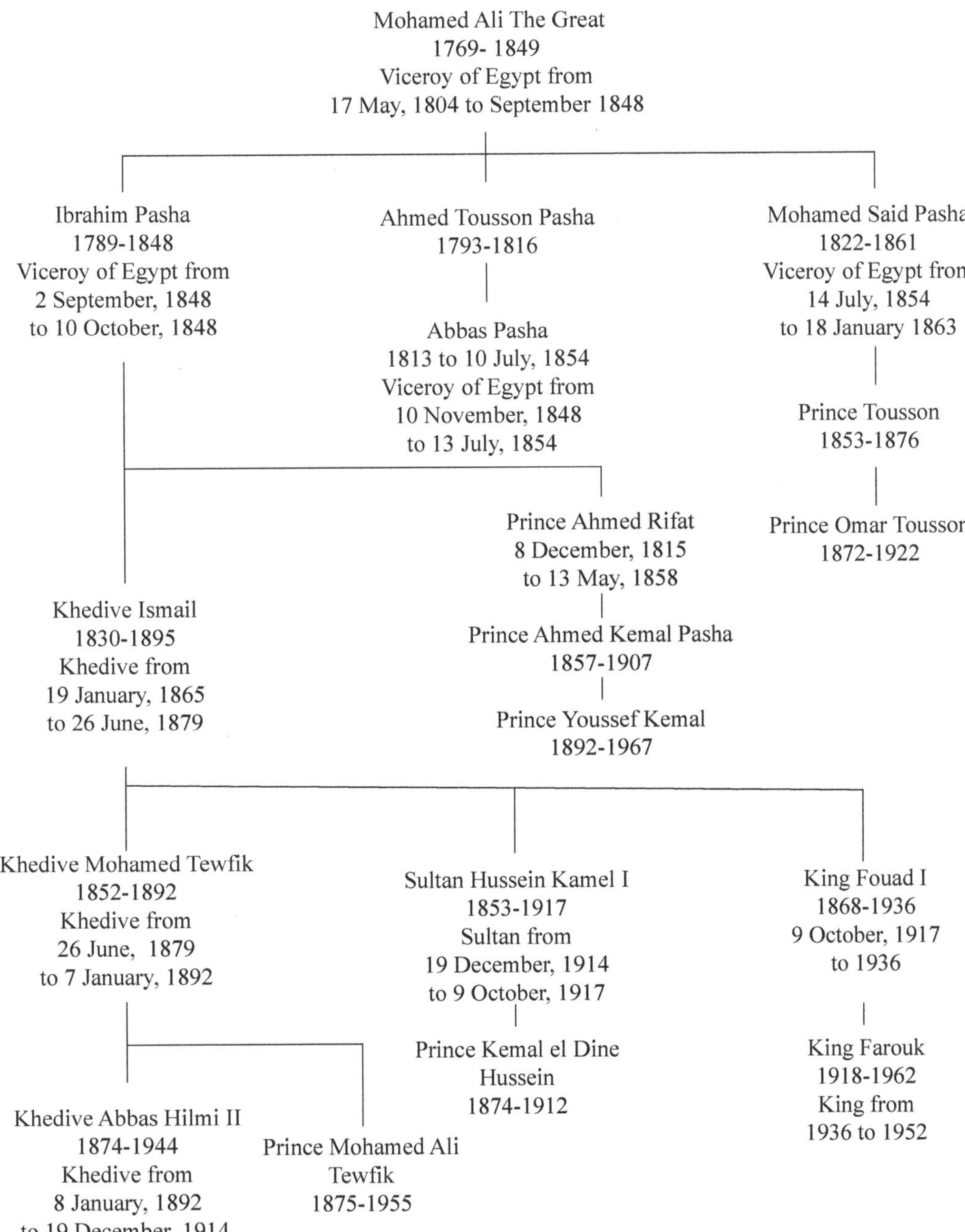
Mohamed Ali The Great
1769- 1849
Viceroy of Egypt from
17 May, 1804 to September 1848
Ibrahim Pasha
1789-1848
Viceroy of Egypt from
2 September, 1848
to 10 October, 1848
Ahmed Tousson Pasha
1793-1816
Mohamed Said Pasha
1822-1861
Viceroy of Egypt from
14 July, 1854
to 18 January 1863
Abbas Pasha
1813 to 10 July, 1854
Viceroy of Egypt from
10 November, 1848
to 13 July, 1854
Prince Tousson
1853-1876
Prince Ahmed Rifat
8 December, 1815
to 13 May, 1858
Prince Omar Tousson
1872-1922
Khedive Ismail
1830-1895
Khedive from
19 January, 1865
to 26 June, 1879
Prince Ahmed Kemal Pasha
1857-1907
Prince Youssef Kemal
1892-1967
Khedive Mohamed Tewfik
1852-1892
Khedive from
26 June, 1879
to 7 January, 1892
Sultan Hussein Kamel I
1853-1917
Sultan from
19 December, 1914
to 9 October, 1917
King Fouad I
1868-1936
9 October, 1917
to 1936
Prince Kemal el Dine
Hussein
1874-1912
King Farouk
1918-1962
King from
1936 to 1952
Khedive Abbas Hilmi II
1874-1944
Khedive from
8 January, 1892
to 19 December, 1914
Prince Mohamed Ali
Tewfik
1875-1955

Figure 12 - *The dynasty of Mohammed Ali the Great of Egypt.*

Figure 13 - *Mohammed Ali, in the 1840 portrait by August Couder.*

The beginning of the dynasty's Arabian horse obsession began with Mohammad Ali the Great himself. Ali kept his collection of Arabian horses at his stud near Shoubra. They were attended by the French veterinarian P. N. Hamont who struggled for years to overcome the disastrous farm conditions that greeted him on his arrival. The horses were diseased and fertility among the mares was very low. The Viceroy gave Hamont full authority to renovate the stable, and to introduce modern breeding conditions. The Turks who ran the stables were displeased, but Hamont's persistence eventually prevailed. Matings were done according to Hamont's judgment, and accurate records of sires, dams, and foals were kept.

By 1842 the stud contained 32 stallions and 450 mares. Although the veterinarian experimented with sires of Dongolian, native Egyptian, English, Syrian and Russian blood, he consistently found that the true Nedji sires were superior. Many of these horses were eventually incorporated into the collection of Abbas Pasha.

The splendor of the Viceroy's horses and palatial stables was known around the world. The grandeur and oriental opulence of Ali's stud was described by the traveler James St. John, who visited the fabled stud in 1832. He described the points of the best horses in the collection as proof of their authentic Nedji origins: "a fine snake head with an expanding and a projecting nostril…a remarkably small pointed ear…an eye expressive of boldness, generosity and alacrity…there is no doubt that he would be elastic, speedy and lasting…the tallest horses were 15 hands high."

The gardens of Shoubra, he wrote, were "among the finest I have seen anywhere". Ali also maintained a menagerie of deer, kangaroos, and giraffes.

Figure 14 - *The Avenue at Shoubra, mid 19th century, from The London Illustrated News.*

Figure 15 - *The Royal Wedding Processional in Cairo for the marriage of the youngest daughter of Mohammad Ali, Zeinab, to Kamil Pasha, one of the functionaries on the staff of the Viceroy, Mohammad Ali. He provided a spectacle for the world to see. The entire population of Cairo celebrated the eight days of festivities and entertainment. As part of Arabic tradition, all of the prisoners in the city jails were released for the occasion. The spectacle was designed to emphasize the power possessed by Mohammad Ali. By the time of this wedding, 1845, Ali had consolidated his power, and accumulated considerable wealth, much of which he lavished on his family, and much of which he lavished on his collection of first-class Arabian horses. Image from The London Illustrated News.*

Following in his father's footstep was the warrior Ibrahim. Ibrahim Pasha kept his collection of Nedji horses at his palace at Kasserling, the site of his palace, on the Nile a short distance from Cairo. About 400 head of horses, captured during the hostilities in central Arabia, were kept in his facilities. The veterinarian Hamont reported that the cruel and barbaric horse management conventions of Mohammed Ai were followed by the managers of Ibrahim Pasha's studs, with foals being shackled by all four legs. The live birth rate for mares was about 50%. Ineffective and primitive Turkish farm managers had absolute control over day to day affairs and Hamont's attempts to introduce modern farm management techniques were doomed by superstitious resistance. Though fond of his asil horses, Ibrahim preferred to ride mules.

Ibrahim once sold some of his land to an Arab by the name of Sheykh Obeyd, who developed it into a pomegranate orchard. The Sheykh, unfortunately, fell into debt and the land reverted to Ibrahim's possession, but the Sheyk's name remained associated with the small plot of land. Lady Anne Blunt, who later bought the grove, was told by the locals that the pomegranates that grew there were originally brought back from Jauf in Arabia by Ibrahim himself, following his invasion of the country decades earlier.

Figure 16 - *Ibrahim Pasha.*

A later line of the Egyptian Royal family tree figures prominently in the narrative of the early days of the Royal Agricultural Society.

Figure 17 - *Ibrahim Pasha, parading the troops in St. James Park London on the occasion of her Majesty's birthday. He is accompanied by Prince Albert and the Duke of Wellington, The London Illustrated News.*

Hussein Kemal (1853-1917) was one of the sons of Khedive Ishmael. While he was not an Arab horse breeder himself, his son was to become an enthusiastic champion and preservationist of the asil desert horse.

Hussein Kemal was placed in the position of Sultan of Egypt in 1914 after the English, exercising their role as "protectors" of Egypt, deposed his nephew Abbas Hilmi II, whose sentiment was anti-British, and more importantly, pro-German.

Sultan Hussein Kemal died in office three years later.

When Sultan Hussein Kemal died in 1917, his son Prince Kemal el Din inherited a number of genuine asil horses. Being an avid sportsman and hunter, his interests took him throughout North Africa in search of big game. He was an adventurer, explorer, and collector of fine Oriental antiquities and art. As a young man, he led an expedition to map the Western Egyptian desert, discovering and naming Gilf Kebir.

Figure 18 - *Khedive Ismail Pasha, (1830-1895) son of Ibrahim Pasha, kept a large stud of about 400 Arabian horses of the finest quality at his stud at Kasserling, where he lived in splendor in his palace on the banks of the Nile River, near Cairo. He ruled as Khedive from 1863 to 1879.*

Figure 19 - *The former Khedive Ismail in 1881, in Vanity Fair, image from Alamy.*

Figure 20 - *The Khedive Ishmail's Country Drive, from The London Illustrated News.*

Figure 21 - *The Khedive Ismael Pasha.*

HSH Prince Kemal El Din Hussein was educated at the Theresian Military Academy in Austria. On reaching maturity, he returned home and in 1904 married Princess Nihmet Allah, (her second marriage, his first) daughter of Khedive Tewfik Pasha. His marriage produced no issue. In 1914 he rose to the rank of Commander in Chief in the Egyptian Army. He died in 1932 in Toulouse.

The prince was heir to the Egyptian throne, but his Anglophobic inclinations led him to despise the English whose tight grip on the ruling family in Egypt was repugnant to him. He abdicated his position as heir to the throne of Egypt, writing to the government that he expressly and voluntarily renounced the succession.

Prince Kemal's fierce resistance to both the Turks and the English were matched by his passion for asil horse breeding. He maintained a select group of authentic *purebred Arabian horses* which were kept under excellent management by Dr. A.E.Branch of the Royal Agricultural Society. In contrast to other members of the royal family, he did not use stock exclusively from his famous horse breeding family. Instead, he used asil Arabian horse stock obtained from Anne and Wilfred Blunt. His horses included Rustem, Serra, Bint Serra, Rasala, Dalal, Bint Dalal, El Zafir, and Bint Zareefa.

Figure 22 - *HSH Prince Kemal ed Dine Hussein (1874-1932). His Arab horse breeding program was significant in that many of his horses were used in the early days of the foundation of the Royal Agricultural Society*

Figure 23 - *Prince Kemal ed Din Hussein.*

Dr. Branch was a key figure in the sale of the Prince's mare Bint Serra to the American Arab horse breeder Henry Babson. The Ibn Rabdan son Ibn Fayda also deserves special mention as a product of Prince Kemal el Din's breeding. The Prince devoted considerable time and energy to the preservation of the *purebred Arabian horse*. He served as president of the Royal Agricultural Society from 1914 to 1932. His judgments regarding the horses were considered beyond reproach and his objectivity in choosing the foundation stock for the RAS was widely respected.

The line of descent from Ibrahim Pasha produced another member of the Ali Dynasty who devoted much of his time and interest to the breeding of the asil desert horses through the Ibrahim Pasha son Ahmed Rifaat (1825-1858). While Rifaat had no interest in breeding horses, his son Prince Ahmed did.

Prince Ahmed Pasha (1858-1907) was an aesthete and a committed breeder, with an eye for beauty and a keen breeder's sense of planned mating. The prince kept several stables and managed large numbers of *purebred* horses obtained from Ali Pasha Sherif as well as several desert-bred horses acquired directly from central Arabia. The Blunts reported on a visit to his farm at the Berkeh (a pond or pool) in 1889 and provided the first eyewitness mention of the influential Egyptian mare

Roda and the stallion Jamil, a Saqlawi Jedran whose foal were considered by the Blunts to be superlative. The prince also had a stallion by the name of Sabbah, a grey Muniqi Sbeyli. The Blunts used him as a sire for some of their mares. Sabbah figures prominently in many of the Egyptian pedigrees seen today.

Figure 24 - *Ahmed Rifaat Pasha, (1825-1858) was a less well-known son of Ibrahim Pasha and Chiwekar Hanem. He died at a young age, killed when his train derailed crossing the Nile at Kafr el-Zayyat. He was married three times and had several sons and daughters. Ahmed Rifaat was the father of Ahmed Kemal Pasha, a man whose passion for breeding the authentic Nedgi horses was perhaps unparalleled.*

Figure 25 - *Prince Ahmed Kemal Pasha. Original rendering from the author's collection.*

Lady Anne was a frequent visitor at Prince Ahmed's farms and she noted that he kept about 58 horses, and grazed them on bersim pastures, a luxury for animals in Egypt.

It is difficult to overestimate the importance of the Arabian horse breeding of Prince Ahmed Pasha Kemal to the Egyptian Arabian horse, particularly because of his fine stallion Rabdan and the notable mare Noura. Prince Ahmed Kemal Pasha also owned or bred the horses Dahman, Bint Roda, Om Shebaka, and Dalal.

The herd that was assembled so carefully and thoughtfully by Prince Ahmed was inherited by his son HH Prince Yusuf Kemal, who was more interested in playing polo than in the family passion for authentic Arabian horse breeding. He held an auction in 1908 and the *purebred* treasures of his father were sold off.

Figure 26 - *Mohammed Said Pasha (1822-1863) was yet another son of Mohammed Ali that would rule Egypt. Said was Viceroy from 1854 to 1863, succeeding Abbas in the dynastic line. Although he was officially under the constraints of the Ottoman, he, in fact, ruled Egypt autonomously. Said was educated in Paris and was fond of all things French. He granted the Frenchman de Lesseps the land concession for the construction of the Suez Canal. Said was also responsible for the armed conquest of Sudan and Ethiopia. This was done in order to obtain slave laborers for Said's expanding military and infrastructure projects. He established the Bank of Egypt and opened the first standard gauge railroad in the country. When Said died, the title passed to his nephew Ismail, his heir presumptive being the deceased Ahmad Rifaat. Said seemed to have had little interest in horses during his 9-year reign as Viceroy of Egypt.*

Little is known of his sons Prince Tousson, Ahmed, and Mahmoud. Said's grandson Prince Omar Tousson, however, played an important role in the preservation of the asil horses. Tousson was instrumental in the formation of the Egyptian government's Horse Commission in 1892. He served as the president of the organization, and under his leadership the first attempt to procure pedigreed *purebred* horses took place. This small government project was to become the RAS, and later the EAO.

Khedive Mohamed Tewfik Pasha (1852-1892 at age 39) was a son of Khedive Ishmael. He was Khedive of Egypt and Sudan from 1879 to 1892. Unlike most of the Egyptian princes, he was not educated in Europe but remained in Egypt throughout his short life. He was unambitious and preferred the quiet life of a gentleman country farmer. His reign was marked by the armed insurrection of the nationalistic firebrand Urabi Pasha, whose passion for freedom from the British is detailed in Wilfred Blunt's book Secret History of the English Occupation of Egypt, published by Alfred Knopf in 1922. The thwarted revolution was aimed at expelling the British from Egypt. Wilfred Blunt played a critical role in supporting Urabi in the uprising.

Figure 27 - *Prince Omar Tousson (1872-1922).*

Figure 28 - *Khedive Mohammad Tewfik Pasha.*

Figure 29 - *Khedive Mohammad Tewfik Pasha with his mounted entourage. Image from the author's collection.*

As Khedive, he was forced to relinquish his title to Sudan at the urging of the British Consul-General Lord Cromer. He maintained a nominal allegiance to the Sultan in Constantinople but cooperated fully with the British, recognizing the importance of their military and financial support to his country.

Khedive Mohammed Tewfik was a significant contributor to Egyptian Arabian horse breeding, not because of his love of horses but because of the interests of his two sons, Khedive Abbas Hilmi II and Prince Mohammed Ali Tewfik.

The Khedive Abbas Pasha Hilmi II (1874 -1944) owned some of the most exquisite Arabian horse stock to be found in Egypt. He also bred some of the horses that would later figure prominently in the creation of the RAS equine breeding program. These horses included Bint El Bahreyn, Bint Yamama, Bint Hadba El Saghira, Bint Gamila, Obeya, and Bint Obeya.

Figure 30 - *Abbas Hilmi II.*

Later in life, Hilmi abandoned breeding Arabian horses in favor of the prevailing Middle Eastern obsession with horse racing and the English Thoroughbred horse. He was ultimately forced to renounce his position as Khedive of Egypt by the British government, whose control over the internal affairs of Egypt was at that time considerable. Hilmi was anti-British in his sentiments. He was replaced by Hussein Kemal, adopting the title of Sultan, a man whose allegiance was to the British overlords. After his deposition, Abbas Hilmi II lived in exile in Geneva, Switzerland until his death in 1944.

HRH Prince Mohammed Ali, brother of the Khedive Abbas Pasha Hilmi II, owned

Figure 31 - *The Khedive Abbas Hilmi II. Image from the collection of the author.*

or bred the mares Negma, Mahroussa, Maaroufa, Farida, Zahra, Saada, and Gamila Manial. The stallions Kawkab II, Mabrouk Manial, and Gamil Manial were also used in his stud at Manial Palace on the island of Roda, in the Nile River at Cairo.

The first man to whom the term "King of Egypt" was applied was Fouad. He was the grandson of Ibrahim Pasha and the brother of Khedive Mohammed Tewfik. Fouad established his personal private Arabian horse breeding stud known as the Royal Khassa – about 23 miles from Cairo. Fouad began his stud with Arabian stock

Figure 32 - *Sultan Hussein was the first to adopt the title of sultan.*

Figure 33 - *King Fouad, born in 1868, and died in 1936. He reigned as Sultan of Egypt from 1917 until 1922, when he issued a declaration making himself King of Egypt and Sovereign of Nubia, the Sudan, Kurdufan, and Darfur. He reigned as King until his death in 1936. He founded the Inshass stud on ascending to the Regency.*

obtained from other members of the royal family, notably Prince Kemal El Din and Prince Mohammed Ali Tewfik. He also acquired horses from the Royal Agricultural Society. The King emphasized racing in his breeding program and one of his early successes was the acquisition of the desert stallion El Deree, a Saklawi Shafi, who was a race champion and a sire of race champions.

The King acquired horses for his Royal Khassa from a variety of sources. From Prince Kemal El Din the stallion Ibn Sara was added. From Prince Mohammad Ali Tewfik the mare Saada (descended from Om Dalal) was obtained, and she produced the mare Raaga by Rasheed. The mare Hagir was gifted to Inshass from the RAS in 1939. The 1937 mare Yaquota, representing the Kuhaylah Rodania matriline, was presented to Inshass by the RAS in 1939. The mare El Samraa was bought by Inshass in 1931. Her descendants were among the most prized in the herd. Her grandson Sameh became a successful sire at Inshass and later at the RAS.

Mabrouka was a gift to Inshass from King Aziz Ibn Saud. She established an extensive line of high-quality performance mares and stallions. Additions to the program included Bint Karima, 1935, bought from Kafr Ibrash Farm. The Inshass stallion El Agouz was the sire of Mubarak and El Saoudia. Nafaa was gifted to Inshass by King Aziz Ibn Saud in1941. The grey mare Hind was given to King Farouk in 1945 as tribute from King Ibn Sa'ud. Her daughter Hanaa produced for Marshall in the U.S.

The Bint El Bahrain bloodlines were used by Inshass through the stallion El Zafir, a 1930 grey out of Bint Dalal of Manial. The family of Radia (representing the Saklawi line of Ghazal) was represented by the grey mare Bint Zareefa, obtained from Prince Kemal El Din. She produced a number of black descendants, notably Ebeda and Emad. Of signal importance from this line was the stallion El Moez.

El Shahbaa (representing the Abbeyah Om Jurays matriline), was purchased from El Haj Mohamed Ibrahim. This line produced Mahdia-Mona-Hanan-Jamill. Probably the most influential member of this line was the mare Mahfouza, grandmother of Magidda, whose fame circumvents description. The Ghazieh I line was present at Inshass by Radia 1922 and was derived from the Bint Yamama stock of Abbas Himi II. This matriline was also present in the form of Bint Zareefa, a 1926 mare by Hadban of HRH Prince Kemal El Din.

El Kahila was added, having been a gift from King Abdul Aziz Ibn Saud. Her descendant Shadia I was transferred to the EAO in 1952 where she was an excellent broodmare. It was the gifts from Shibly Bisharat Bey that scintillates with quality. Bashar El Ashkar, Badria, and Ward were his gifts to the King. Bashar El Ashkar was bred to Badria, resulting in the stallion Badr (1946). Badr was bred to Mahdia, of El Shahbaa descent, producing the mare Mona, mother of Hanan. The Badria line did not persist in matriline. The mare Ward had no lasting impact on Egyptian breeding.

The royal stud was inherited by the young Farouk. The name was changed to the Inshass Stud. The breeding population expanded as the result of gift horses given to the King by regional governments, potentates, and dignitaries. The stud was managed by Dr. Muhammad Rasheed. One of his roles was to re-home the gift horses from foreign dignitaries that did not measure up to royal standards. The King's stud manager kept careful records of the pedigrees of its horses in a stud book referred to as the Inshass Original Herd Book or IOHB. Study of the horses at Inshass requires attention to detail since several of the horses there had the same name, similar names, or multiple names.

Figure 34 - *Queen Fawzia, the daughter of King Fouad, became the Empress Consort of Iran by marrying the crown prince Mohammad Reza Pahlavi, later the Shah of Iran The marriage was arranged for political reasons and lasted only 9 years.While the common man in Egypt lived in squalor and desperate poverty, the Royal family adopted the glamorous and decadent lifestyle of the West, spending enormous sums of money on personal acquisitions.*

Figure 35 - *King Farouk (1920-1965) Farouk inherited the title from his father, ruling from 1936 until 1952 when he was overthrown during Egyptian Revolution.*

The stud was eliminated following the Egyptian Revolution of 1952, and many its horses were merged with the RAS stock to form the Egyptian Agricultural Organization stud at El Zahraa.

The nearly complete loss of the *purebred Arabian horse* had begun with the internecine wars of the peninsular Bedouin in the 19th century and was accelerated by the efforts of Ibn Saud to unify the country in the 20th century.

The "slow emergency" had run its course, and following the Revolution of 1952, the future of the *purebred Arab horse* lay in the resuscitative hands of the Egyptian Agricultural Organization. From a very few surviving bloodlines, representing the last vestiges of the once wild and *purebred*, genetically unique, horses of the Arabian Desert, the EAO continued the bold experiment to reconstitute the once prolific and proud race of horses, synthesizing the few remaining strands of the asil heritage into a carefully protected core of authentic desert Bedouin horses.

Figure 36 - *King Farouk with his first wife Safinaz Zulficar (Re-named Farida)and her daughter Ferial.After the 1952 Revolution and the forced exile of the King, Farouk married the commoner, Narriman Sadek. Photo from the Library of Congress.*

Figure 37 - *A Bedouin elder of the early 20th century, in Beersheba. Photograph from the Library of Congress.*

Figure 38 - *Bedouin horsemen, from The Raswan Archives.*

**More remains hidden than has been revealed.*

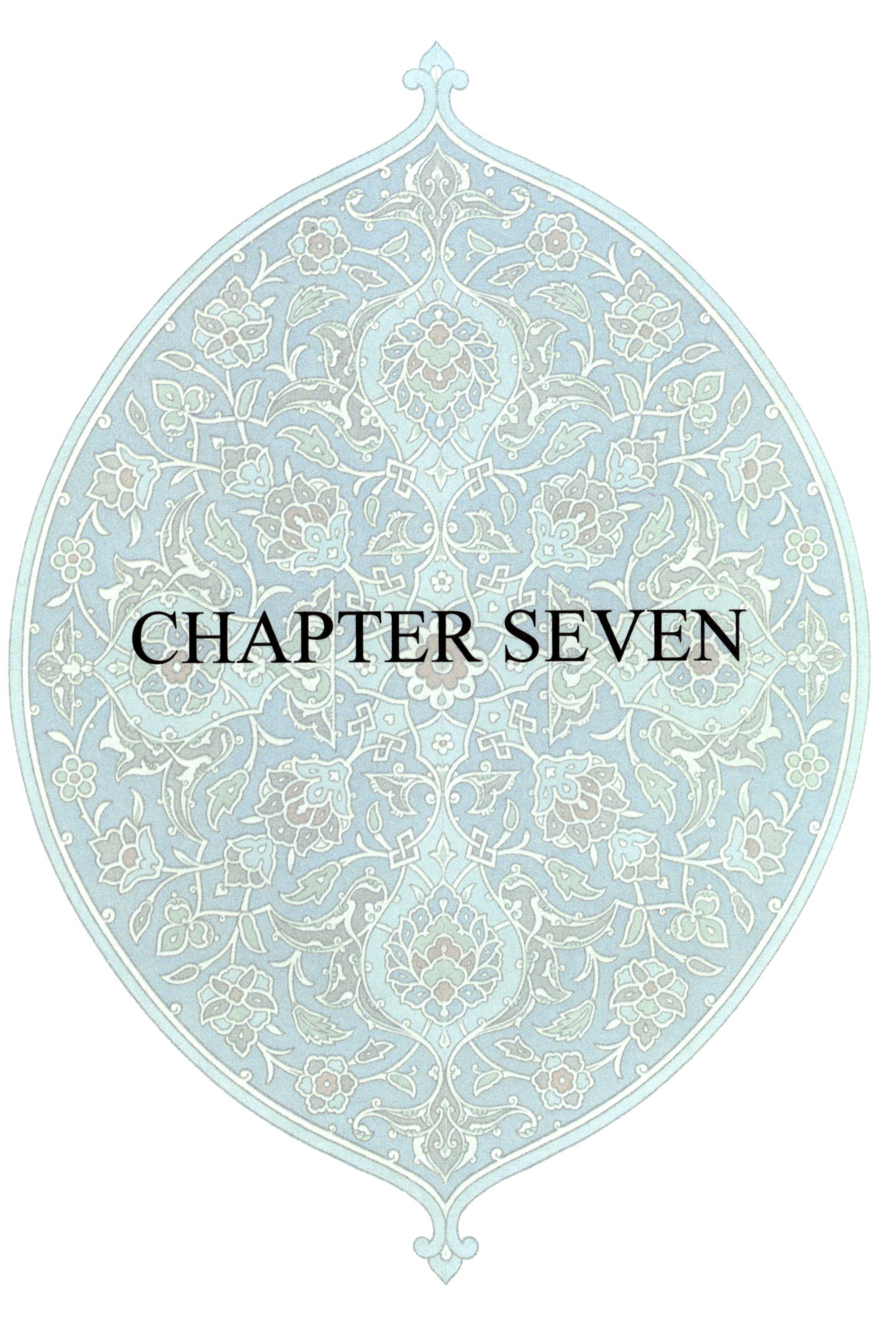

CHAPTER SEVEN

CHAPTER SEVEN

WESTERN ARABIAN HORSE BREED ORGANIZATIONS

In order to understand the significance of the original Egyptian Arabian horse mitochondrial DNA information presented in Chapter Eleven, an awareness of the modern day usage of the term *purebred Arabian horse* is needed. This involves a review of the past history of this term as well as the origins, evolution and current operations of the various Arabian horse registration organizations that use this term.

The subject of Arabian horse matrilines and their pedigrees has occupied researchers for decades. In that this book focuses on the matrilineage of Egyptian Arabian horses, an investigation into the pedigrees of the horses involved is needed. Since animal breed organization exist principally to maintain records of the pedigrees of the horses under their purview, the origins and the methodology of the various Arabian horse breed organizations is essential information for the novice Arabian horse breeders. The subject is complex, and there is no single reference source to consult.

In the preceding chapters, the nature of the Egyptian Arabian horse, its sources in the Arabian Deserts and environs, and the matter of pedigrees were discussed. The conditions pertaining to Arab horse breeding on the Arabian Peninsula in the 19th and 20th centuries were examined. The process by which the original horses were selected at the inception of the Royal Agricultural Society, forerunner of the EAO, were reviewed. In these chapters, the term "*purebred Arabian horse*" was used without an explanation of its meaning. The recurrent usage of the term may leave the reader with the impression that the term has a universally recognized and accepted meaning. This is not the case.

Figure 1 - *Charles Darwin, from the collection of the Library of Congress in Washington D.C.*

The fact is that the term itself has a long and tortuous history and warrants close critical scrutiny. This chapter will serve as a roadmap that can be followed in order to understand how this term originated and how it has come to be used in Arabian horse commerce today.

> "False facts are highly injurious to progress, for they often endure long; but false views, if supported by little evidence, do little harm, for everyone takes a salutary pleasure in proving their falseness."
>
> Charles Darwin

In 1900, Western Arab horse breeders generally assumed that the term "*purebred Arabian horse*" referred to a genetically unique and homogeneous race of horses, free from non-Arabian breeding that had lived in a state of geographic isolation deep in the heart of the Arabian Peninsula, having been present there in a feral state long before the Bedouin nomads existed. This understanding was self-evident. It had been the generally accepted definition for centuries.

Horsemen of the time assumed that the documented and registered *purebred Arabian horse* as it was found outside of Arabia was a genetically pure entity. The definition of a *purebred Arabian horse* was simple and straightforward: a "*purebred Arabian horse*" was one imported from the Bedouin tribes of the central Arabian Deserts whose ancestors all came from the central Arabian Deserts. This definition implied that these animals were all wholly derived from genetically unique and uncontaminated stock that had existed in the central Arabian Peninsula from extreme posterity.

This understanding was part of the allure and romance of the Arabian horse that attracted the interest of horsemen in the Western world during the 19th century. Later, as these horses were exported from Arabia and reproduced new generations of offspring in the West, the definition of a "*purebred Arabian horse*" became "a horse descended entirely from *purebred Arabian horses* originally imported from the Arabian Desert". While the public may have been naïve to accept such a fanciful intimation, it was nonetheless widely embraced. It was an inherently quixotic notion that had great appeal for the sensitive aesthetes, the dreamers, and the romantics. The 19th century Western world was awash in these notions.

This definition continued to be considered a "gold standard" during the first half of the 20th century, but by the year 2000, this view was largely discredited. The idea had become a very vexed question.

By 1880, Arabian horses were being imported from the Middle East to Europe, Russia, and America in significantly increasing numbers. The majority of these horses were bought from desert tribes and horse dealers located far from the Nejd: Egypt, Syria, the Levant, Turkey, and Baghdad. The implications of the dislocation of the *purebred Arab horse* from the Nejd were recognized by European horse buyers, and this created an atmosphere of distrust and suspicion regarding their genetic purity. The ability to document and verify the genetic purity of any particular Arabian horse became even more problematic.

The idea of national Arabian horse registries grew out of the problem of controlling fraud in the buying and selling of horses. Dishonest horse traders were known to falsify pedigree documents to increase the value of an otherwise unpedigreed average horse. Fraud and dishonesty are problems that afflict all forms of horse buying and selling, and it is an age-old problem. The first attempt at forming a breed-specific, legally binding, reliable, and trustworthy horse registry took place in England.

A brief review of the history of the General Stud Book of England is useful to the student of Arabian horse breeding, as it was the prototype of all future horse registration schemes. This model would later be applied to the Arabian horse, and its application of these principles would result in the creation of the World Arabian Horse Association. The parallels are striking.

English "running horses" were the descendants of a diverse collection of interbred horses, beginning with the Royal mares of the King of England Charles II in 1661. The mares were from North Africa, had no pedigrees and can only be described as of unknown origin. The use of several Eastern stallions on these mares led to the formation of what the English nobility came to refer to as the "thoroughly bred" horse. They could run.

Various combinations of mares and stallions were tried and the offspring culled to select only those horses that excelled at the "sprint", the short distance turf race. These horses were produced, generation after generation with only one purpose in mind; speed on the turf. And speed on the turf had one intention; gambling.

Tregonwell Frampton was credited with being "the father of the turf" in that he came into the Thoroughbred horse racing business when it was still in its infancy and races were still being conducted informally. He turned Thoroughbred horse racing into a phenomenon.

By using selective breeding, the Thoroughbred horse had developed throughout the reign of the English King Charles II, and then continued to evolve through the reign of his successor James II. Frampton came onto the scene in 1695 with the succession to the throne of William III and Mary II, Queen of England, when he was appointed "keepers of the running horses" to the King.

He stabled and trained the Royal horses at Newmarket, employing a staff of 10, and brought excitement, glamour and unabashed showmanship to the sport as he made matches for the horses of Royal Stables. When not racing horses, he was also a master of hawking, hare coursing, and cock-fighting. His only superior in England was the Royal Master of the Horse. Otherwise, he was a law unto himself. Newmarket became the epicenter of horse racing for the entire country.

Figure 2 - *Tregonwell Frampton, (1641-1727) portrait used with the permission of the British Museum.*

His cruelty to animals was well known. Once his fastest stallion won a private match against the mare of a wealthy nobleman. The mare's owner, incensed, extended a bet that no gelding could beat his mare. Frampton gelded his stallion that night, and he ran the race against the same mare the next day. Frampton's horse crossed the finish line first and promptly fell to the ground, dead.

Frampton was an inveterate gambler. He would bet on anything. He was called, in print, a pimp. He was bold, self-assured, and immodest. He was called the cunningness jockey in England. He was a shameless and a well-known cheat. He sought out and bought fast horses for his own use. They were often presented to the public with falsified pedigrees. He was a chronic cozener with no sense of decency.

His favorite scheme was to arrange a match with a wealthy opponent, and then, feigning indecision and hesitation, would suggest that a trial run be made first, before the actual race. He would then conceal weights on the person of his rider, and as a result, his horse would finish the trial race in second place. The opponent, now emboldened by his apparent superiority, would agree to even higher stakes. At the match, Frampton's horse would, of course, be much lighter and faster and take the purse. He made large sums and his inappropriately brazen attitude toward his Royal patrons, which he kept until his death, earned him the name "father of the turf".

His talent for arranging high-stakes matches was legendary. He could create tremendous suspense and excitement by spreading rumors among his not inconsiderable network of touts. He arranged the famous Merlin match which attracted the interest of the English nobility, and entire magisterial estates were put up as wagers. And, as expected, many entire estates were lost on that one match.

The disgrace was so profound, so far-reaching, and so shocking, that Parliament was forced to enact a bill prohibiting the legal recovery of any bet made on a horse race that was over 10 pounds. Such was the power of Tregonwell Frampton, the weaver of spells.

Not coincidentally, turf authorities began to keep carefully written accounts of all race matches. These men were disinterested third parties and became known as "keepers of the match book". The many episodes in which Tregonwell Frampton falsified the pedigree of a horse alerted the turf community that something must be done. Records must be legitimized and reliable pedigrees must be placed in print before the public.

In 1727 one such keeper of the match book by the name of John Cheny published An Historical List of All Horse Matches Run. Cheny also published the first recorded pedigrees of the original Thoroughbreds. He began the world's first stud book.

Concerning the origins of the sport under Charles II, he wrote:

> "King Charles the Second sent abroad, the Master of the Horse, (Which some say was a late Sir Christopher Wyvil; others, the late

Sir John Fenwick) in order to procure a Number of Foreign High bred Horses and Mares for breeding; and the mares, thus procured by the said King's interest and brought to England (as also many of their off-spring) have, for that reason, been called Royal mares; one of which was the dam of Hautboy."

Figure 3 - *Betting at a Horse Race, from the collection of the British Museum.*

Cheny also wrote his own defense:

> "With regard to the authenticity of the pedigrees, nothing is here inferred for which there is not some authority; as little as possible is hazarded on uncertainty, and nothing upon conjecture."

The following example is typical of the type of authority that he cites in his decision to include certain horses in his book as being Thoroughbred. Cheny wrote:

> " I have been assured by a person of Rank and Great Honor, that the Horse called by sportsmen the Byerly Turk which horse (as observed) was the Grandsire of Bucephalus, was in fact an Arabian."

Figure 4 - *The Godolphin Arabian, one of the foundation sires of the English Thoroughbred horse.*

The English populace became fanatically enthusiastic about horse racing. The popularity of the sport was based on the popularity of is associated activities; gambling, excitement, danger and suspense, Spectators were thrilled at the sights and sounds of accidents and injuries, the agony of spilled jockeys and horses with broken legs and necks dispatched on the spot with a bullet. Nearly everybody in England, from the members of Parliament to the cooks and coachmen of Georgian England, frequently had a "small flutter" at the racetrack.

Figure 5 - *The Royal Mares, 1636, by J. Sibrechts.*

Figure 6 - *The Royal Mares, 1636, by J. Sibrechts.*

The British Jockey Club, an 18th century English gentleman's club, became the organization that informally controlled all aspects of the Thoroughbred business and racing in England. But the Jockey Club was having a problem.

People were cheating.

AN

Hiſtorical LIST,

OR

ACCOUNT

Of all the

Horſe-Matches Run,

And of all the

PLATES and PRIZES

run for in *England* (of the Value of Ten Pounds or upwards) in 1727.

CONTAINING

The Names of the Owners of the HORSES, &c. that have Run, as above, and the Names and Colours of the HORSES alſo.

WITH

The Winner diſtinguiſhed of every MATCH, PLATE, PRIZE or STAKES: The Conditions of Running, as to Weight, Age, Size, &c. and the Places in which the loſing HORSES have come in.

With a LIST alſo of the Principal COCK-MATCHES of the Kingdom in the Year above, and who were the Winners and Loſers of them.

LONDON,

Printed in the Year M.DCC.XXVII.

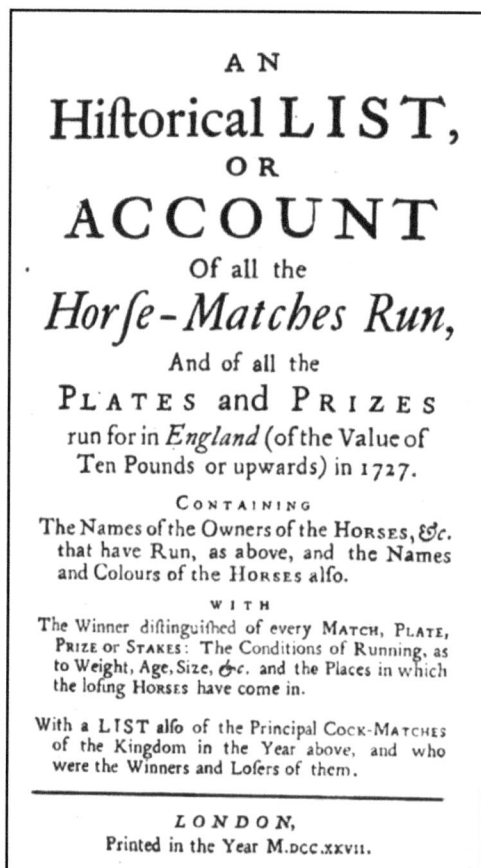

Figure 7 - *The Cheny document: A list of horses that competed as racing Thoroughbreds in England in 1727. Weatherby used this document as a starting point to begin compiling horse names, owner's names and pedigrees for his registry.*

To bring order and dignity to this morass of misbehavior, The Jockey Club, an "association of noblemen and gentlemen", took upon itself the position of supreme authority over the sport and became the official governing body for all matters related to turf racing in England. The Rules of Racing were drawn up, codified, and distributed to the members. A monitoring, enforcement and penalty system was set in place.

But there was still no official stud book.

James Weatherby, secretary for the group, had begun compiling a list of authentic Thoroughbreds for the Jockey Club in 1770. Building on the Cheny model he compiled a studbook in an attempt to combat the increasing frequency with which non-Thoroughbreds were being falsely presented at the major stakes races as true Thoroughbreds. The motivation for this dishonesty was financial; a win at a major stakes race led to the promotion of the horse and monetary gain for the owner. If an unscrupulous individual came into possession of a horse with exceptional speed but of unknown pedigree, it could be passed off as an authentic Thoroughbred. Much money could be made. Horse racing and cheating were synonyms, even in 18th century England.

The English Thoroughbred was known to have been created from an acknowledged mixture of several genetic sources, so genetic purity was not an issue in creating a formal stud book; cataloging the correct horses and verifying the correct pedigrees were the issues. The definition of the *purebred* English Thoroughbred was well known to Jockey Club members at the time; it was any horse descended in all lines from a generally recognized and widely acknowledged group of foundation mares and stallions, dating to the time of the English King Charles II. Everyone involved in turf matters knew this, but the details had never before been documented and formalized.

Weatherby compiled a database of all living horses, with complete pedigrees, that he considered being valid Thoroughbred horses. He worked directly with owners and was careful to investigate, validate and verify conflicting records or suspicious individuals. The Weatherby database became the Registry Office, and all records, pedigrees, and registrations were handled there. No foal could be registered without thorough scrutiny by the Registry Office staff.

The names of some of the first entries into the database give some idea of how arbitrary the initial pedigree validation system must have been. There were horses named Sister to Miss Partner, Gower Stallion mare, and Brother to Fearnought Mare.

Weatherby saw that it was not possible to unequivocally verify the pedigrees of all living and potentially registrable Thoroughbred horses under review, so he made the decision to include or exclude each horse under review based on his best judgment.

He knew that he would probably include horses that should not be included, and he knew that it was likely that he would exclude horses that were truly Thoroughbred. There were no foolproof criteria to establish the unimpeachable truth.

Despite the fact that errors would inevitably be made, Weatherby knew that an arbitrary database was better than no database at all. It was regrettable but necessary. He saw the need to bring order and uniformity in record keeping to the business, for the sake of the business. Cheaters were ruining horse racing. Weatherby created a registry of horses that would allow reliable and verifiable identification of living horses. With a fixed database in place, the accuracy in registering future foals resulting from the breeding of horses in the database could be assured. Applications for registration of new foals were cross-checked against the Service Records that owners were required to submit to the Club officials. They cross-checked proposed names of foals submitted for registration; they would not allow the same name to be used twice. From this time forward, fraudulent horse dealers would find it very difficult to register a foal with a manufactured pedigree.

The past was past, decided Weatherby, and it could not be changed. Fraud and deception had certainly occurred, but that could not be corrected. It was regrettable, he thought, but he decided it was best to make a fresh start based on the prevailing general consensus of which horses were truly authentic and verifiable *purebred* Thoroughbreds. Working in conjunction with his nephew, a pedigree specialist, he published the first General Stud Book in 1791. Its purpose was clear; to prevent future fraud and deception in the English Thoroughbred business. Its decisions were final; any horse not registered with the Registry Office was not a Thoroughbred horse. Today, the affairs of the Club are now directed by three stewards who monitor and enforce the rules.

The parallels between these events and the evolution of the World Arabian Horse Organization are evident.

In general, horse breed registries are primarily founded for the purpose of facilitating the buying and selling of horses, and they are inherently only marginally concerned with the issue of genetic purity. They are created as a way to keep the competition honest. The secondary purpose of any animal registry is to provide a sheen of authority and legitimacy for its breeding activities since the appearance of legitimacy is so important in cultivating the interest and trust of each new generation of customers – the newcomers. Providing "papers" with an animal is a time-honored and effective marketing tool for gaining the confidence of the neophyte or the beginner.

All horse breeding operations depend on the continuous and uninterrupted flow of naïve newcomers (potential customers) onto the scene, providing the veteran breeders with a group of inexperienced people to whom they can sell the produce of their farms. The steady inflow of newcomers and the anticipated outflow of disillusioned customers several years later has always been, and always will be, the key elements of the business model for horse breeders.

To deal with the inevitable and growing problem of pedigree authenticity and verification, the concept of the national Arab horse registry came into being in several Western countries. From the beginning, the national registries were organized and controlled by powerful and wealthy Arab horse-owning enthusiasts. The registries were small and elite organizations. The question of actual blood purity was not approached in an orderly and systematic manner because the documentation of actual blood purity was not foremost in the minds of importers. Authenticity was the issue, and the early Arabian horse importers were clear on what that meant; an authentic *purebred Arabian horse* was defined as a horse which was imported from Arabia. It was very simple. It was self-evident. However, registries were afflicted from the beginning with discord and disagreements over questions of which horses to register and which horses to reject. Personalities came into conflict, and accusations of fraud, deception, and misrepresentation were common. Documentation of a horse's provenance and the reliability of the records was a source of dispute from the very beginning.

In 1881, the General Stud Book in England opened a new section to record and register Eastern or Arabian horses. In 1918, the Arab Horse Society was established in Great Britain for the purpose of providing a central registry in England of the pedigrees and provenance of the growing numbers of Arabian horses now outside the Arabian Peninsula. The idea was put forward by H. V. Musgrave Clark, and a society was formed with 30 council members. W. S. Blunt, a man of independent wealth and owner of the world famous Crabbet stud, was the first president. In 1921, the General Stud Book closed its Arabian section; its last registered Arabian horse was the Polish stallion Skowronek. The Arab Horse Society had meanwhile started its own *purebred* stud book, and its definition of the *purebred Arabian horse* was "a horse in whose pedigree there is none than pure Arab blood". This definition is noteworthy.

Figure 8 - *H. V. Musgrave Clark.*

Figure 9 - *Wilfred Scawen Blunt.*

In America, The Arabian Stud Book of America (ASBA) was established in the late 19th century. It did not survive long, crippled by internal factional disputes over the validity of the documentation of a number of the horses presented for registration. It was later replaced by the Arabian Horse Club Registry of America (AHCR).

The AHCR was created by a small group of individuals who were disturbed by the arbitrary refusal of the ASBA to register a number of eligible and qualified Arabian horses imported into America. The ASBA's refusal to register these horses was based on a claim that the horses were of inferior quality and were improperly documented. The ASBA's attempts to force its own judgments concerning purity onto the public were viewed as biased and prejudiced. As a result, the AHCR was formed as an alternative registry and was enfranchised by the United States Department of Agriculture to be the sole national registry for the Arabian breed. This fact is noteworthy.

Matters of blood purity and disputes over the definition of blood purity were central to this organization, and controversy was rife, even at the very beginning. The first stud book was published in 1909. Homer Davenport was influential in the early stages of its formation, having become enraptured by the beauty of the Arabian horses that he saw at the 1893 Chicago World Fair.

Figure 10 - *Arabian horses at the 1893World's Fair, the white stallion Obeyran on the left and Nedjme on the right. Used with the permission of the Raswan Archives.*

The first definition of qualified imported horses, or a *purebred Arabian horse*, for the Arabian Horse Club Registry was "All horses of pure Arab blood furnished with certificate of registration in any recognized European stud book". Personal differences soon began to emerge. Due to objections from some of the members of the Board of Directors, a list of recognized countries from which *purebred Arabian horses* could be imported was created and included only Britain, France, and Australia. One of the first horses registered was Gouneiad #21, a "Turkomen bred Takhtaravan" owned by the AHCR treasurer. Objections to this inclusion were raised, and in 1918 the power of the Board was expanded to give them complete discretionary power to accept or reject individual imported Eastern horses.

Figure 11 - *Homer Davenport (1864-912) was a widely syndicated political cartoonist and satirist, and was to become the first large scale importer of Eastern or Arabian horses into the U.S. He was a prime mover in the formation of the new Arabian horse registration body. Photo used with the permission of the Library of Congress.*

By 1930 the AHCR was operated from offices in the Woolworth Building in New York City. H.S. Gregory of Berlin, New Hampshire, was the Secretary for the group and he managed all applications for registration of *purebred Arabian horses* and foals. The Rules for registration were rather broad:

1. Any horse of pure Arabian blood, born in Arabia, when accompanied by a certificate bearing the seal of a Tribal Sheykh affirming that the horse is a *pure blooded Arabian horse.*

2. Any horse of pure Arabian blood previously registered with the General Stud Book of England, The Australian Studbook, Le Stud-Book Francaise *or any other foreign stud books approved by and accepted by the Board of Directors.*

3. Any horse registered in the Arabian section of The American Stud Book.

4. Any offspring born to parents that qualified under rule 1, 2, or 3.

In its infancy, the AHCR began to experience difficulties with the problem of international reciprocity. In matters of registration among the various nations of origin, there were disagreements over what constituted a *purebred Arabian horse*. The AHCR sought to improve relationships with foreign registries and attempted to standardize criteria for evaluating imported Arabians that were presented for registration. In 1937-38, the stud books of Poland were added as acceptable sources of *purebred Arabians*. These additions were made to accommodate the desire of two wealthy and influential AHCR Directors Col. Dickinson and Henry Babson, who wished to import and register Arabian horses from Poland. In 1939, the AHCR and the Arab Horse Society of England signed an agreement of reciprocal registration.

The first Polish imports to America were the Crabbet-bred sons and daughters of the Polish Arabian stallion Skowronek. Their acceptance by the American registry required a considerable degree of accommodation since the foundation stock of the Polish studs included more than a few non-Arabian mares. W.R. Brown, American Registry president from 1918-1939, advised the first importer of Polish Skowronek-bred Arabians, W. K. Kellogg, that "Skowronek had a lot of doubtful ancestors". Brown was correct is his assertion; Kellogg was not concerned about the implications of the statement.

The AHCR soon came under fire from the rank and file members when some animals presented for registration were refused admission. By policy, each horse was examined by a registry-appointed inspector who viewed the animal and its papers of provenance. The inspector had full authority to accept or reject the animal. This was not a rubber stamp operation; many horses were rejected. The AHCR began to claim that their purpose was to admit only those animals that were clearly above average in quality, as they wished to improve the overall quality of Arabian horses in the U.S. The policy was generally regarded as motivated by the desire of the very wealthy members of the Board of Directors to limit competition in the U.S. This would enhance the reputation of their own horses.

It did not go down well. Critics such as Miss Ott of Blue Book fame saw to that. As she put it "one man's ideal horse is another man's dogmeat."

The AHCR had its detractors. A rival registry known as The International Arabian Horse Registry of North America (IAHRONA) was formed. This group was created to redress unanswered grievances put before the AHCR, including the cost of registration, methods of identification of horses, and the use of artificial insemination. IAHRONA used a deceptively low-tech method of Permanent Identification of horses at the time of registration: photographs of the callosities of the inner sides of the hocks and forearms of the animals were used. The registry believed that the patterns and size of the callosities were sufficiently unique to a given horse that they could be used to definitively establish a horses' identity for future reference. This did not prove to be workable.

Until the 1960s, the AHCR continued to not only inspect import documentation, but it also employed its own importation inspectors to personally inspect each horse presented by its importer for registration. In this way, the registry sought to insure the authenticity and validity of the Arabian horse's documentation. The registry expected the horse owner to pay for all expense in the process. Fees for registering mares was $1000 and for stallions, $2500.

The AHCR ironically was accused of exclusiveness and favoritism, refusing to register some imports because they were considered, in the opinion of AHCR, to be of inferior quality. Controversies resulted.

By the mid-1960s, the registration policy for imported Arabian horses had changed again. Only imported horse from England, Poland, and Egypt were accepted without question and without inspection. In addition, horses from Spain and Germany could be considered for registration, but only if an AHCR inspector went to the country of origin and personally inspected both the horse and the stud book in which the horse was registered. The client had to pay for this service whether the horse was accepted or not. Naturally, this policy did not last, due to complaints from the public and charges of bias and fecklessness on the part of the inspectors. The AHCR was eventually dissolved and was reorganized in 1969 as the Arabian Horse Registry of America. The headquarters were moved from Chicago to Colorado.

The AHRA continued to promote its idea of purity and was in the vanguard of horse registries by adopting biological testing for parentage verification. The first attempt at pedigree verification was the introduction of stallion blood typing in 1976. Full parentage identification of foals using blood typing began in 1991. DNA parentage verification was initiated in 2000. From its inception, the AHRA had been committed to the preservation of the "blood of the *purebred Arabian horse*". The use of the term "*purebred*" in its bylaws was to have significant future implications. The AHRA asserted that all of its registered horses were *purebred Arabians* in that they traced in all lines of descent to the Arabian Desert Bedouin, with no contamination by foreign or non-Arabian blood. The AHRA began to formulate a database of horses that it considered "*purebred Arabian horses*".

The database that the AHRA began to compile in 1969 was based on records held by the previous organizations but was heavily influenced by the work of Carl Raswan and his book The Raswan Index, published in 1957. This book contained information concerning the provenance of Arabian horses from around the world, information that could be found nowhere else.

CARL RASWAN AND THE RASWAN INDEX

At about the same time that Dr. Mabrouk of the Royal Agricultural Society of Egypt was scouring the deserts of the Middle East for new Arab horse blood, a young German adventurer named Carl Schmidt (later Raswan) was immersed in his passion for the Arab horse that would change the way in which the Western world viewed the Arabian horse pedigree and the Arabian horse itself.

Raswan was a figure of immense energy and scholarship, obtaining records of the pedigrees of thousands of Arabian horses throughout the world as a consequence of his first-hand and intimate contact with the Bedouin tribes of the desert of Arabia.

He was born Carl Reinhardt Schmidt in 1893 in Tolkewitz-Laubegast, near Dresden Germany. His father Martin Schmidt was a doctor. His mother was a native of Hungary. According to Dr. Johannes Flade, Carl first became enthralled by the Arabian horse when, early one dawn, he happened to see the young prince Ernst-Heinrich of Wettin, son of King Friedrich August III riding his grey Arabian horse into the mist-covered lake at Moritzburg Castle. The stallion's beauty and vivacity deeply affected the young Carl. It was an epiphanal moment.

As a young boy, Carl received equitation training at the Royal Wettin Gymnasium, learning to ride in the European fashion using only leg and voice commands. In 1910, he came across a copy of the book by Lady Anne Blunt, A Pilgrimage to Nejd, and he was infused with the romantic ideas of the perfection and intoxicating beauty of the horses of the Arabian Bedouin.

This was the beginning of a lifelong obsession in pursuit of the ancient horse. Carl was now on a mission.

Beginning in 1912, at age 18, he spent many years tirelessly journeyed among the horse-breeding desert tribes of Arabia. His immersion in the oriental way of life began modestly enough when he took a position as a manager at the Santa Stefano Rach in Ramle, east of Alexandria in Egypt. Chance encounters with the local Bedouin fueled his fascination with their free way of life and independent spirit. His spirit had found its brothers in the desert dwellers. He was enraptured.

Over a period of 34 years, he made 13 forays into the desert, studying the horses of 19 of the indigenous pastoral Bedouin tribes as they followed their annual migration patterns. He studied the Arabic language and became fluent before he ever set foot in the desert. He wrote: "I did not have to depend on word-of-mouth reports of wandering Bedouins, but was able to check my findings with their educated men."

Figure 12 - *Raswan, second man from the right, among the Ruala Bedu chieftains, from the Raswan Archives.*

Schmidt was 21 years old when the First World War broke out. He served in the German Cavalry in the 18th Hussars, first in Russia and later in the deserts of the Middle East. He fought at the Battle of Suval Bay in 1915. As an advisor to the Ottomans in Arabia, he was once fired on by Bedouin rebels led by the British adventurer T. E. Lawrence.

He joined the army of General Kress von Kressenstein, Chief of Staff to Jemal Pasha IVth Turkish Army, and saw action in Sinai and Northern Arabia.

He fought in the battles at Gallipoli and the Dardanelles. He later saw action in battles at Suez. Attached to the 4th Turkish Brigade, he fell victim to spotted typhus.

Figure 13 - *Dresden was the capital of the German state of Saxony, with a long history as the capital and royal residences of the Electors and Kings of Saxony. Known as the "Jewel Box" of Europe the city was filled with baroque art and architecture. It was a sophisticated cultural landmark.*

Typhus was the most feared disease of the First World War. It is caused by the parasitic bacterium *Rickettsia* and has a mortality rate in excess of 40 %. Surviving this potentially fatal disease, he was sent to Russia in the spring of 1918 to work in the prisoner clearing camps. Soldiers too debilitated to fight were sent there for rest and recuperation. When the Great War ended, Schmidt was in the Ukraine.

His time in the desert had prepared him for the hardships and privation of war. His exposure to the Bedouin had solidified the essential elements of his character.

He emerged with an insatiable curiosity and an affinity for the cause of Arabian political independence.

He had come to know the principle that governed life in the Arabian Desert. "The Eternal law of nature was being fulfilled; death to the weak, the maimed and the forsaken. The strong trampled over the weak, gaining from their victim's fresh strength and endurance to push on."

Following the War, Carl immigrated to America, joining his mother who had left Germany and settled in Coachella Valley near Oakland California. The year was 1918, and Carl was frail and weak from his deprivations and illnesses suffered in the army. He did not recover his vibrancy for several years.

He worked for a time at the W. K. Kellogg ranch in California, where his charisma and personal charm made him popular with the horsemen of the area. A close observer remembered him as "rather short, strongly built, with high cheekbones and a prominent nose. His hair was heavy for a man of his age, and combed straight back from the forehead . . . he was keen, highly intellectual, humorous, God-fearing, and encompassed with a fine humility of spirit."

In America, Raswan became convinced that it was his calling to promote the breeding of absolutely *purebred* noble Arabian horses. "It is not a mechanical accomplishment to breed Arabs, but it is, in my opinion, the appreciation of the most wonderful creature that left the hand of God on the day of creation." Raswan was not speaking fancifully when he wrote this. He meant it literally.

He was very specific about his conception of and definition of the *purebred Arabian horse*:

> The wild mare of Arabia was an Ultimate achievement of Nature, a mature and perfect creature . . . Upon an unseen but not totally abstract pattern of beauty and perfection the spirit of God has created a harmonious animal and endowed it with a gentle and intelligent soul which has the capacity to understand his mission in this world as a companion to man.

<u>The Raswan Index</u>

In 1927 he returned to the Bedouin of the Inner Desert of Arabia, and he was shocked at the differences that he saw there. "Since the World War, the last romance and ideals of the Bedouin are crumbling. Automatic rifles and machine guns, and now automobiles, destroy hundreds of horses today in battles that used to be fought with spears and primitive weapons…causing only non-fatal wounds, the passions tempered by knightly virtues and laws. In October 1927 I witnessed a case with the Fid' an Anaza Bedouins, who lost 135 mares in one day."

During his long career of travel and research, he visited major Arabian breeding centers outside of Arabia including Egypt, the flourishing European studs, England, and America. Throughout these years, he kept notes, hand written on index cards, of virtually everything he saw and heard.

Near the end of his life, he compiled this mountain of information into a book. His encyclopedic work, <u>The Raswan Index</u>, was published in a series of volumes, beginning in 1957. The entire project was hand typed by his wife Esperanza Raswan. 380 copies of each serialized volume were printed as a limited edition, and the plates were then destroyed.

Esperanza Raswan issued a new edition of <u>The Index</u> in 1990. In her introduction, she singles out Miss Jane Ott for special recognition as her central supporter and close friend. Of Ott, Esperanza wrote, "Without her dedicated work, Carl's purpose of conveying his study of the Arabian horse could not have again been perpetuated as Carl had intended it to be done."

The book contained pedigree research which was to profoundly influence the world of Arabian horse breeding. Raswan's book was notable in that the author was not primarily a horse owner or breeder, but rather he was a researcher. He was unbiased in his writing and sought only to establish the facts. He spoke, wrote, and read Arabic. The opinions that he expressed possessed an admirable degree of candor and objectivity, free from bias and prejudice. He was not writing to promote himself, nor was he writing in an attempt to sell horses. The book was not a self-advertisement.

While the data and conclusions found in the book are at times not unimpeachable, The Raswan Index was respected as a sincere and unbiased attempt to bring together in one place the surviving historical data regarding Arabian horse pedigrees. It was in many ways an inflammatory work, for some of Raswan's conclusions were contrary to public opinion and contradicted many pedigrees attested to in Arabian horse registries around the world. Raswan wrote with integrity, sincerity, and gravity. He was unique.But for Raswan, purity was the issue. Absolute purity.

Figure 14 - *Carl Raswan with Amir Nuri Ash Sha'lan of the Ruala, The Raswan Index.*

The publication in 1957 of Raswan's seminal work The Raswan Index marked a high point in the history of Arabian horse pedigree scholarship. The Index was comprehensive and authoritative; promoting the idea that pedigree information could be collected, sorted and analyzed. The goal of his research was to arrive at accurate reconstructions of the Arab horse's true genealogy. Raswan's goal was to define and identify the 100% *purebred Arabian horse* of the Bedouin tribes of the Arabian Desert. His motivation came from his growing awareness that the true full-blooded Arab had disappeared from the desert and was rapidly disappearing from the breeding programs that existed outside of Arabia.

The book was subtitled *A Handbook for Arabian Breeders*, and Raswan viewed the book this way; not as a stately tome, but as a practical resource to be consulted regularly and repeatedly by breeders.

> The Index will be of great importance to those horse breeders who always wanted to trace the pedigrees and ancient descent of their Arabians to their very source among the Bedouin and "unearth" any hidden faults and flaws which may exist anywhere in between those most distant ancestors of Desert Arabia and those imported to Europe and/or America. It is in European stud farms where many so-called Near Eastern horses of "unknown" or doubtful Arabian origin turn up. We find Turkoman, Turkish, Syrian, Dongolian, Barb, and other foreign blood among them.
>
> Carl Raswan, The Raswan Index

Raswan focused on the idea that while some Arabian horses were genuinely 100% *purebred*, most generic Arabian horses were actually part-bred Arabian horses, containing genetic material from non-Arabian horses.

For Raswan, the definition of the *purebred* horse was simple and, at the time, universally accepted:

> The *purebred Arabian horse* was one which traced in all lines of descent to the Bedouin tribes of the deserts of Arabia.

There was the general impression among Arabian horse enthusiasts that horse pedigrees could be effectively researched, and through research, the truly *purebred Arabian horses* could be identified and separated from the part-breds. There was a general impression that blood purity mattered in breeding Arabian horses, and many breeders believed that blood purity was a matter of unsurpassed critical importance.

Many pedigree researchers were motivated by the conviction that the ability to identify those horses that were truly 100% *purebred Arabian* Desert Bedouin-sourced horses was a race against time. Many authorities considered this research to be essential; many felt that its success or failure would determine whether or not the Arab horse could continue to exist in its classic and timeless form. Raswan believed that once the last genetically pure Arab horse was gone, the remaining part-bred Arabs would inevitably decline in quality and deviate from type. Raswan believed, as the

Bedouin believed, that the presence of even a small percentage of non-Arabian blood in any Arabian horse breeding population would eventually lead to significant and irreparable deleterious effects on the quality of the horses that were produced.

Raswan wrote:

> The most important ancestors of an Arabian horse (as far as these powerful tendencies are concerned which transmit good qualities and faults as well) are the eight great-grandparents. They hold the key to the size, weight, shape, conformation, type, Arabian characteristics, and temperament of the horse.
>
> <u>The Raswan Index</u>

But these eight individuals must all be 100% desert Arab horses, free from mongrelizing impurity. "If any one of them was not pure," said Raswan, "alterations in the quality of subsequent generations would reveal this fact."

Raswan pointed to the detrimental influence of non-Arab blood in a pedigree.

> "The purer we breed the more simplified becomes our task, and the stronger, more beautiful, more distinctive, and more perfect and harmonious the offspring."
>
> "Why am I so fanatic about the pedigrees of our Arabians? The answer is that the authentic (pure) Arabian horse is dying out and many Arabian breeders are desperately trying to protect and save the few doubtlessly pure Arabians which are left."
>
> <u>The Raswan Index</u>

The Raswan Index was the spark that lit the flame of "the era of definitions". This book provided much of the pedigree material that fueled the enthusiasm and scholarship in the search for the truly *purebred Arabian horse*. The Index also brought the concepts of Bedouin horse breeding theory and practice to the attention of horse breeders in the Western world. It was the first work to establish an intense and sustained interest in the accuracy of the traditionally accepted pedigrees of Arabian horses. It was scholarly and minutely detailed, based on data that Raswan had for the most part collected first-hand. This interest would lead to a broad-based movement among Arab horse enthusiasts to re-evaluate all that was known and uncritically accepted about the Arab horse. The sheer volume of information coupled with its sincerity, enthusiasm and authenticity has not yet been equaled.

Figure 15 - *Carl Raswan in native dress, The Raswan Index.*

<u>The Raswan Index</u> contained an extended pedigree of the Polish stallion Skowronek, one of the most heavily used and influential stallions of the 1940-1950 period. Even today, 90% of the horses registered with the Arabian Horse Association in the U.S., *purebred Arabian horses* contain Skowronek somewhere in the pedigree. This ancestral disclosure revealed for the first time the non-Arabian elements in the stallion's pedigree. It is an impressive fact.

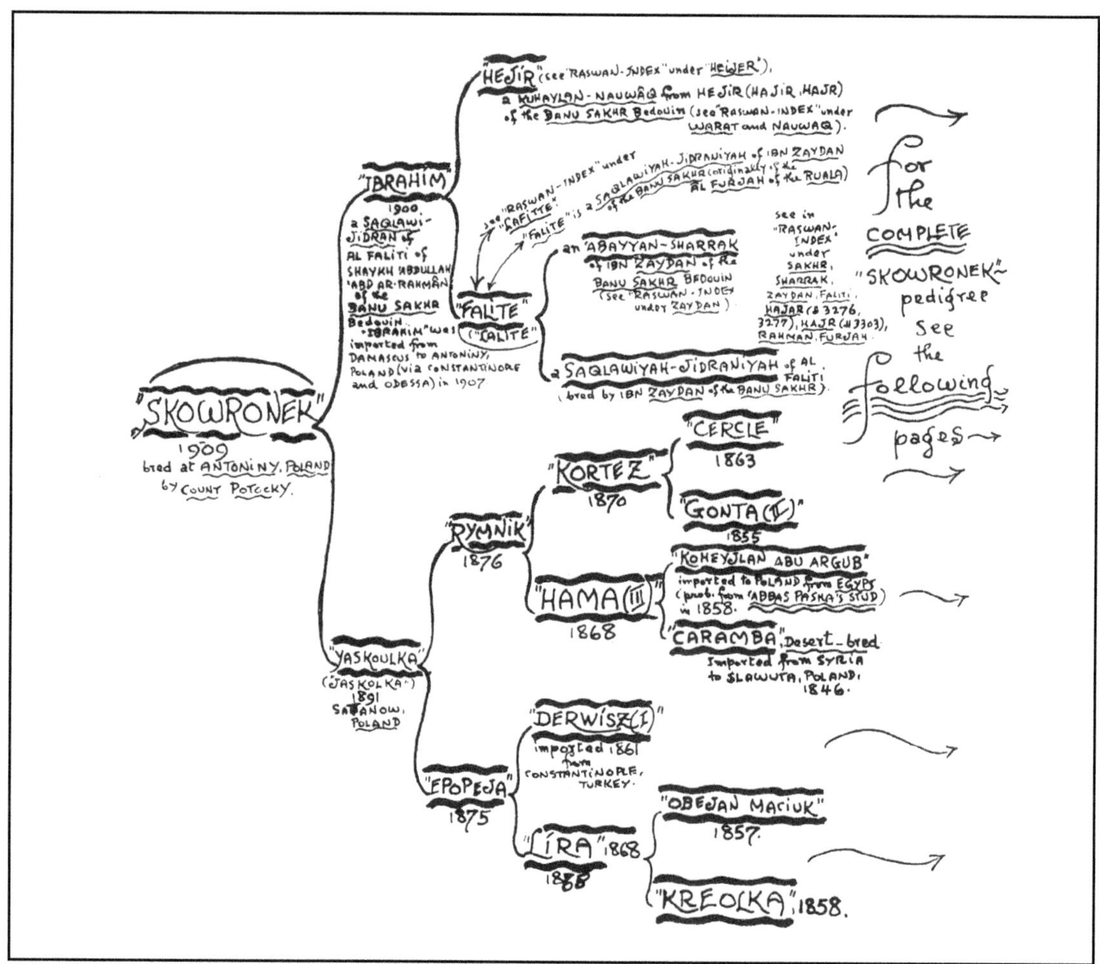

Figure 16 - *The pedigree of Skowronek as published in* <u>*The Raswan Index,*</u> *showing the first five generations of his ancestry. The non-Arabian ancestors are found much farther back in the pedigree.*

The pedigree that Raswan published contradicted the accepted pedigree promulgated by Skowronek's owner and champion of the Crabbet Stud, Lady Wentworth. The Skowronek discussion provides a good example of the problems encountered by researchers who wished to demonstrate the direct connections of all lines of all Arabian horses to the Arabian Desert. Raswan utilized existing historical records, but he also sought out and personally interviewed those living individuals who had intimate knowledge of these matters.

The pedigree of Skowronek contained many outstanding and unimpeachably *purebred Arabian horses* directly imported from the inner deserts of Arabia. But the pedigree of Skowronek as it was published in The Index (Figure 16) showed several crosses to the stallion Szumka (1824), a Polish horse out of Polka I (1808), who was a domestic English mare. In addition, a number of the female lines in the pedigree were founded by mares with no provenance. These mares were most likely local domestic mares, bred as work animals by the local farmers who domesticated them from the Konic breed, a dun-colored small tarpan-like feral horse.

But there was much more to be found in the Skowronek pedigree.

The Antoniny stud that bred Skowronek had used the formula of breeding desert Arabians to part-bred locally sourced mares for over a century.

The equine expert Dr. Lukomski described the pattern of breeding at Antoniny following his visit to the stud:

> "Anglo-Arabian mares are mated with English Thoroughbred stallions and the Arabian mares, depending on their size and constitution. The smaller type of Arabian mares were bred to English Thoroughbred sires, the larger Arabian mares to Arabian sires. The Anglo-Arabian mares did not differ much in their appearance from the Arabian mares. The larger mares sold easier."

Skowronek's sire Ibrahim was shown to have no verifiable pedigree. Potocki's agents bought the horse from traders in Constantinople who claimed the horse was bred in the desert of Arabia but gave no pedigree. Potocki recorded in his private stud book that the stallion was a desert bred original Arab. Raswan provided a pedigree for the stallion, shown in Figure 16.

The purity of Ibrahim, however, registered in the Polish records merely as "desert bred", was not the essential problem with the pedigree of Skowronek. The fact was that many of his ancestors on his mother's side were simply not Arabian horses.

But Skowronek's mother Yaskoulka was to become a problem.

The Poles were aware of the fact that a beautiful Arabian horse could result from the breeding of part-bred parents, and they were aware that *purebred* parents did not always produce offspring that were appealing. The Polish Count Sangusko remarked that Szumka II was a magnificent and outstanding individual, pure black with no white markings, but that he never produced quality in his offspring. Since Szumka II was 50% English, this was not unexpected.

The Count opined that:

> "a breeder should not be misled simply by the magnificent appearance of a stallion. I could mention here many a plain, small, and faulty Arabian stallion who left us an excellent progeny only for the one reason that he descended from doubtless good and noble parents."

The presence of the stallion Cercle in the pedigree extension (Figure 17) is noteworthy. He was the sire of Kortez, who was the sire of Rymnik, who was the sire of Jaskolka, the highly regarded Polish broodmare. Cercle (1863) was out of the Polish mare Krytyka (1858), who was herself out of Kokietka (1853). Kokietka's pedigree contains two lines to the black stallion Szumka II, whose dam was an unidentified English Thoroughbred mare.

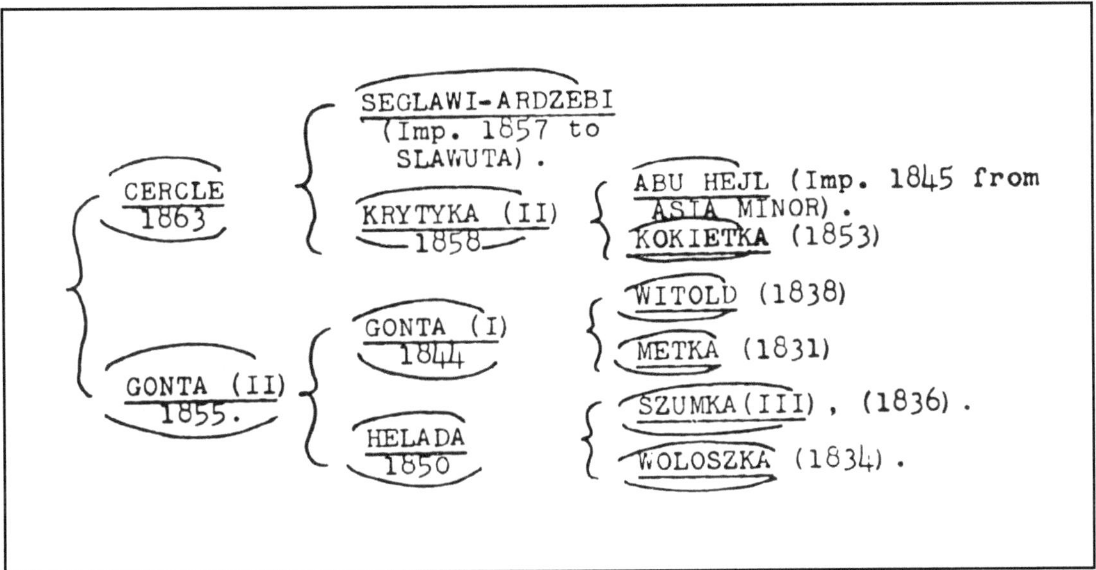

Figure 17

The following portion of the pedigree traces the mare Kreolka to the Polish mare Iliniecka. Here the record of ancestry ends: the Sangusko family records contain no mention of the parents of this mare.

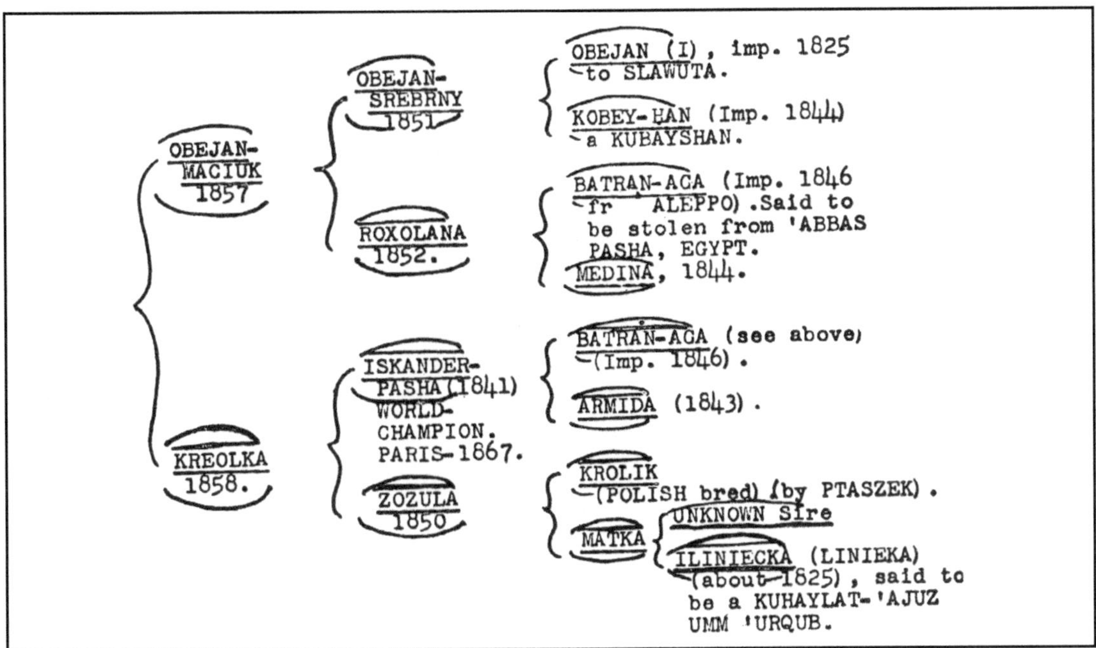

Figure 18

The next four segments of the pedigree below illustrate that Szumka II was a half-Arab whose mother Polka was an English mare. Polka is 10 generations removed from Skowronek.

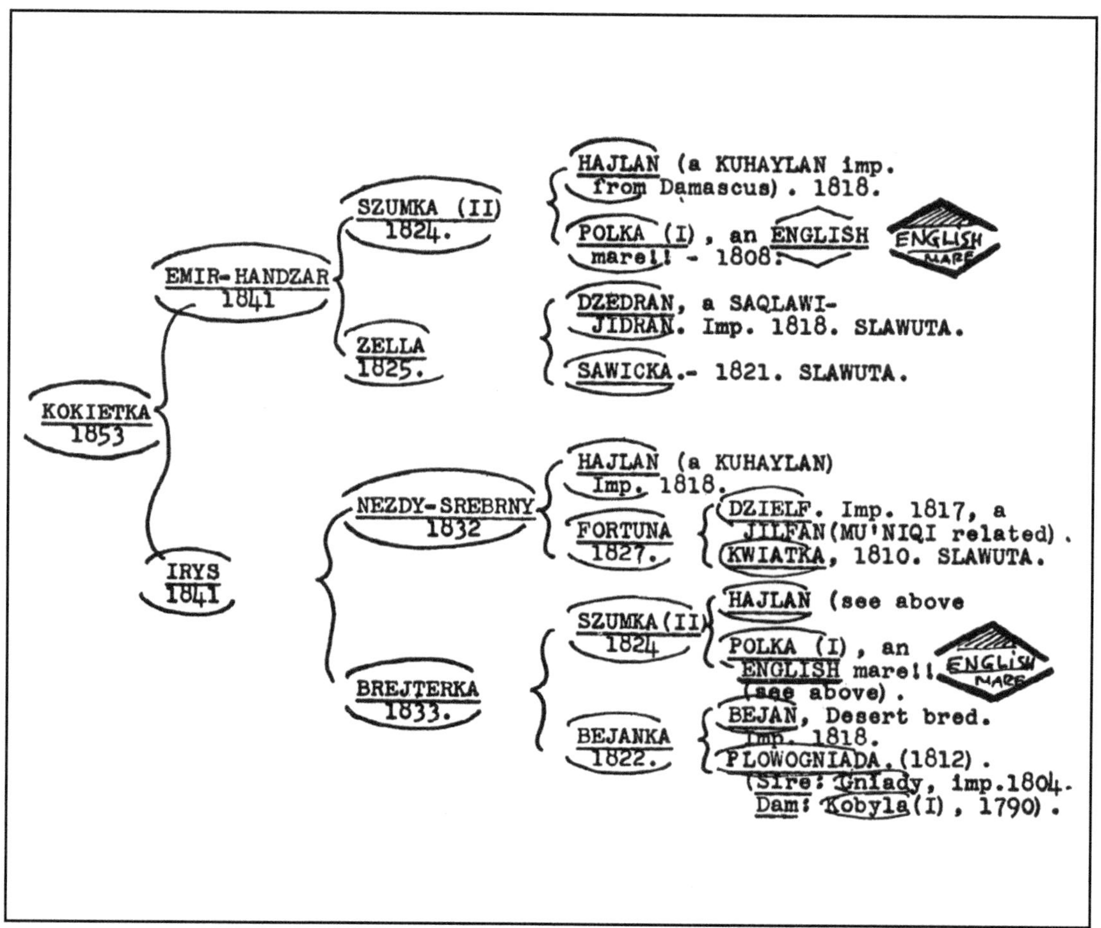

Figure 19

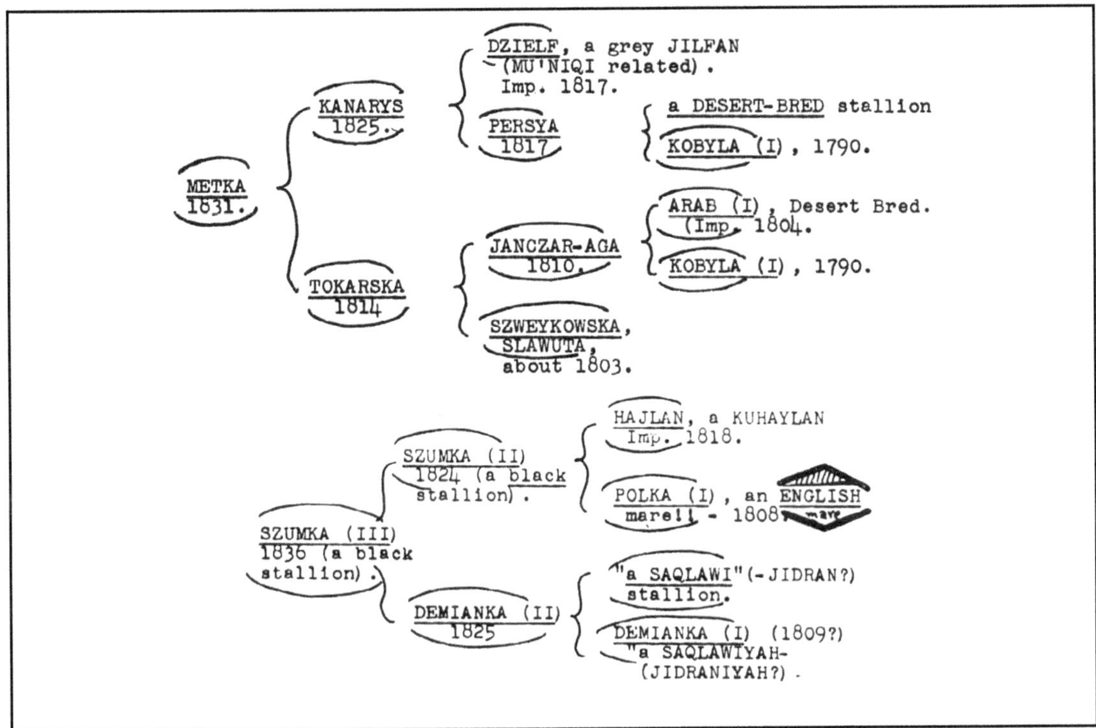

Figure 20

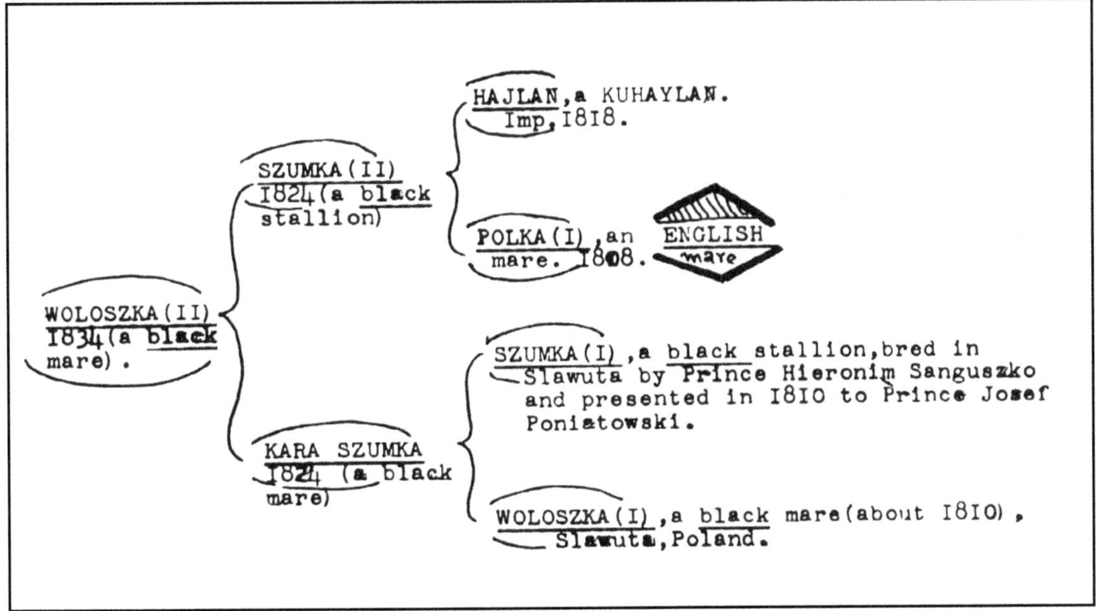

Figure 21

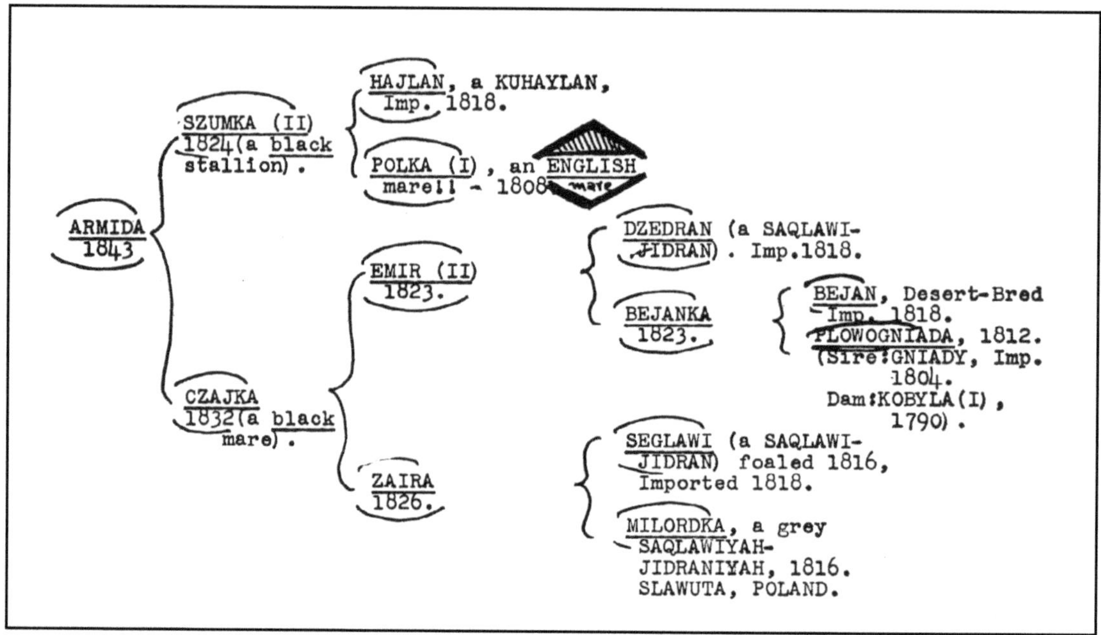

Figure 22

The following pedigree segment introduces several of the undocumented females in the extended pedigree of Skowronek.

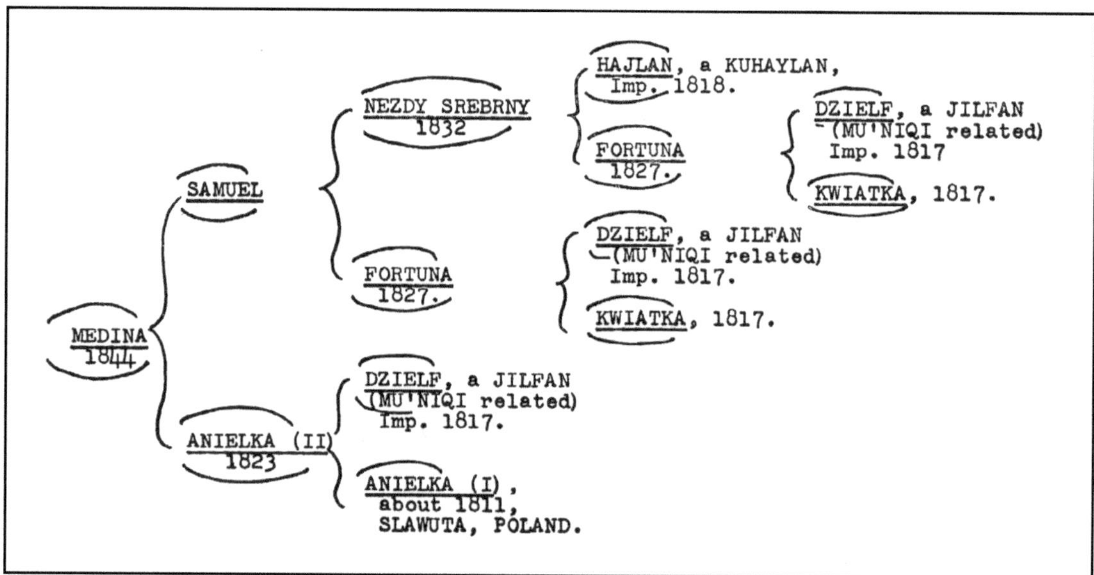

Figure 23

The names of the non-Arabian foundation elements of Skowronek's pedigree are enclosed in boxes in Figure 25. Those horses that are enclosed in a red box were local Polish mares with no known pedigree. The stallion Derwisz is close up in the pedigree. He has no pedigree and was imported from Turkey in 1861. He was probably a Turkoman horse.

The names of the locally bred mares were generally common Polish family names. Anielka means "Angel". The word "Kwiatka" does not exist in the Polish language, and appears to be derived from the Polish "kwiat" meaning "flower". However, "kwiat" is a masculine form and would not be appropriate for the name of a mare. The mare's name was probably created by the owner by adding the traditional "ka" suffix to create a feminine word to be used as a name.

Czajka is a Polish common name for the Northern Lapwing, *Vanellus vanellus*.

Sawicka is a common and widely encountered family name, a surname,among Poles.

Kobyla, a Turkish loan word to the Slavic languages, means "filly or mare". The names Milordka and Malikarda do not exist in the Polish language and appear to have been cobbled together by their breeder using English and Polish elements. Woloszka, Sawicka, Withold, and Szweykowska are common Polish family names.

Plowogniada is a Polish word that means "a buckskin horse". Gniady is a proto-Slavic Polish word that describes the coat color known in English as "sorrel" or "bay". It is a popular name given to unpedigreed farm animals. Notably, the Konik breed of horses is predominantly buckskin colored.

The matriline of Skowronek deserves special attention. The presence of the mare Ilinecka, which is properly spelled Linecka, has caused some confusion. Her name in Polish means "from Lince". It is not a name but a description of the place where she was born. She was born in 1815 at the Sangusko stud in Lince, one of the many scattered Sangusko family breeding farms. She was transferred to the Slawuta stud in 1822. She was therefore not desert bred. Lince (or Linieck) is a small rural village in modern-day Ukraine. Records show the sale in 1833 of the Lince property by Prince Sangusko, along with several villages contained on the property. The bill of sale indicates that the name of the property had formerly been Ilince. This accounts for the spelling of the mare's name as Ilinecka. The correct name for this mare is Liniecka (a mare from Linieck, or Lince, as modern Polish renders it).

HEJIR

IBRAHIM

FALITE

SKOWRONEK

RYMNIK

YASKOULKA

EPOPEJA

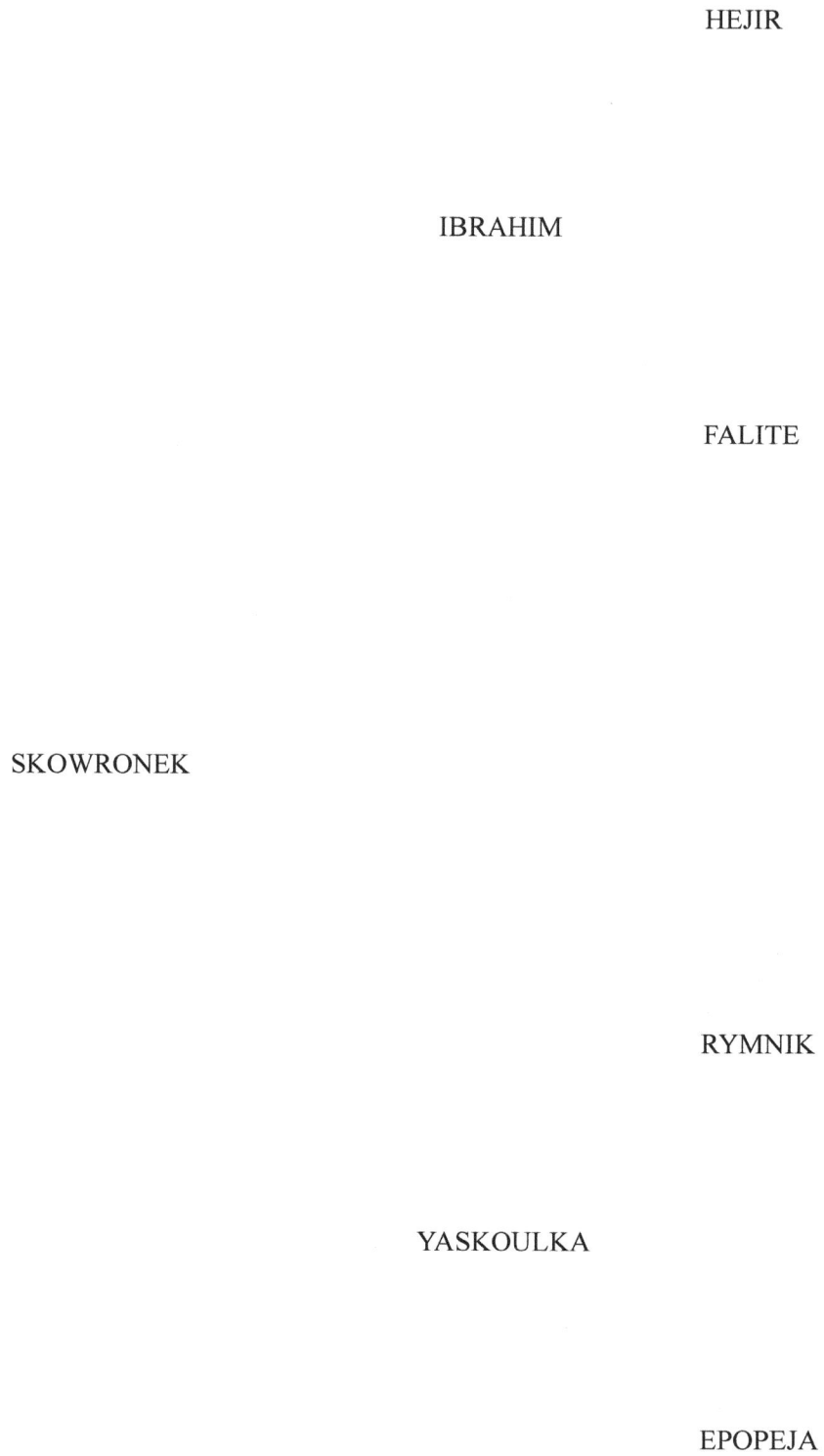

Figure 24 - *Skowronek's pedigree.*

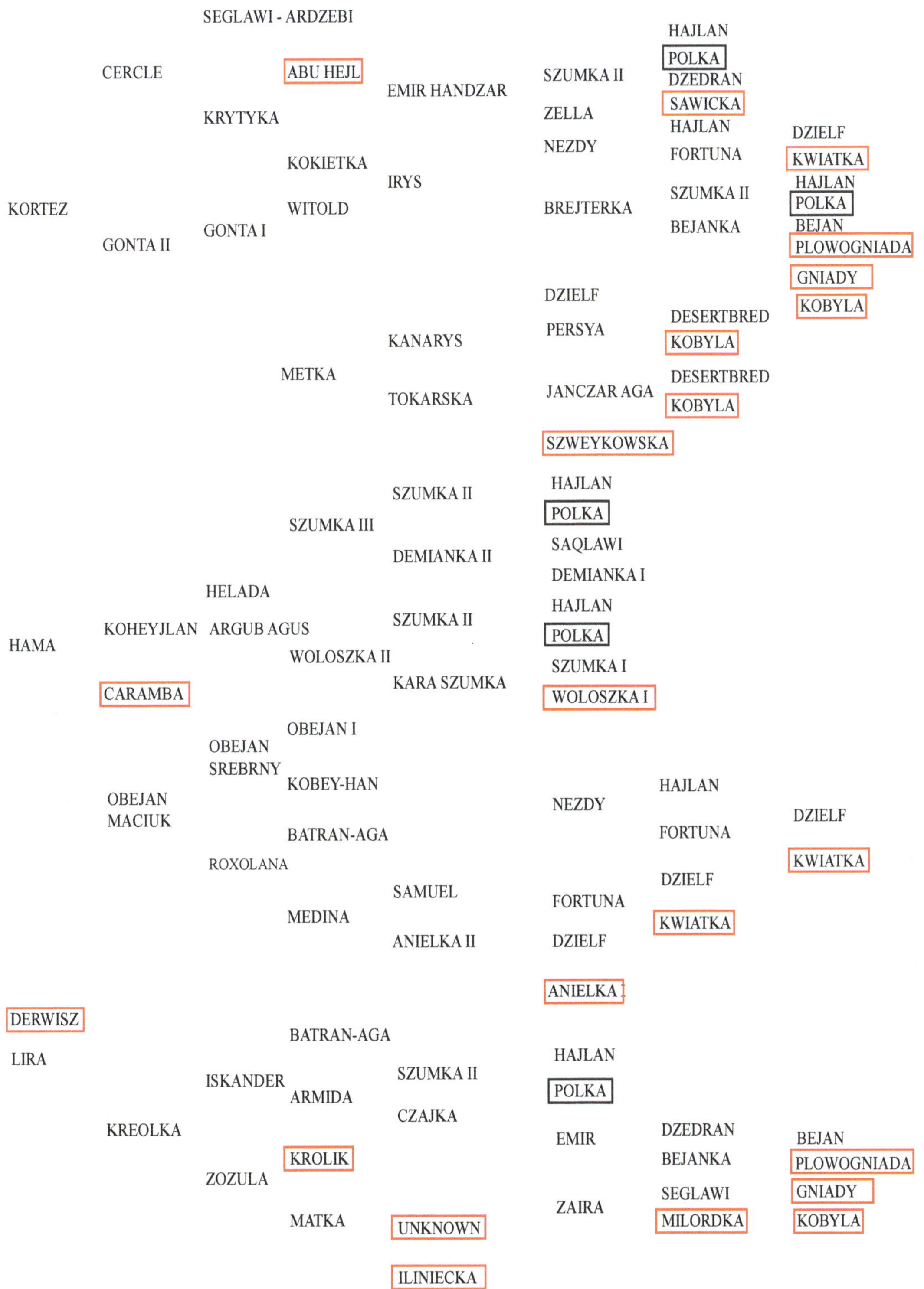

Figure 25 - *Skowronek's pedigree.*

The name of the English Thoroughbred mare Polka is shown in the pedigree enclosed in a black box (Figure 25). She was a bay mare, born in 1808. Her pedigree is unknown. Her source is unknown. She is present in the pedigree four times. The identity of Polka 1808 as a Thoroughbred mare was confirmed by a series of highly regarded Arab horse specialists; Dr. Lukomskiski in Das Arabische Pferd in Slawuta, written in 1906, Dr. J.E. Flade in Die Arabishe Vollblut in Polen, Prince Wladyslaw Sangusko in his book published in Krakow in 1850, and Dr. Edward Skorkowski.

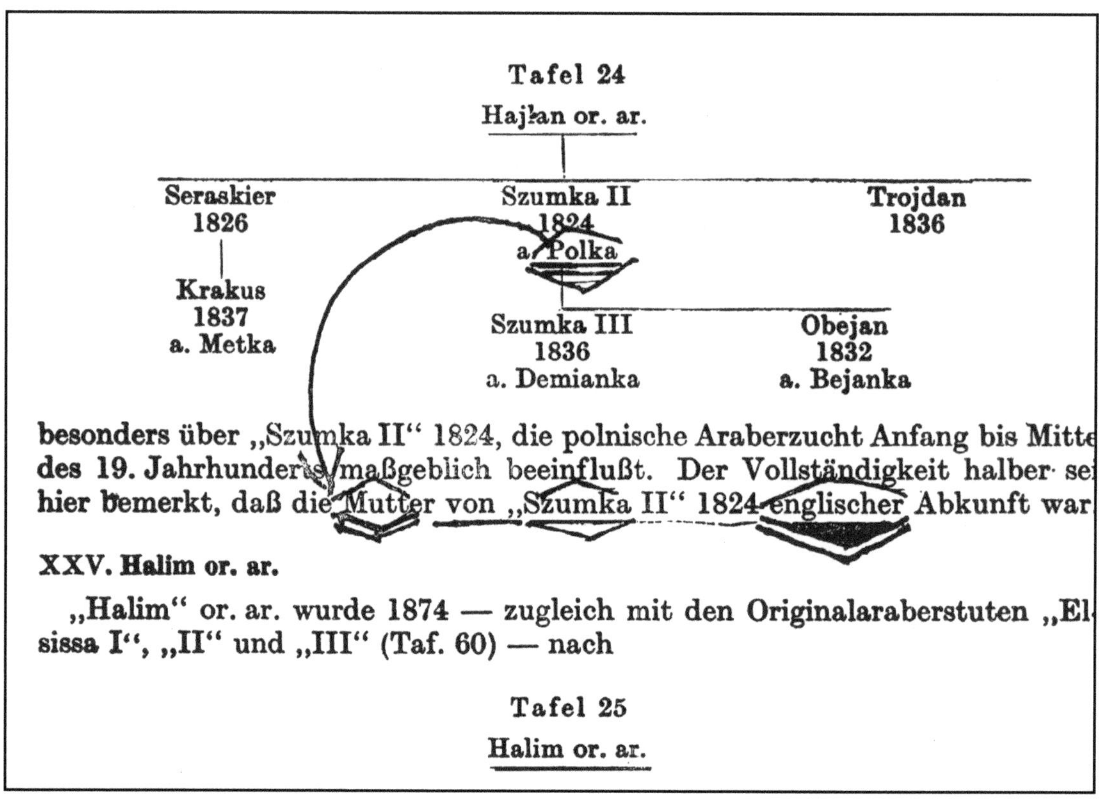

Tafel 24

Hajłan or. ar.

Seraskier **1826**	**Szumka II** **1824** **a. Polka**	**Trojdan** **1836**
Krakus **1837** **a. Metka**	**Szumka III** **1836** **a. Demianka**	**Obejan** **1832** **a. Bejanka**

besonders über „Szumka II" 1824, die polnische Araberzucht Anfang bis Mitt des 19. Jahrhunderts maßgeblich beeinflußt. Der Vollständigkeit halber se hier bemerkt, daß die Mutter von „Szumka II" 1824 englischer Abkunft war

XXV. Halim or. ar.

„Halim" or. ar. wurde 1874 — zugleich mit den Originalaraberstuten „El sissa I", „II" und „III" (Taf. 60) — nach

Tafel 25

Halim or. ar.

Figure 26 - *A pedigree indicating that the mare Polka is the mother of the stallion Szumka II, from the book Die Arabische Vollblut-zucht in Polen, by Dr. Johannes Flade, published in Berlin in 1958, photostatic copy from The Raswan Index. The text explains that the stallion Szumka II is a son of the English mare Polka.*

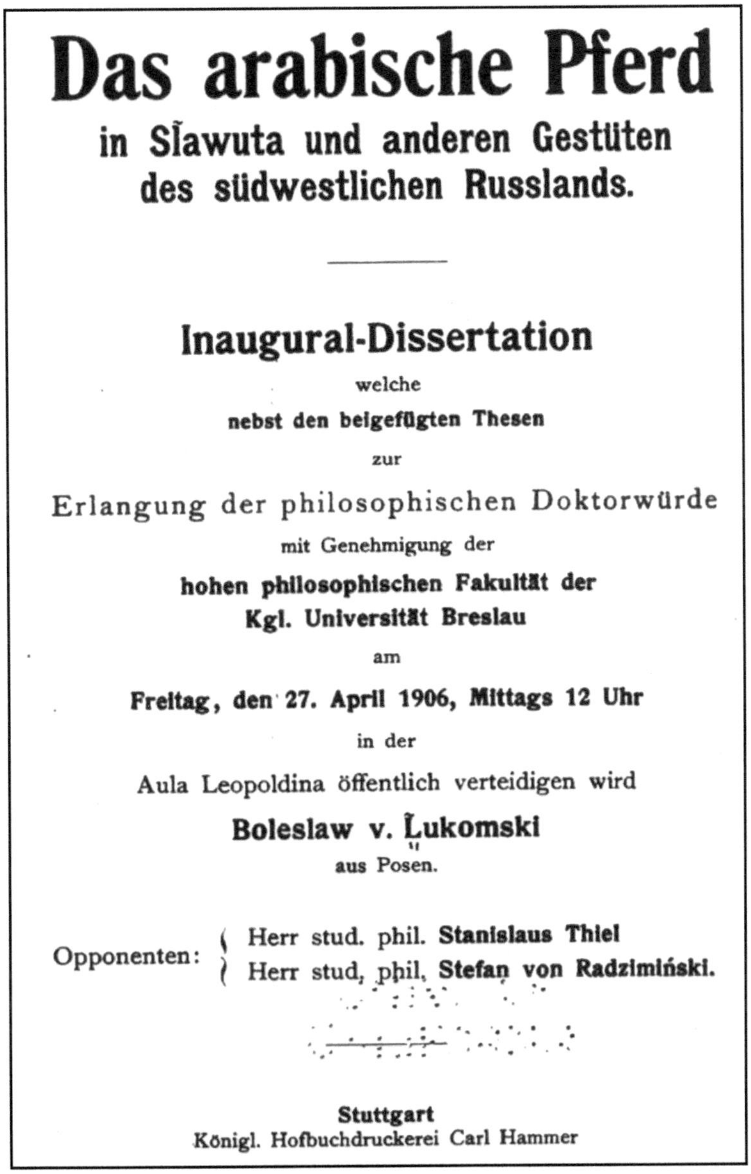

Das arabische Pferd
in Sławuta und anderen Gestüten des südwestlichen Russlands.

Inaugural-Dissertation

welche

nebst den beigefügten Thesen

zur

Erlangung der philosophischen Doktorwürde

mit Genehmigung der

hohen philosophischen Fakultät der Kgl. Universität Breslau

am

Freitag, den 27. April 1906, Mittags 12 Uhr

in der

Aula Leopoldina öffentlich verteidigen wird

Boleslaw v. Łukomski

aus Posen.

Opponenten: Herr stud. phil. **Stanislaus Thiel**
Herr stud. phil. **Stefan von Radzimiński.**

Stuttgart

Königl. Hofbuchdruckerei Carl Hammer

Figure 27 - *The announcement of the 1906 doctoral dissertation of Boleslaw v. Lukomski concerning the Arabian horses of the breeding farms of Poland. Lukomski, a Pole, did extensive research on the subject and was personally familiar with the farms, the breeders and the horses themselves, spending many months personally examining each of the farms and horses in his study. The dissertation contains the pedigree of Szumka II, giving as his mother the mare Polka, an English Thoroughbred.*

There was another female horse named Polka in Poland, born in 1843 at the Chrestowka stud in Poland. She was not connected to the Skowronek pedigree. Her mother was Persia, a matrilineal descendant of the Polish mare Kobyla, another of the many locally sourced Polish horses that were bred into the Arabian stock in the early 19th century.

Szweykowska (1803), Woloska (1810), and Milordka (1816) were foundation mares at the Sangusko studs.

Leaving aside the issue of the origin of Skowronek's sire Ibrahim, the pedigree of Yaskoulka contains many non-Arabian horses:

1. Over 50% of the foundation stock in the pedigree of Skowronek is non-Arabian.

2. The English Thoroughbred mare Polka appears five times.

3. The unpedigreed Polish mare Kobyla appears four times.

4. A case of intense inbreeding is noted. Fortuna was bred to her son to produce Samuel, who was then bred to Fortuna's half-sister to produce Medina.

5. Two stallions were imported from Turkey: Derwisz and Abu Hejl. There is no evidence that they were bred in Arabia.

6. Caramba was acquired in Syria. There is no evidence that she was from Arabia.

7. There is no evidence that Skowronek's matriline originates with an Arabian mare. The evidence suggests that she was not.

Raswan is not read much these days. His legacy recedes more with each passing decade. His ideas about blood purity have lost much of their appeal, given the success of the Arabian horse breed relativists. Yet his work remains. The terms that he used to refer to desert horses, such as *purebred Arabian horse* and O.A. (Original Arabian) now seem antiquated.

Arabian horse authority Erika Schiele wrote:

> "It was not until Carl Raswan, a German who spent nine years in all of his twentieth-century life among nineteen different Bedouin tribes, came on the scene, speaking and reading Arabic as well as his mother tongue, that he was privileged to reveal to the West the broad outlines of Arabian horse breeding. To him we owe the knowledge that all strains and families are classifiable under three biotypes, all issuing from the old "Kuhylan Ajus" (old pure blood).

His critics were numerous. One stated that his book is 'cryptic, garbled, ambiguous, incoherent, confusing, and downright contradictory'. But then, prophets and pioneers often are misunderstood.

ARABIAN HORSE BREEDING IN CENTRAL EUROPE IN THE 18TH AND 19TH CENTURIES

Skowronek's pedigree is constructed in much the same way that the English Thoroughbred was created: by breeding imported desert Eastern stallions to local non-pedigreed mares in order to create a race of horses that differed from any of its individual components. The Polish Arabian horses of today bear a passing resemblance to the authentic Arabian horse precisely for this reason; many *purebred Arabian horses* were used in its creation. The English Thoroughbred of today most closely resembles the Turkish and Turkoman desert horses from which it was created, and less from its Arabian antecedents.

This method of breeding horses was followed by most of the governmental and private studs of Europe during the 19th century. It was standard policy for horsemen of the time to regard them as "*purebred Arabian horses*". There was nothing controversial about it. Not at the time.

There were sporadic and isolated attempts to create 100% pure desert sourced Arabian horse studs in Europe. The wisdom in doing so was certainly appreciated by the European breeders who most admired the true desert horse. The will to make the attempt was certainly present. But obtaining a sufficient number of high-quality desert mares to create a viable breeding program outside of Arabia was nearly impossible. Good stallions could be found on the peninsula, and the Arabs would sell them. And there was the problem of the money needed to finance a venture of this type.

The Bedouin would not part easily with a first-rate mare, at any price. Men such as de Portes, King Wilhelm I of Wurtemberg, Prince Puckler, Count Rzewuski, Count Dzieduszycki, and Colonel Brudermann tried, but they all failed to leave a durable legacy.

King Wilhelm established his own private Arabian breeding stud near Weil, Germany, beginning with importations from the desert in 1812. When he died in 1864, the entire herd was willed to the Marbach stud.

The Frenchman Monsieur de Portes made an attempt at founding a *purebred Arabian* breeding program in Pau, France. In 1820 he secured 39 aristocratically bred stallions from Syria and brought them to France. The most highly regarded of the group was Massoud, bred by the Fid'an Bedouin. He was regarded as "a rare beauty of perfection". The group, however, did not persist intact, apparently due to the lack of sufficient numbers of quality desert mares. The French were not fond of the de Portes horses, judging them to be too small to be of any use.

The German Prince Hermann von Puckler-Muskau was able to purchase a number of asil horses from the Roala tribe in Arabia. The stallion Zarif, bought from the tribe in the Hamad Desert of Northern Arabia, was a superior specimen. But Puckler was a nobleman and an artist, and he lacked the commitment to sustain his interest in horses.

He was a peculiar man, with a consuming interest in landscape gardening. He traveled extensively in the Middle East. During a trip to Cairo in 1837, he happened to notice an exotic Ethiopian slave girl of about 14 years of age on the auction block. He promptly bought her and named her Mahbuba, the beloved. He returned to Vienna with the girl, displaying her to all of the Viennese high society as his consort. Sadly she contracted tuberculosis and died in 1840.

Figure 28 - *Zarife, by Emile Volkers.*

On Puckler's death in 1871, Zarif was given to the Weil stud where his contributions to Arabian horse breeding and Anglo-Arab horse breeding were significant.

Count Waclaw Rzewuski's acquisitions of desert horses were more numerous, but as a group, they left less of an impression on European breeding. He was a wealthy Pole and was passionate about the Arab horse. In 1820, after spending three years collecting Bedouin sourced horses in Arabia, he sailed for Europe with 89 stallions and 45 mares. He settled the horses on his family estate near Sawran, Poland, and he lived there with them for eleven years. At his death, the horses were auctioned and dispersed.

The Polish Count Dzieduszycki imported 23 stallions and 4 mares from Arabia, breeding them on the family estates at Jarczowce, Niesluchow, and Jezupol. The project was notable for the acquisition of three particular mares; Gazella, Mlecha, and Sahara. These mares of impeccable credentials form the core of Polish Arabian horse breeding to this day.

Colonel Brudermann was an Austrian, and a military man, head of the Babolna stud farm in Hungary. In 1856 he brought 16 stallions and 50 mares to Europe from the Arabian Desert, all horses of the finest quality and of unquestionable purity. But he was a cavalryman, and his horses went to studs that used them to breed with non-Arabians, for the purpose of improving the quality of the stock. They were not kept intact as a *purebred Arab* breeding program. His horses became important elements in the breeding programs of Lipizza and Babolna.

The Blunts of Crabbet Stud spent a great deal of time with the Polish Arabian breeders, and they were familiar with the Arab horse breeding methods and philosophies used in Poland. Lady Anne noted that the early stud books of the farm were kept under lock and key and were shown to no one. Count Jozef showed the later breeding records to the Blunts, but would not reveal the names of the foundation horses of Antoniny.

Figure 29 - *Lady Anne Blunt and Wilfred Blunt, from Greely, Arabian Exodus, J.A. Allen.*

Lady Anne Blunt and her husband Wilfred worked closely with Count Potocki during the 1890s and visited his farm in Poland several times. They had trouble following the pedigrees as Potocki translated for them out of the Polish Stud Book, but Lady Anne made the general statement that "the discussions were clear to show that there is not one without a flaw." "It seems that most of them (the pedigrees) are faulty having a flaw some generations back."

As for Potocki himself, her opinion was unreserved; "though he does not know a good horse when he sees it he ought to know how to behave."

"Polka," she wrote, "sounds like a Polish mare and Shumka II is a doubtful sire." The horses that she saw convinced her that they were mostly "not pure". She noted that the presence of English Thoroughbred blood in the background of the horses was evident "in the head if one knows it".

Figure 30 - *Prince Roman Sangusko at the head of the stallion Achmet.*

11

to z sobą podobieństwo: iż nikt nie byłby zdołał znaleść w nich jakiéjkolwiek wady, a nawet, (śmiem powiedzieć) dopatrzyć się braku przymiotu w którymkolwiek z nich.

Nigdy nie widziałem silniejszego i dzielniejszego konia od *Zbója*, a *Szumka* jest do téj pory najpiękniejszym koniem jakiego zdarzyło mi się w długim moim życiu widzieć. *Szumki* ojciec był synem konia arabskiego, matka zaś jego, była kobyła angielska a sam urodzony u dziada mego w Sławucie. Pochodzenie *Zbója* było całkiem niewiadome, gdyż Turek co na nim jeździł poległ w bitwie, a koń był schwytany od żołnierzy ruskich. Jednakowoż dla znawców niewątpliwą było rzeczą: iż byłto Turkomański koń, rodem z mniejszéj Azyi, i że pochodził z najlepszych zawodów tamecznych. Obydwa te ogiery były używane w tém samém stadzie, przez więcéj jak piętnaście lat, i mimo, że im dawano najcelniejsze kobyły, nic dobrego po nich się nie zostało. Po *Zbója* bez wyjątku wszystko musiało być brakowane, a po *Szumce*, którego ród był więcéj do reszty stada przybliżonym, ledwie parę kobył o tyle się udało, że nie poszły w brak, i że je do stada używano.

Chcąc daléj przykładami moje zdanie popierać, mógłbym tu nie jednego ogiera brzydkiego,

SZTUCE
KONI
I UTRZYMANIU STADA
PRZEZ
Władysława księcia Sanguszkę.
(Nakładem Autora).
KRAKÓW 1850
W ZAKŁADZIE WY. DZ. KA. PRZY ULICY FLORYAŃSKIEJ.
N. 503.

Figure 31 - *Wladyslawa Sangusko wrote a book about the state of Arabian horse breeding in Poland, and it was published in Krakow. Above is a photostatic reproduction of a portion of the 1850 Sangusko book.*

The following is a translation from Polish of the text on the left side of the image (Figure 31). It contains a clear and authoritative discussion on the state of Arab horse breeding among the Polish aristocracy during the 18th and 19th centuries.

ON THE ART OF REARING HORSES

AND MAINTAINING A HERD

By

Prince Władysław Sanguszko

Edition by the author

KRAKÓW 1850

At the Wy. Dz. Ka. Workshop by Ulica Floryańska No. 503

(Page 2)

> (Continued from previous page) similarity with it: so nobody would be able to find any fault, and even (I dare say) see the lack of attribute in any of them.
>
> I have never seen a horse stronger and braver than Zbój (Bandit), and Szumka (female name derived from 'to hum') is until now the most beautiful horse I have ever seen in my long life. Szumka's father was a son of an Arab horse, and his mother was an English mare, and he himself was born at my grandfather's in Sławuta. Zbój's origins were quite unknown, because the Turk who rode it died in a battle, and the horse has been captured from the Russian soldiers. However, it was clear to all experts: that he was a Turkish horse, straight from Asia Minor and that he came from the best contests there. Both these stallions were used in the same herd, for more than fifteen years and even though they were given the best mares, nothing good was left after them. After Zbój, everything without an exception had to be defective, and after Szumka, whose breed was closer to the rest of the herd, as little as a few mares were fine and did not go defective and they were used in the herd.
>
> In a desire to further support my opinion with examples, I could not find one ugly stallion here.

At the Branicki Stud, her observations were similar; "Branicki has abandoned pure Arab horse breeding through lack of stallions, and the English Thoroughbred is invading everything."

The Braniki Stud at Szamrajoka was very old, founded in 1778. The Branikis had originally imported a number of important desert bred horses, and some of them are found in pedigrees today. The stallion Van-Dyke 1898 was a Braniki horse.

Although the Counts Braniki were avid supporters of desert horse breeding, they were most heavily involved in cross breeding experiments, mating Arabian stallions to a variety of locally sourced Russian horse breeds. Only briefly did they breed desert horse to desert horse, despite the fact that they imported more desert stallions from Arabia to Poland than any other family. The Don horses of the Russian Cossacks were derived largely from Arabian stock that the Braniki's gave them with the goal of producing a superior cavalry horse, one capable of warfare, hardy and durable.

The Sangusko family bred Arabian horses at the studs of Slawuta, Chrestowka, and Gumniska beginning in the early 19th century. The policy of breeding first quality desert Arabian stallions to local mares was standard policy. The idea of a truly pure Arabian breeding program was considered impracticable due to the difficulty in obtaining *purebred* desert mares of high quality from the Bedouin. The Sangusko contributions to Arabian breeding include the stallions Kuhailan Hafi and Kuhailan Afaz.

The Blunts impressions of Count Sangusko were markedly different from those of Potocki; Lady Anne wrote in her journal that "he comes off as a breeder of exceptional knowledge and discrimination." Unlike his fellow Polish breeders, Sangusko took the matter of blood purity seriously. When Sangusko was pressed regarding the question of what constituted a "*purebred Arabian horse*", he admitted that he could not say that any of his horses were pure-blooded Arabian. It was for him a matter of honor and scruples. Concerning the Count's own collection of imported desert Arabian horses, he wrote: "des lacunes dans le pedigrees chez tous, ce que je declare de bonne foi." Sangusko, like the Blunts, did not believe that a question mark in a pedigree could be expunged by the mere passage of time.

In 1909, the stallion Skowronek was born at the Antoniny Stud of Count Potocki. The Count did not think much of the little horse, and he was sold to an English painter by the names of Walter Wynans who used him as a model. The horse then passed to Mr. Webb Wares who used him as a hack. The horse was ultimately acquired by H.V. Musgrave Clark. The stallion was, through intermediaries, obtained by Lady Judith Wentworth, and his achievements at Crabbet Stud in England would change the face of Arabian horse breeding around the world forever.

Skowronek, when bred to the authentic desert Arabs of Crabbet, produced a new type of horse. His sons and daughters were sold throughout the world, and the Skowronek sons Raseyn and Raffles came to dominate American Arabian horse breeding for decades. The Skowronek son Naseem was sold to the Tersk State Stud in Russia and his issue proved to be very popular. The sale of Naseem sons and daughters further saturated the world's Arab horse breeding population with the blood of Skowronek.

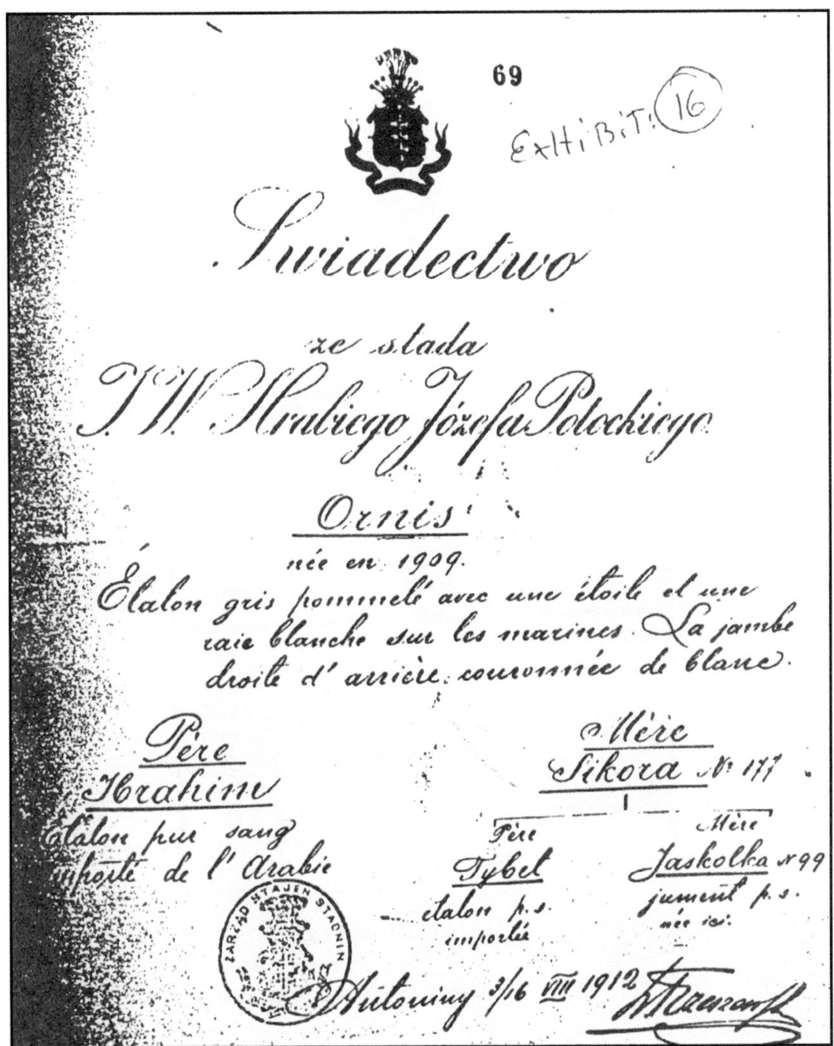

69

Exhibit 16

Świadectwo

ze stada

J.W. Hrabiego Józefa Potockiego

Ornis

née en 1909.

Étalon gris pommelé avec une étoile et une raie blanche sur les narines. La jambe droite d'arrière couronnée de blanc.

Père	Mère	
Ibrahim	Sikora № 171	
	Père	Mère
Étalon pur sang importé de l'Arabie	Tybet	Jaskolka № 99
	étalon p.s. importé	jument p.s. née ici.

ZARZĄD STAJEN STADNIN

Antoniny 3/16 VIII 1912

Figure 32 - *Count Potocki's method of record keeping at Antoniny was ornate. This document regards Ornis who was sold to the Spanish government. Ornis was out of Sikora who was sired by Tybet and was out of Jaskoulka, Skowronek's dam. Tybet, the document indicates, was imported to Antoniny, not bred there. Subsequent research has shown that Tybet's pedigree contains non-Arabian blood. Sikora, therefore, had non-Arabian ancestral elements from both the mother and the father. This document was to figure prominently in the conflict between the Spanish Government's Arabian horse registry and WAHO. From the* <u>PINDEX</u>, *used with the permission of Hansi Heck.*

The countries with the most intensive use of Skowronek blood included Argentina, Spain, Poland, Australia, and South Africa, and the U.S.

Skowronek died in 1930 at age 22. His skeleton was donated to the British Museum.

In 1954, the argument that had begun with the observations of Lady Anne Blunt in 1895 was continued in Cairo.

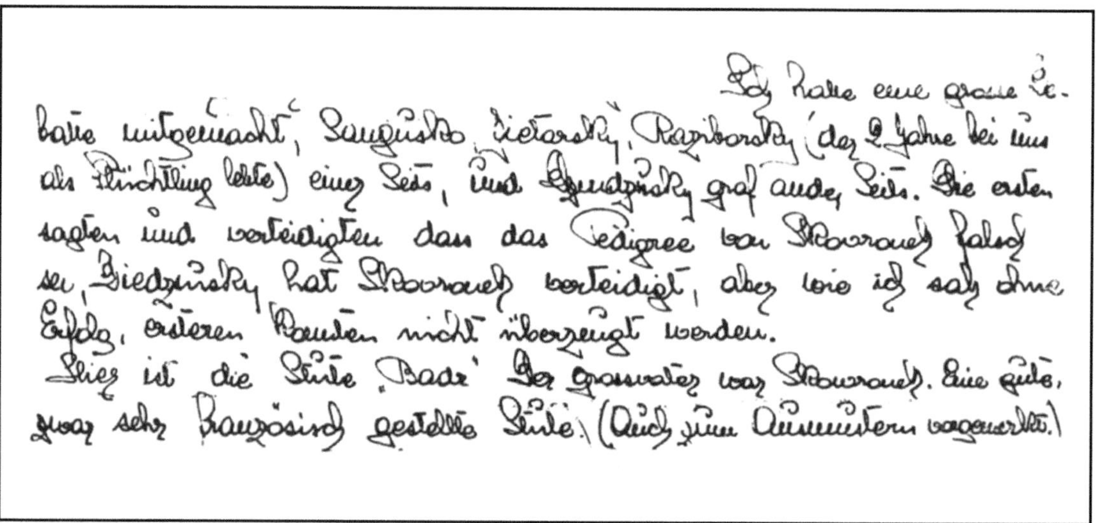

Ich habe eine grosse Debatte mitgemacht, Sanguszko, Zietarsky, Raziborsky (der 2 Jahre bei uns als Flüchtling lebte) einer Seits, und Dziedzinsky graf ander Seits. Die ersten sagten und verteidigten dass das Pedigree von Skowronek falsch sei, Dziedzinsky hat Skowronek verteidigt, aber wie ich sah ohne Erfolg, ersteren konnten nicht überzeugt werden.
Dies ist die Stute „Badr" der Grossvater war Skowronek. Eine gute, zwar sehr französisch gestellte Stute. (Auch zum Ausmustern vorgemerkt.)

Figure 33 - *A letter from General Tibor von Szandtner at the EAO to Carl Raswan in 1954.*

The following is a translation of the letter in Figure 33:

> I had taken part in a lengthy debate, which included on one side Prince Roman Sanguszko, Colonel Bogdan Zietarski, and Raziborsky, and on the other side Prince Dziedzusky. The first group said and defended that the pedigree of Skowronek (as presented by Lady Wentworth) was "falsch", wrong. Dziedzusky defended Skowronek, but I noticed without success since the others could not be convinced.

This debate, which took place among noted authorities on the Egyptian Arabian horse, capsulized the future debates between the skeptics and the purists, the absolutists and the relativists. The hotly debated issues of pedigree analysis and the meaning of the word "*purebred*" were gaining momentum. It was a debate regarding definitions.

The Polish breeders never claimed that their horses were 100% desert bred Arabians. Their definition of a *purebred Arabian horse* was one produced by breeding a highly qualified true desert stallion to a locally sourced non-Arabian mare. This disagreement was based entirely on semantic differences and was the *forme fruste* of the debates that would take place between the pedigree purists and the skeptics of the future. It represented the nucleus of the debate that was to occupy the minds of prominent Arab horse breeders of the 1960s, leading to the development of the WAHO doctrine of genetic dilution and pedigree relativism: the "definitions".

Figure 34 - *General Tibor von Szandtner and Moniet el Nefous at the EAO, The Raswan Index.*

Dzieduszycki, the sole defender of the *purebred* status of Skowronek, was a member of a distinguished Polish family that had maintained an Arabian breeding program at the Jezupol Stud in Poland for several generations. The Dzieduszycki family had imported the desert bred mares Gazella, Mlecha, and Sahara in 1845, and the Indian stallion Zulejma in 1910. The blood of these mares was later incorporated into the Polish Government breeding program. Count Alexander Dzieduszycki who took part in the debate with von Szandtner, was President of the Arab Horse Breeding Society of Poland from 1925 to 1945.

Figure 35 - *Carl Raswan, used with the permission of The Raswan Archives.*

This debate occurred three years before the death of Lady Wentworth and prior to the revelation of 1957 that the pedigree of Skowronek, which she had artfully constructed and with vigor promoted, was in places false and in places incomplete.

Horsemen of this time were much less focused on the absolute truth of a pedigree, and were much more focused on the qualities of the horse itself. This was natural. These men rode horses for pleasure and for hunting, and they worked astride their horses. They rode these horses in battle. They did not simply admire them. Europeans horsemen bred horses for improved utility and enhanced performance characteristics, and they knew from experience that this was best accomplished by breeding authentic desert Arab sires to mares of mixed European domestic stock. They took advantage of the animal breeding law of hybrid vigor. This fact is to their credit. The fact that they often glossed over matters of pedigree accuracy is regrettable but understandable, given that they never imagined that the supply of high caste authentic desert horses from central Arabia would ever come to an end. They were not yet aware of the emerging crisis of the failure of horse breeding in the Arab desert, and consequently, they were able to appreciate the part-bred Arab, as well as the *purebred Arab*, with a more naive and less cynical enthusiasm.

Raswan had high regard for all Arabians – *purebred* and the part-bred alike. He rode. Raswan was not strictly a purist or a skeptic; he was a horseman, through and through. While it is true that he pursued the quest for blood purity with unflagging energy, this was not the full extent of his interest. He was just as enthusiastic and devoted to the part-bred Arabian. His encounters with the part-bred Arabian stallion Skowronek and Lady Wentworth at Crabbet Stud in England read like the impressions of a poet entranced and transported by a feeling of deep emotion produced by a vision of extreme beauty. His heartfelt attraction to the Skowronek offspring resulted in his acquisition of the stallion Raswan, a story of love and betrayal that he tells in The Raswan Index with great feeling and sensitivity.

To horsemen in the first half of the 20th century, the word "part-bred" had not yet acquired the pejorative sense that it would acquire in the latter half of the 20th century. These early Western breeders were not preservationists, they were cavalrymen and pioneers. The awareness of the importance of preservation simply did not exist then. These early breeders would have been puzzled by the pedigree wars that were taking shape; to them, a good horse was a useful properly conformed tractable horse. The importance of a good horse to a cavalryman engaged in the heat of battle was no less significant than the importance of a good horse was to a Bedouin fleeing from his enemies across the endless flat desert in Arabia.

The American horse breeder Henry Babson was an eclectic breeder. He was not averse to using "part-breds", mixing many different types of bloodlines in his breeding program. In the spirit of continuous experimentation, he included Polish blood and Saudi bloodlines into his herd, although he found from long experience that these non-Egyptian additions did not prove in practice to have a beneficial influence on his breeding program, and he eventually eliminated them from his herd.

Despite its flaws and inconsistencies, and in spite of those breeders critical of his research methods and conclusions, The Raswan Index set a new standard of rigorous pedigree research and verification. More importantly, it created a sense among breeders that the issue of blood purity was an issue worthy of serious attention. For Raswan, purity was the issue. He saw first-hand the preference in Europe and later in America for the part-bred Arab horse that was taller and stouter than the authentic desert Arab that he was so familiar with from his desert experiences. He regarded this Westernization of the Arab horse to be a real threat to the survival of the truly *purebred* antique Arab horse, distracting the horse public from the importance of breeding the antique horse in a straight line of genetically stainless individuals. He foresaw the day when this race of horses might become extinct.

The Purists (some people call them fanatics) among Arabian Horse breeders are not tempted by these new conditions, they ignore the demand for more size, bone and muscle and continue to preserve the original ("antique") Desert-type (an Arabian that even in a starved condition still looks magnificent) and whose finer bone the purists compares to fine steel.

Carl Raswan, The Raswan Index.

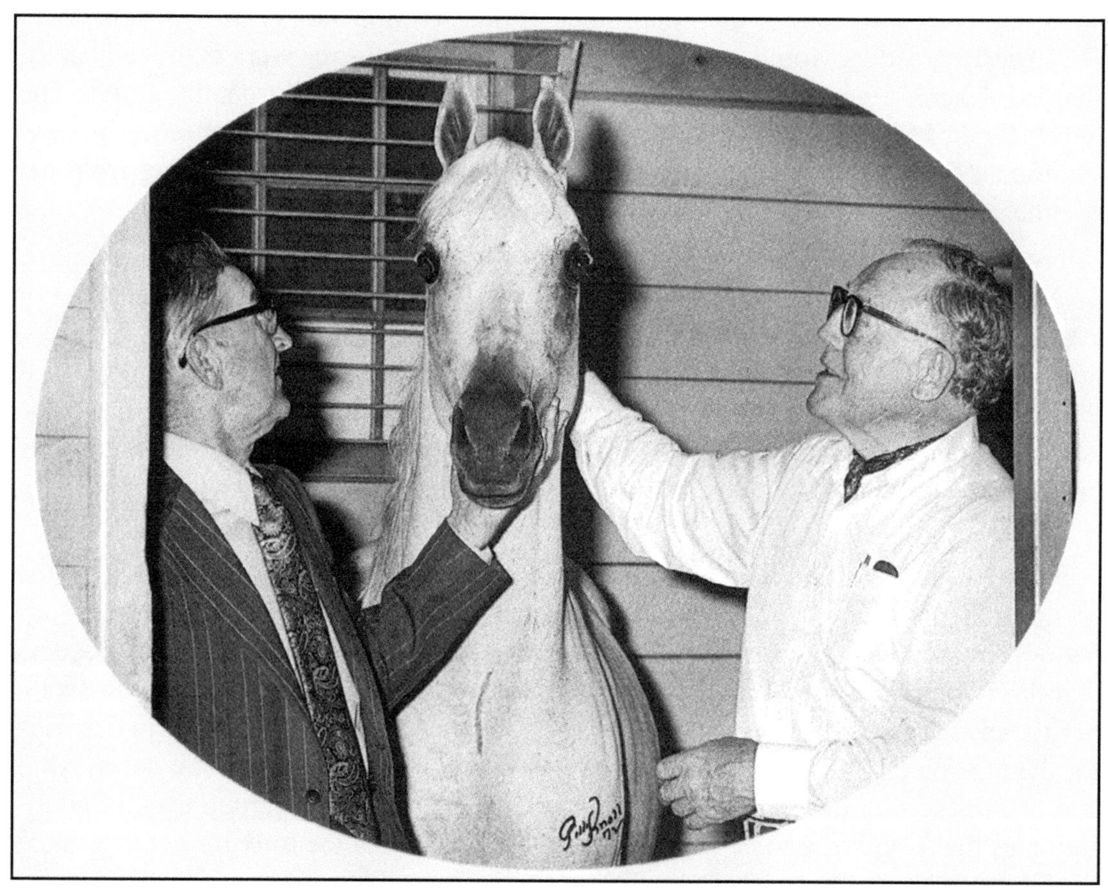

Figure 36 - *Frank McCoy, Ferzon, Daniel Gainey. Ferzon, a popular and widely used stallion in the U.S. was born in 1952 at the farm of Frank McCoy. He was purchased by Daniel Gainey for $10,000 and went on to create a dynasty, producing 251 offspring. He was the grandsire of 7,161 foals. He exemplified the degree to which Skowronek blood had penetrated the American market. His pedigree contained three crosses to Skowronek. Photograph by Polly Knoll.*

His work drew widespread public attention to the fact that the pure desert horse was rapidly disappearing from the Arabian Peninsula. He sounded the alarm that there would be no more journeys into Arabia with the intention of procuring classic Nejd horses of clear provenance and high caste. It was his opinion in 1957 that only a small number of truly 100% pure-blooded desert Arabian horses were still in existence anywhere in the world. His call to action created the first glimmerings that interested parties could and should identify these horses and collect them into a central genetic storehouse, to be preserved in perpetuity. This work established the notion among breeders that most of the world's registered Arabian horses did, in fact, contain elements of the blood of non-Arabian horses. There were also some horses, Raswan opined, that were recognized as Arabian that were, in fact, authentic genetically unsullied *purebred Arabian horses*. To Raswan, this meant a horse entirely descended from 100% desert stock and unadmixed with foreign blood. And that is all that it meant.

Skowronek was a problem; his blood had saturated the world of Arabian horse breeding to an extent that has never since been equaled. From this time forward, the serious pedigree researcher could regard very few horses registered with the Arabian Horse Registry of America or the Arab Horse Society as unequivocally 100% *purebred Arabian* bloodstock. Researchers in subsequent decades were inspired by the scholarly example of Raswan, and this enthusiasm led to a variety of efforts to re-analyze the historic data in an attempt to arrive at a set of verifiable and universally acceptable pedigrees of the rootstock of the true authentic genetically unadmixed desert Arab horse. There was a sense in the air that the final search for the Holy Grail of Arabian horse pedigree authenticity was underway. There was also the sense that time was running out. The work of Raswan established indelibly in the minds of breed enthusiasts the notion that the truly *purebred Arabian horse* was endangered, threatened on all sides by those who found the claims of the Bedouin conception of blood purity to be singularly unconvincing.

Raswan influenced many of the Arabian horse breeders of his day including Richard Pritzlaff, (1902-1997) a Harvard educated American horseman from Rancho San Ignacio in New Mexico. Pritzlaff was by training and inclination an artist. He was a friend of the American Southwest artist Georgia O'Keefe and was active in the National Endowment for the Arts. He loved the Arabian horse for its beauty and usefulness under saddle.

Richard Pritzlaff was born in May of 1902 in Milwaukee, Wisconsin, U.S.A.

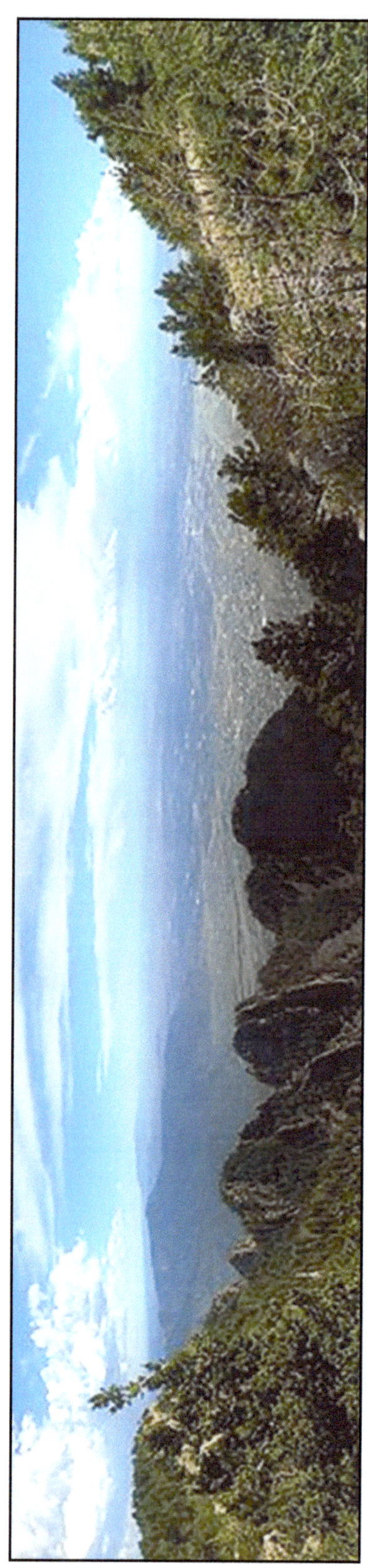

Figure 37 - *Sandia Mountains, New Mexico, the view from Sandia Crest.*

During the 1940s, Pritzlaff lived alone in a cabin high in the New Mexico Mountains. He became impressed with the functional value of the Egyptian Arabian horses, having ridden daily with the high country cowboys to check on the cattle that were pastured there. He saw the stallion *Zarife at the Van Vleet Ranch in Colorado. He also visited the Babson Farm the in Illinois and was again impressed with the functional durability of the Egyptian horses.

Figure 38 - *Bint Moniet El Nefous.*

A chance meeting with Raswan in New Mexico resulted in a lifelong friendship. After recovering his health, Raswan established a farm in the Sandia Mountains east of Albuquerque. It was only 75 miles from Pritzlaff's sanctuary high in the mountains around Sapello, New Mexico, near Hermit's Peak. Raswan began a modest breeding program of his own.

Figure 39 - *Richard Pritzlaff and Dymoniet RSI. Richard remained active into his 80s. He was fond of telling visitors that he decried the modern market in Egyptian horses, as the breeders had been "corrupted by money". He died at the age of 94.*

Richard Pritzlaff was a man of independent means and he became Raswan's benefactor during the 1950s as Carl worked year after year to complete his Index. The two friends often rode together, and Pritzlaff's fondness for the Arabian horse led him to obtain his own Arabians for use at the New Mexico ranch.

Richard was impressed by Raswan's command of the body of knowledge concerning Arabian horses, and he asked Raswan for advice on buying Arabians. Raswan recommended that he buy the weanling filly Rabanna, bred by Delma Gallager of California. This purchase was to have significant effects for Pritzlaff in the future.

When Rabanna matured, he bred her to the Doyle stallion Ghadaf in 1955 and she produced the filly Kualoha in 1956. This decision too was to have future repercussions. But Richard was a confirmed Egyptian advocate by this point. He was caught up in the spirit of the Egyptian excitement.

At Raswan's urging, Pritzlaff went to Egypt and sought out the EAO director General Tibor von Szandtner for advice about importing horses from Egypt. Five horses were selected, four of whom were for sale. These were Rashad Ibn Nazeer, Bint El Bataa, Bint Nefisa, and Bint Dahma. But the EAO management was very reluctant to let go of the fifth horse, Bint Moniet el Nefous. However, pressure was applied by the Nasser government, who demanded that the EAO must remain economically self-sufficient. This was in 1958, only six years after the Egyptian Revolution, and conditions in Egypt were desperate. The General himself knew that he could not stay in Egypt much longer, living under such dire conditions.

"I made an offer on the five horses, and although the E.A.O. did not want to lose *Bint Moniet el Nefous, Egypt needed the exchange of American money for medical supplies and other necessities. So the Government prevailed and my offer was accepted."

The Pritzlaff importation of 1958 represented the most significant introduction of Egyptian horses into America since the 1932 Babson importations.

Raswan died in 1966 from silicosis, a form of pulmonary failure caused by the long-term inhalation of airborne sand particles. His many years in the desert had come at a very high price.

Nearing death, he wrote: "The old times are passing too in Arabia and the pure absolutely pure strains are disappearing fast the farther north you go (away from the Nafud Desert) the less you find which are considered absolutely pure."

There was much praise from his friends and colleagues, who remembered him as a man of magnanimity and enthusiasm.

> We who did him justice while he was living have nothing to say about him now that he is dead. His work will outlive any words of friends, false friends, or open enemies. The Original Breed of Arabian Horse is his living monument and will continue to exist now as long as there are men who love it with a fraction of his ardor.
>
> Ott, The Blue Arabian Horse Catalogue, 1967

Figure 40 - *Carl Raswan died on October 14, 1966 and was buried at Calvary Cemetery in Santa Barbara California.*

The process of pedigree research accelerated after Raswan's death. Many serious and tenacious researchers devoted a lifetime of work to this cause. Methods varied. In Egypt, the validation of pedigrees was less problematic than in other countries, owing to the uninterrupted records kept by the RAS/EAO. Very few questions of pedigree validity needed to be addressed and very few errors in pedigree have subsequently come to light. The problems were more daunting in cases of horses originating in the older established studs of Europe, where horses had been bred more or less continuously for 300 years by the nobility as well as by governments such as those in Hungary, Austria, and Germany. Pedigree research involved thorough investigation of the "paper trail" that existed in the stud books of Europe and Egypt, and when the trail ended or became contradictory, researchers sought out and interviewed those still living who had personal but unpublished relevant knowledge.

The Index of Partbred Arabians, by Hansi Heck-Melnick, provides facsimiles of some of these very old records and gives an overview of the problems encountered by researchers as a result of the condition of the documents themselves: language differences, record availability, poor record legibility, and record inconsistency.

This was the high water mark of the enthusiasm of the pedigree purists, but it would not last long. The Raswanian spirit of tireless investigation into questions of pedigrees and the purity of Arabian horses came slowly to a halt. Many came to regard The Index as the work of a fabulist, and his theories were often met with derision. However, The Raswan Index was by then published and available to anyone who wished to use it as a research tool. Its subtitle was A Handbook for Arabian Breeders, suggesting that Raswan's intent was to provide guidance for future breeders.

Figure 41 - *Hansi Heck, pedigree specialist and untiring advocate for the preservation of purebred Arabian horses.*

The arrival of the 1960s would see the introduction of commercial factors into the Arabian horse business that would alter forever the way in which breeders and the Arabian horse public regarded the importance and validity of the Arabian horse pedigree, as well as the importance of blood purity itself.

The 1960s ushered in the era of private Arabian horse breed organizations. A number of groups composed of private individuals were formed, each motivated by a nascent appreciation of the special value of the horse of the Bedu. These horses were now largely in the hands of Western horsemen and were present in significant numbers. This was to be the decade of "definitions", a period of time in which individuals and groups of individuals responded to the growing awareness of the dangers of indiscriminate Arab horse breeding practices which many felt posed a threat to the *purebred* horse. There was a general appreciation among breeders that many of the world's Arabian horses possessed pedigrees that were not complete or correct, but some breeders were convinced that a small number of Arab horses were indeed 100% genetically *purebred*. There was a growing awareness of the importance of determining which ones were *purebred* and which ones were part-bred.

Until the 1960s the term "*purebred Arabian horse*" was used to refer to any horse registered with a recognized national registry, such as the AHRA or the Arab Horse Society in Britain. Now all of this changed. Opinions on the subject of genetic purity were varied, and disagreements became the rule, often resulting in highly animated discussions and irreconcilable conflicts. It became problematic that during this decade of "definitions" so many intelligent, sincere and well-meaning people were unable to determine with any degree of certainty a generally acceptable definition of what constituted a *purebred Arabian horse*. By the end of the 1960s, a number of organizations and fraternal entities had codified and were promoting their own widely disparate versions of the definition of the *purebred* horse of the Arabian Peninsula. A matter that for centuries had seemed marginally significant became a matter of central importance. Fractious debate among and between the groups became the norm.

The American breeder Mrs. Kathleen Llewellyn Wright Ott, born in rural Virginia in 1894, was among the first to be influenced by the work of Carl Raswan, who she first met in 1937. An Arab horse enthusiast from an early age, she had been studying Arabian horse pedigrees since before World War I and was greatly impacted by Raswan's assessment of the dwindling stock of *purebred* horses left in Arabia. From her home in New England, she was able to personally inspect many of the registered desert-bred Arabians that had been imported into the U.S. up to that time. After World War II she started her own breeding program in earnest, beginning with Babson stock and later adding imported original Arabian Desert bred stock.

Mrs. Ott and her daughter Miss Jane Ott set for themselves the goal of preserving blood purity of the Arabian horse as it was originally found among the Bedouin tribes of Arabia.

Aware that the term "*purebred*" was being misappropriated and misused by breed organizations, she abandoned the use of this term altogether and forged her own system of Arabian horse classification.

> We realized that the vast majority would not care for the original types and that the recent evolution of a bastard type . . . was a prime necessity to provide the public with what it wanted . . . it must be understood that none of the tribes ever allowed a good mare to leave the country, and so all mares we began this program with were

culls . . . to Fadl and Nasr we are indebted for the modified dishes that crop up from time to time in our BLUE STARS . . . we would prefer that it never appeared at all since it is not an attribute of the genuine *purebred Arabian*, but a throwback to the Abbas Pasha early selections from the desert, which we have been given to understand consisted only of what he could find that had dished faces . . .

Originally we could depend on the Egyptians to be at least Blue List; that is, varying amounts of Muniqi on the *purebred* stock, but not now. The EAO has bred some of the Skowronek and Basilisk stock they got with their Blunts . . . and now "Straight Egyptian" no longer means a Blue List as it once did.

Mr. Harris probably did more to popularize the Arabian horse in America but the kind of Arabian he admired was rarely of any lasting importance to the breed . . . General Dickinson bred some of the finest Arabians ever seen in the country but again we have been unable to see what he is trying to do other than produce beautiful and saleable horses . . . Mr. Babson's interest in Arabians centers on what he likes to look at. His present experiment is running but we anticipate the usual culling action at the end.

Mrs. John Ekern Ott in Arabiana.

The Otts developed and promoted a conception of Arabian horse blood purity which was termed "the Blue List", defined as those horses in America which their research indicated were originally found in the possession of the desert Bedouin tribes of Arabia, unadmixed with non-Arabian blood. She further defined a more restrictive sub-list of these horses which she called "Blue Star", consisting of those Blue List Arabian horses which contained no Muniqi blood.

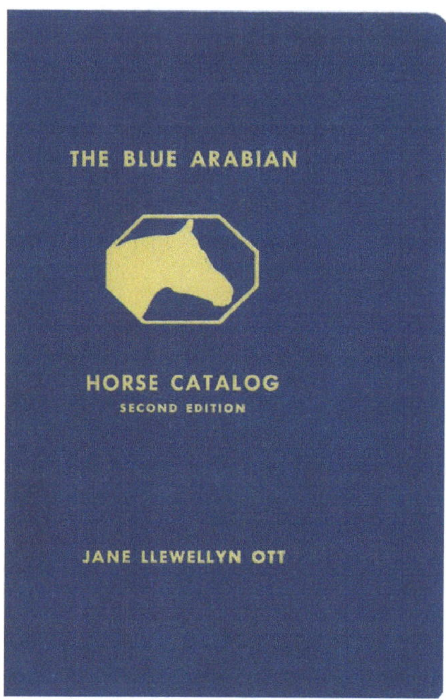

Figure 42 - *The Blue Arabian Horse Catalogue.*

Ott stated that the Muniqi strain was developed when the Salqa Bedouins bred Muniqiyat mares to Turkoman stallions, about three hundred years ago. This produced a taller, more angular animal lacking much of the quality of the pure Arabian but possessing greater speed. She singled out the Muniqi strain for special censure because of her conviction that the presence of this blood in an otherwise pure Arabian pedigree led to loss of quality. She does not provide her reader with references or data in support of her contentions.

Miss Jane Ott began compiling a list of these horses which she kept in a binder with a blue cover. This work would later lead to the publication of The Blue Arabian Horse Catalogue, and the popularity of the terms Blue List and Blue Star. She was especially motivated to formalize her research when she discovered that in 1956 only 10 Blue List foals were born in the U.S.

The Blue concept was promoted by the Otts informally among like-minded breeders, and its first appearance in print was in the April 1958 issue of the *Arab Horse Journal*. In 1961, The Blue Arabian Horse Catalogue, (written by Miss Jane Ott), was published, in which she listed those horses which she determined, based on her research, met her criteria for authenticity. She decried the fact that so many authentic Arabian mares were being bred to Arabian stallions with "contaminated" pedigrees. The task as she saw it was educational, and she set out to put the facts before the American Arabian breeder. She states the motivation behind her preservational movement: ". . . in the modern craze for extra size and other characteristics foreign to the original Arabian horse, many breeders are overlooking the serious loss of original type." The Blue List contained the names and pedigrees of the foundation Arabians that Miss Ott considered *purebred,* and all descendants of these horses were considered "Blue List". Miss Ott was well aware of the pedigree controversy that was raging in both the U.S. and Europe; the conflict between the pedigree purists and the pedigree skeptics, the absolutists, and the relativists, was being actively debated.

In 1964 she wrote:

> . . . The Poles themselves admit that their "*purebred Arabians*" are not *purebred*. They have always been incredibly naïve about the breeding of the "Arabians" that early polish breeders imported with and without pedigree from all over the Near and Middle East, but they have never denied the part played by captured 17th century Turkish cavalry horses and native Polish mares, only claiming that the impurity is not enough to matter. . . there is also much undercover strife between those who want to try saving whatever is left and those who think the infiltration has gone so far that there is nothing left to do but hush it up and hope the stock will eventually stabilize somewhere . . .

In 2011, Jane Llewellyn Ott was interviewed by the author of this book on the subject of her organization.

She contributed the following note for publication in this book, declining to comment further.

> The Blue Arabian Horse Catalogue was published by and for breeders who prefer the original type to all the others that have invaded the stock outside Arabia. We have abandoned the term "*purebred Arabian*" to those who value that name for its own sake, and it is now just a synonym for registered Arabian. If you want to know what Blue Arabians are like, the editor of The Blue Arabian Horse Catalogue, P.O. Box 224, Canton, TX., 75103 can supply illustrated literature and directions to Blue breeders in this country. Don't assume you want Blue Arabians until you see them. Most people prefer the new types. That's why the original one so nearly became extinct.

When asked in the interview about the difference between Blue Star and Blue List, she replied that "the less said about that, the better."

The Blue Arabian Stock as formulated by Ott consisted of the following groups: Huntington, Hamadie, Davenport, Egyptian, Blunt, Crabbet, New Egyptian, Ibn Saoud, Crane, Ayerza, Ibn Jiluwi, and Bahreyn. She included the individuals: Leopard, Shabaka, Nejdran, Sunshine, King John, and Mirage.

The Ott phenomenon became the prototype of all subsequent Egyptian Arabian horse organizations. The features included an organizational structure, definitions, membership, member breeders updates, newsletters, yearly foal and import update, and promotional publications.

Douglas Marshall, the importer of Morafic, was also influenced by Raswan. Marshall had seen the horses of the RAS during his time in Egypt with the U.S. Military during World War II. It was a chance sighting, but he described his first visit to the RAS as being in "heaven".

Figure 43 - *The Pyramid Society.*

Back home, Marshall acquired the Egyptian stallion Moftakhar, (a 1946 RAS Egyptian stallion by Enzahi and out of Kateefa). He also struck up an acquaintance with Carl Raswan, known for his expertise regarding Arabian horses, their pedigrees, and their sources. Marshall had a vision of obtaining *purebred* desert horses and bringing them to his Texas farm to breed. Raswan mapped out a Middle East journey for Marshall to begin the search for horses, but Raswan died suddenly, and the project was halted.

Marshall's visit to the EAO followed several years later, and his fascination with the Egyptian horse was complete.

In 1969, the idea of an organization devoted to preserving a nucleus of Egyptian horses was conceived by Douglas Marshall and Judith Forbis during an afternoon visit at Gleannloch Farm, viewing the Egyptian Arabian horses at Marshall's farm at Spring, Texas. Marshall recommended the name Pyramid Society. Eventually, a group of like-minded breeders became associated with each other, all favoring the formation of this breeder's group.

Those breeders of Egyptian horses interested in preservationist ideals included Douglas B. Marshall, Judith Forbis, B. D. Heck, Richard Pritzlaff, Wilbur D. Winter, Dr. Thomas Atkinson, and Willis Flick.

When the Pyramid Society became an official organization with Articles of Incorporation, the founding members were Douglas Marshall, James M. Cline, Judith E. Forbis, Willis H. Flick, and Bradford Heck.

Douglas Marshall wrote a historical review in 1997 concerning the Pyramid Society in which he stated that "as living founders, it is our position that the Pyramid Society was founded as a result of that suggestion and that meeting between us in 1969."

Hansi Heck recalled the early days of large-scale importations:

> "A major importation of 23 straight Egyptian stallions and mares from Egypt took place November 2, 1968. The horses arrived in 20 below zero at the Toronto Airport and the Canadian TV stations, newspaper reporters, Canadian government officials . . . were there to welcome this priceless shipment of authentic Arabians . . . most of these were selected by Douglas B. Marshall...the importation included * Serenity Sonbolah . . . * Sakr . . . * Magidaa . . . and * Soufian."

In 1970, the Egyptian war with Israel threatened to engulf the EAO in Cairo. Marshall and others joined together to acquire additional EAO horses that would have been unavailable to them except for the threat from the advancing war front. A shipment of 32 horses was arranged, at great peril and expense to those involved. This group included *Ibn Hafiza and *Zaghloul.

The next two decades were to be a golden age of breeding for the Egyptian Arabian horse in America. The Pyramid Society chose to use the term "Straight Egyptian", a term familiar from Miss Ott's <u>The Blue Arabian Horse Catalogue</u>, to designate those horses of Egypt and the EAO which in their opinion were especially worthy of preservation. In its first handbook published in 1973, the Society defined the straight Egyptian or Pyramid Arabian as "one who traces in every line of his pedigree to horses listed in Egyptian stud books and thence to Arabia Deserta". The handbook explains that there really are no "pure Egyptian" or "straight Egyptian" horses at all in the strict sense of the words since many horses in the EAO stud book trace to horses that were bought in Arabia by the Blunts and sent directly to England. They were bred there for three generations at which point several of them were sold to the EAO and incorporated there into its breeding program.

The "Crabbet detour" in the EAO pedigrees caused considerable debate and dissention during the formative years of the Pyramid Society. The problem was the "definition".

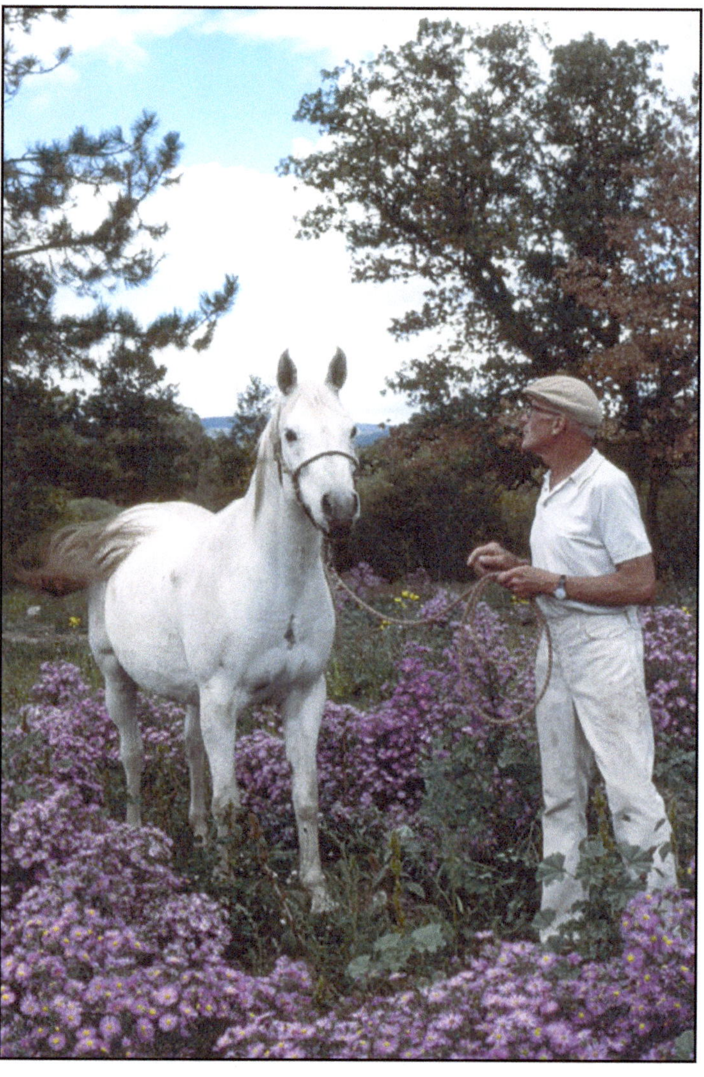

Figure 44 - *Richard Pritzlaff and Tatu (John Doyle x * Bint Moniet El Nefous). John Doyle was the son of Ghadaf, and out of Rabanna. Despite the fact that John Doyle was of pure EAO breeding, Tatu was not, again owing to the pedigree of his mother.*

The founders of the Pyramid were aware of the fate of earlier importations into the U.S. from Egypt and wanted to avoid the mistakes of the past. They wanted to be fair to all horse owners and breeders concerned, but they firmly believed that a line must be drawn and adhered to, for the sake of the Egyptian Arab horse. Their motives were admirable; they did not want the fate of dilution to befall the "New Egyptians". A clear and unambiguous definition was required. A line was drawn in the sand, but the choice of sand as a medium had clear metaphorical implications.

Figure 45 - *W.R. Brown.*

The Board never claimed that Pyramid horses were better quality horses than others, nor did they claim that they were *purebred Arabian horses*. They made only one claim: Pyramid horses were a special group that deserved protection and preservation as a closed genetic pool.

Henry Babson's use of imported desert horses from King Ibn Saud of Saudi Arabia in his herd of first-rate EAO horses provided an example of what the Pyramid Society wished to avoid in the future.

Henry Babson (1875-1970) is remembered today as one of the premier American breeders of Egyptian horses, but he did not just breed Egyptians. Miss Ott wrote in the Blue Book that "Babson would breed anything that came down the pike", referring to the many experiments that Babson conducted with Arabian blood. His core Egyptians were Bint Bint Sabbah, Bint Bint Durra, Bint Saada, Bint Serra, Maaroufa, and the stallions Fadl and Metsur.

While Babson as a rule bred and then inbred these animals as a closed pool, he imported the Polish stallion Sulejman, who became a popular sire. Babson also acquired the Saudi stallion Aldebar, purchased from the Prince of Wales E. R. Ranch in Alberta Canada in 1938. Dr. Zaher, breeding consultant of the EAO, described the horse as "off-type". In 1958 Babson imported the Skowronek-bred stallion Nimrod (Champurrado x Nautch Girl) from Musgrave Clark in England. Nimrod was used on the Egyptian mares briefly and then sold.

Babson bred these stallions to his Polish mares, but he also bred them to some of his Egyptian mares. The results were considered unsuccessful. The experimental cross-bred progeny were, in the prophetic words of Miss Ott, culled. The produce of this experiment meant the loss of those breedings to the Straight Egyptian pool.

Babson also acquired and used the Saudi mare Turfa at his farm. She was a grey mare from the stud of King Ibn Saud, given to King George VI as a gift. Turfa was sent to the King's farm in Canada during World War II, and it was there that he saw her. Babson bought Turfa and bred the mare to Fadl for several years, producing a number of foals. Ibn Turfa was considered the best of the lot and he was retained. He was bred to some of the straight Egyptian mares, but the resulting foals were not considered desirable. Eventually, all animals containing Turfa blood were sold. The Babson farm returned to its policy of breeding straight EAO bloodstock exclusively.

Figure 46 - *Henry Babson with an unidentified horse.*

The case of William Robinson Brown (1875-1955) was similar. Brown established the Maynesboro Stud in Berlin New Hampshire in 1912, and at its peak, it was the largest Arabian breeding farm in the U.S. His foundation stock was largely Crabbet blood. However, in 1932 Brown purchased 6 horses of the finest Egyptian Arabian breeding from Prince Mohammed Ali Tewfik. He bought 2 stallions, Nasr and Zarife. In addition, he bought 4 mares; Aziza, H.H. Mohamed Ali's Hamama, H.H. Mohamed's Hamida, and Roda. Brown was, however, in no sense a preservationist. He was an agent of the U.S. Remount Board and believed in crossbreeding Arabians to non-Arabians to produce animals fit for endurance riding. By the time Brown dispersed the herd and sold the farm during the Great Depression of 1933, the Egyptian horses that he owned were dispersed. They went to General J.M. Dickinson, William Randolph

Hearst, Kellogg, and Roger Selby. They were eventually lost in pure form by dilution due to the prevailing supremacy of Skowronek and his male descendants.

While breeding an Egyptian stallion to a non-Egyptian mare presents no threat to the EAO stock, breeding a non-Egyptian stallion to an EAO mare mean that there is one less straight EAO foal in the world. With such a small number of straight EAO mares, to begin with, each foal is precious. Each foal represents a unique genetic opportunity to advance the goal of protecting the purity of the straight EAO population. This was the reason that the Pyramid Society Board stood firmly by its position as stated in the definition.

The problem of the "Crabbet detour" posed a different dilemma.

Richard Pritzlaff had begun breeding Arabian horses with the purchase in 1947 of the mare Rabanna (Rasik x Banna). He bred the mare under the guidance of Carl Raswan and by the time of his 1957 importation of the EAO horses, he had a large number of Rabanna-bred horses, both male and female. He used Rabanna-bred sires on some of the straight Egyptian mares.

By 1969, the year that he became involved with the Pyramid Society, Pritzlaff had been breeding his horses in this way for 12 years. He assumed that all of his horses qualified as Pyramid horses. His assumption was incorrect.

As the Pyramid movement began to expand and prosper, Pritzlaff was made aware by his colleagues that his mare Rabanna did not qualify as a Pyramid horse because of the presence in her pedigree of the Blunt horses Kars and Jeroboam. These were two of the many horses that the Blunt's had obtained in the Syrian Desert and shipped directly to their Crabbet Stud in England. Neither Kars nor Jeroboam were ever in Egypt.

Pritzlaff, being an early member of the Pyramid Society, was displeased by this ruling and made a concerted and vigorous attempt to challenge it. He argued that the Pyramid-approved Blunt mare Rodania had also never been in Egypt, noting that this fact alone did not disqualify her for inclusion as a Pyramid horse.

He further asserted that the non-Pyramid elements in Rabanna's pedigree were, in fact, all represented at the EAO by virtue of the ancestors of the horses that were present in the large purchase of Crabbet stock by the EAO in 1920. This group included 16 stallions, 2 geldings, and 2 fillies.

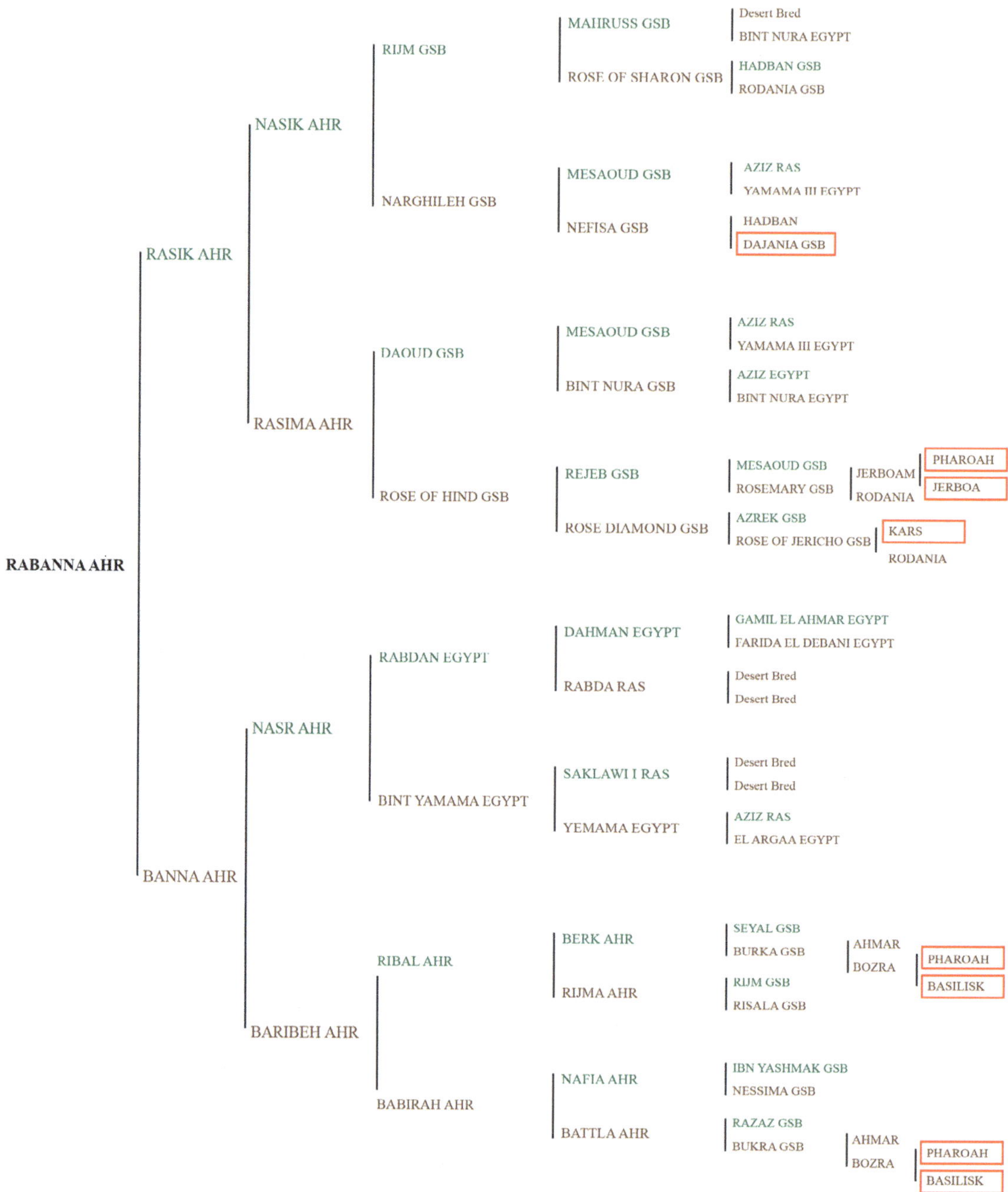

Figure 47 - *The pedigree of Rabanna. The names in red boxes represent horses that the Blunts collected from the Syrian Desert during their early attempts at pure Arab horse breeding. These horses were discarded by the Blunts after a few years and the experiment, called the "First Attempt", was abandoned by Lady Anne, who began again by purchasing stock that was specifically derived from the horses of Abbas Pasha and Ali Pasha Sherif.*

Note especially the two fillies that made the trip from Crabbet to Cairo in 1920. They became and have remained a major female line of EAO breeding, that of Rodania. Note also that these horses were not bought from Lady Anne or Wilfred, both by then deceased, but they were bought from their daughter Lady Wentworth, who was by this time deeply involved in her Skowronek breeding program.

While the Pritzlaff argument was technically correct, most of the stallions in this group were considered "off-type", and, after they arrived in Egypt at the EAO, they were, with a few notable exceptions, never heard from again.

The Pyramid Society Committee to Investigate Doyle/Pritzlaff horses was created to look into this matter.

Figure 48 - *Richard Pritzlaff and Rashad Ibn Nazeer.*

The horses of Dr. J.L. Doyle presented the Society with a problem that was identical to Pritzlaff's dilemma. Dr. Doyle had assembled a small group of horses noted for the high percentage of Abbas Pasha/Ali Pasha Sherif blood that they contained. Although Dr. Doyle had died in 1957, his family bred the group of horses and made the case for them to be denominated as Pyramid qualified horses. Like Rabanna, all three of the Doyle foundation horses, Nusi, Ghadaf, and Gulida had one or more lines to Blunt desert imports (directly imported from the Arabian Desert to Crabbet Stud in the U.K.) that had never been used for breeding in an Egyptian program. These included Queen of Sheba, Pharoah, Basilisk, Kars, Jerboa, and Dajania.

The Society board wrote their assessment:

> "... The horses bought from Crabbet by the Egyptians are listed in their RAS Studbook prior to page 45, were for the most part not used in the subsequent RAS *purebred* breeding program. Four of the horses were thought so little of (Zeidan, Samir, Besheir, and Rayyan) that their pedigrees weren't even listed...Hamran II and Rustem were the two main horses that were used.
>
> While the Society has made exceptions for the specific individuals; Julep and Ghadaf (but not necessarily their immediate ancestors)...it is my personal recommendation that no further stock be acknowledged."

This large importation was to be the cause of considerable debate among Egyptian Arabian horse breeders in the ensuing years. The names of the stallions from the 1920 Blunt purchase as recorded in the Egyptian records are Besheir, Bustan, El Borak, Kamar, Karun, Keslan, Nawab, Raseed, Ras el Mal,Rayyan, Razaz, Samir, Solajan, Zeidan, Mabrouk Blunt, Ibn Yashmak, Rustem, Kazmeen, and Hamran. Ibn Yashmak was the sire of the key 1920 mare Bint Rissala while still at Crabbet. Zeidan produced three foals recorded in the Inshass herd book, Rustem produced 9 foals at the RAS, Kazmeen produced 13 foals and many grandchildren at the RAS. Hamran produced ten recorded offspring. Mabrouk Blunt produced one registered daughter at the RAS. Kazmeen and Rustem formed key elements that led to the creation of the Egyptian horse as it exists today. That is to say, many of the horses in the 1920 sale from Crabbet to Cairo were essential to the progress of the RAS and high-quality reproducers, whose contributions to RAS breeding were immeasurable. It was a difficult decision. The results of the Committee's findings, however, did not favor Pritzlaff's position. The Pyramid Society Board of Directors did not accept the Pritzlaff argument. A modification in the rules was needed.

In 1976 the Pyramid Board voted to introduce the words "used to create and maintain the RAS/EAO breeding programs" into its definition of the straight Egyptian horse, thus dealing with Pritzlaff and Doyle, and eliminating from eligibility Rabanna along with a host of other horses as well.

Board President Jarrell McCracken wrote to Pritzlaff in 1977 with the board's final decision:

> "I am very sorry that this report is not favorable from the standpoint of accepting these bloodlines as conforming to the Pyramid Society definition of the Straight Egyptian. There is enough question existing in the pedigrees to reflect unfavorably upon the integrity of the Pyramid Society... for them to be accepted. I truly wish it was otherwise....I also wish it from the standpoint of my personal friendship and feeling for you."

Pritzlaff responded to the decision by publishing the details of both sides of the argument in the July 1978 edition of the *Arabian Horse World.* He then resigned from the Pyramid Society, never to return.

The 1979 edition of the Pyramid Society reference handbook addressed the Pritzlaff controversy by clarifying in detail the exact sources of the stock which it considered to be Pyramid horses or Straight Egyptian. The components were: a horse that traces in every line of its pedigree to one or more of the following categories: a) owned or bred by Abbas Pasha I or Ali Pasha Sherif, b) used to create and maintain the RAS/EAO breeding programs, with the exclusion of Registan and Sharkasi, and c) a horse which was a lineal ancestor of a horse described above, and d) a horse (other than a horse having Registan or Sharkasi as a lineal ancestor) which was conceived and born in a private stud program in Egypt and which was imported directly from Egypt to the United States and registered with the AHRA.

Later the Pyramid Society again amended the definition of a Pyramid Straight Egyptian horse to include "...Other than those excluded above, a horse conceived and born in a private stud program in Egypt and imported directly to the United States and registered by the Arabian Horse Registry of America prior to the extension of the EAO's supervision of private Egyptian stud programs as reflected in Volume 4 of the EAO's stud book." Volume 4, published in 1975, was the first stud book to include private farms of Egypt. There were several entries in Volume 4, including Al Badeia Stud Farm, Shams El Aseel Arab horse Stud, and Ahmad Cherif Stud Farm.

Item b) was important in clarifying the dilemma posed by the mare Rodania, since she was never in Egypt. The Blunts had obtained her in the Syrian Desert and shipped her directly to Crabbet in England. By using this criteria, the Society specifically rejected the prevailing opinion that the straight Egyptian had to come out of Arabia and thence directly to an accepted breeding stud in Egypt. Some Arabian horse breeders still expounded the importance of the "straight from Arabia, directly into Egypt" credo.

Pritzlaff continued to take issue with the growing trend to compartmentalize and label Arabian horses based on their place of origin or place of birth. He objected to the fad of nomenclature which he viewed as a marketing technique. The definition of an Egyptian Arabian horse had for many years been one whose parents had been incorporated into the RAS breeding program from the Arabian Desert. He pointed out that the 1920 RAS purchase of Blunt Crabbet stock, which included the Rodania descendants Bint Risala and Bint Riyala, made the designation of types of Arabians problematic. Rodania was, after all, taken by the Blunts directly from the Arabian Desert to the stud at Crabbet in England. Did this make the Rodania descendants English Crabbet Arabians? When they returned to Egypt, did they then become Egyptian Arabian horses?

> There are now only a few Arabians that could be termed pure, straight, or Pyramid straight Egyptian. Of the Gleannloch, Ansata, Bentwood, Schimanski and Pritzlaff Arabians listed in the Al Khamsa Directory I, only the Inshass stallion *Ibn Hafiza is without Crabbet blood.

> Only *Ibn Hafiza, he notes, is free of Rodania blood. This was a remarkable insight.

Regarding the use of Egyptian Arabian stallions in America, he wrote: ". . . why should twenty or so be considered so superior that they should saturate the breed? Are we getting away from the true Bedouin horse through fads and artificiality? Is the Arabian horse, through artificiality, losing his true character?

The stated purpose of the Pyramid Society was to preserve and perpetuate these Egyptian bloodlines as a nucleus of outcross blood for other breeds of horses. In its series of handbooks, the Society published the pedigrees of those foundation stocks that fall within its definition, along with the provenance of each horse. The Pyramid Society is the most exclusionary of the major breed groups. Many of the Al Khamsa horses do not qualify as "Straight Egyptian" under the Society's criteria. The Pyramid Society did not make assertions regarding the presence or absence of genetic impurity in the horses which it terms "Straight Egyptian", maintaining only that they are of "desert origin" and that they are a "proven source of highly desirable characteristics". What these characteristics might be is not enumerated. The Society specifically avoided using the term "*purebred*" in its definition, deferring to the Arabian Horse Registry of America as the ultimate authority for defining, categorizing and registering the Arabian horse. As the handbook states, the Arabian Horse Registry of America had a position of eminence among the world's registries that was a source of pride to all.

The Pyramid Society does not register or maintain lists of Straight Egyptian horses; the Society handbook states that such a registry would be of academic interest only since such a list has no legally binding consequences and would only be of interest to members themselves.

The Pyramid Society stands out from other definitional groups by limiting its approved foundation stock to a specific national breeding program. Its success is in large part attributable to its worldwide appeal. The Pyramid Society encouraged a spirit of fraternity with breeders around the world who were in sympathy with their mission and who bred their horses according to the Society definition. Horses from the EAO have been dispersed internationally, and the Pyramid Society definition applies to all of them, including all of the progeny born from two straight Egyptian parents.

During the 1970s a group of preservationists saw the need to foster the Arabian horse in a manner that was broadly inclusive. Al Khamsa Inc. was a private member-supported American breed organization, established in 1974 by a group of concerned individuals who are interested in the preservational breeding of the authentic antique horse of the Arabian Desert, which it termed The Al Khamsa Arabian Horse. The organization limited its scope from the beginning to horses in North America. This restriction was the result of practical considerations, particularly the need for personal and physical inspection of the horses in question.

As early as 1973, a group of American breeders, inspired by the spirit of The Blue Arabian Horse Catalogue, sought to carry forward the work begun by the Otts. These individuals ultimately coalesced into the North American Arabian horse organization Al Khamsa. Many Arab horse enthusiasts participated in its formation including Diana Marston, Charles and Jeanne Craver, Walter Schimanski, Carol Lyons, Jackson Hensley, Don Austin, James Bullard, Nan Burket, H.B. Stubbs, Darrell and Ruby Perdue, Carol Schultz, Dr. and Mrs. Fred Mimmack, Barbara Baird, and Jim Brown.

Figure 49 - *Al Khamsa Arabian.*

With the publication of its 1976 directory, the enterprise was launched. Al Khamsa Arabian horses were defined as: "horses in North America that can reasonably be assumed to descend entirely from Bedouin Arabian horses bred by horse-breeding Bedouin tribes of the deserts of the Arabian Peninsula without admixture from sources unacceptable to Al Khamsa". The organization did not claim that its horses were "*purebred*". In fact, it did not even use the term "*purebred*" in its deliberations. This designation was left in the hands of the Arabian Horse Registry.

Al Khamsa has published extensive historical information on the Arabian horse as well as a list of Al Khamsa foundation stock – those horses that it considered met their definition. The analysis of the pedigrees of the foundation horses and the reasoning used by the Al Khamsa founders in including each one are available in its serial reference books called Al Khamsa Arabian.

In contrast to the Pyramid Society, Al Khamsa is a more inclusive organization, representing a broader region of the Arabian cradle of origin than Egypt alone. Its by-laws allow for the future addition of foundation horses whenever further information becomes available. It also publishes updates periodically documenting the descendants and progeny of this original nucleus of horses. The ancestral elements accepted by Al Khamsa are represented by benefactors, importers or specific importations such as Mrs. Connie Cobb, the Ayerza horses of South America, Blunt (Wilfred and Anne), The Charles Crane imports, the Homer Davenport imports, most but not all of the Egyptian horses, the Hamidie Society (Arabian horses imported to the U.S. in 1893 under the authorization of the Ottoman Sultan Abdul Hamid II), Randolph Huntington's horses, Inshass, (the Royal Stud of King Fouad and King Farouk of Egypt, horses bred by Amir Saud Ibn Abdullah Ibn Jiluwi, the horses of the ruling Al Khalifa family of Bahrain, horses from the stud of Ibn Saud in Saudi Arabia, horses imported by Major Roger Upton, and a number of individual horses of varied provenance).

Hence the Al Khamsa root stock includes many horses whose pedigrees are not within the Pyramid Society definition.

Figure 50 - *Asil Club.*

The same spirit of idealism and dedication to the recognition and preservation of the authentic asil desert horse of Arabia was present in Europe, and during the late 1960s, a group of concerned individuals began to discuss the importance of taking action to prevent the loss of asil blood entirely. The Asil Club was founded in Germany in 1974 after many years of informal association between a small and highly dedicated group of like-minded and interested individuals. Their goal was to promote the ancient horse breeding techniques of the desert Bedouin and to guard the asil bloodlines of Arabian horses that were then present in Europe. The club took its name from the Arabic word for purity and nobility. It defined the *purebred* or asil horse as "a horse whose pedigree is exclusively based on Bedouin breeding of the Arabian Peninsula, without any crossbreeding with non-Arabian horse at any time."

Dr. Georg Wenzler, Landoberstallmeister and Dr. W. Uppenborn, Landstallmeister Directors of the Marbach Stud, Germany's oldest horse breeding institution, were among the most vigorous of the Asil Club's early exponents. W. Georg Olms, Georg Thierer, and Dr. F. Bonno Klynstra were instrumental in organizing the Club and publishing the group's Asil Araber, in which the principles and documentation of the Asil Club are presented.

Dr. Olms first became involved in the *purebred Arab horse* movement in 1939, following his experience with the Arabian horse Yanik. A former Polish Army horse, Yanik was descended from the famous stables of Count Waclaw Rzewuski. Olms found that under saddle, "It was so magnificent that I felt something had to be done to preserve such a breed." The Asil Club was centered in Hildesheim, Germany. The gravitas of its founders lends the Asil Club considerable legitimacy and authority.

The Asil Club's philosophy with regard to the Asil horses was characterized by Dr. Olms:

> "They deserve to be specially cared for as they are a 'gene-pool' of immense importance for all breeds of *purebred Arabians* as well as for horse breeding as a whole; they must not be lost in the total stock."
>
> W. Georg Olms, Asil Arabians, 1985

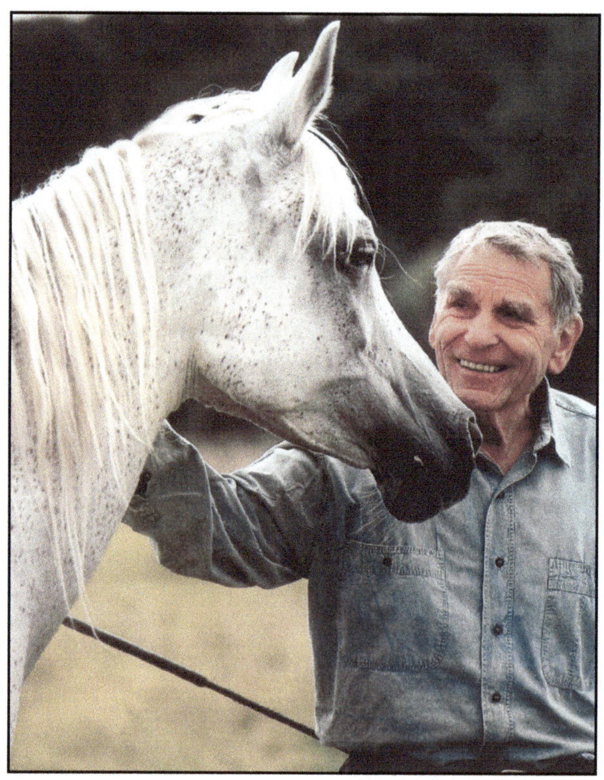

Figure 51 - *Dr. W. Georg Olms, founder of the Asil Club e.V., with his foundation mare Nafteta (by Kaisoon OA). Photograph used with the permission of Dr. Olms.*

Figure 52 - *Dr. George Wenzler.*

Dr. Wenzler added the following:

> "The Age of automation has not only embellished our lives but it has also taught us to be afraid. Ruthless exploitation of unspoiled nature, loss of traditions and civilizations, a decrease in human dignity and in responsibility towards future generations, are the menaces of today."
>
> Dr. Georg Wenzler, Asil Araber III

Dr. Wenzler spoke from a position of considerable authority, having been the Landoberstallmeister of Weil-Marbach Stud in Germany from 1949 to 1974. He was instrumental in bringing outstanding Egyptian Arabians from the EAO to join the program at Weil-Marbach, notably Hadban Enzahi and Jasir. The mare Sanacht, producer of champions, was born at Weil during Wenzler's directorship.

The Asil Club was unique among preservationist groups because of its primary insistence on the importance of recognizing and promoting the athletic ability in the Arab horse.

Dr. Wenzler wrote:

> "The Asil Club refuses to give way to sudden fashions and to worship grace and beauty for themselves alone, so as to make the horse a mere object for the show ring. The members understand the Arabian horse to be primarily an excellent riding horse for sport and recreational purposes and must be bred for performance otherwise he will degenerate . . . Continuous performance tests and strict selection are thus indispensable."

The Asil Club philosophy does not subscribe to the dictum of "art for art's sake". For the Asil Club, the horse must perform its "*telos*", its right function, in order to have any legitimate value.

Right action, not abstract delight, is the proper criterion to be used in evaluating the worth of a work of art. British author W. S. Maugham wrote:

> It seems to me that the philosophers were right who claimed that the value of art lies in its effects and from this drew the corollary that its value lies not in beauty, but in right action. For an effect is idle unless it is effective. If art is no more than a pleasure, no matter how spiritual, it is of no great consequence; it is like the sculpture on the capitals of columns that support a mighty arch; they delight the eye by their grace and variety, but serve no functional purpose. Art, unless it leads to right action, is no more than the opium of the intelligentsia.
>
> W.S. Maugham, A Writer's Notebook.

The Asil Club's goal in promoting the preservation of the Asil horse was not that it should be preserved in an artificial state of suspended animation, but that it should be used as it had been for centuries, for the improvement of other horse breeds. Dr. G. Wenzler wrote: ". . . this eldest breed of horse is justly considered a regenerating force for all breeds."

Figure 53 - *Dr.Wilhelm Uppenborn.*

The Asil Club has an active outreach program designed to promote their vision of the Arab horse, participating in horse shows and educational events around the world. Emphasis is placed on endurance riding and performance events, and a strong commitment is made to therapeutic uses of the Egyptian horse in youth and disabled individuals. The Asil Club is governed by a Board of Directors which consists of a chairman, a vice-chairman, and ten members elected from the general membership.

Dr. Wilhelm Uppenborn, (1904-1988) was an avid proponent of the principle, philosophy, and activities of the Asil Club from its inception. Dr. Uppenborn was a scholar in agriculture, an author (<u>Pferdezucht and Pferdehaltung</u>) and a specialist in the fields of equine management and breeding. As he wrote in Asil Araber III, "...those beautiful Arabian horses must never be an end in themselves...there is no more honest horse, no better or more versatile horse than the Asil Arabian. Thanks to its excellent disposition for performance it has decisively influenced almost all important breeds of horses. This disposition is to be conserved and encouraged."

The Asil Club operates on the basis of the following principles:

1. The Asil Arabian is defined as a horse whose pedigree is exclusively based on Bedouin breeding, without any crossbreeding with non-Arabian horses at any time.

2. Horses may be considered Asil if their parents are registered in the stock list of the Asil Club.

3. The horses Registan, Sharkasi, and Ibn Galababi and their progeny are specifically excluded from the Asil Club definition of asalah.

Over the years, the Asil Club has remained unapologetic about its position regarding the authentic antique desert horse. Commenting on the terminologic controversy that is still contested, the Asil Club makes the following point; "The term 'asil' is today no longer identical with the terms '*purebred*' *Arabian,* Arabian, and Arabian horse, because internationally the original definition of purity - a *purebred Arabian* is a horse that must trace in all of its ancestors to the Bedouin breeding of the Arabian Peninsula cannot be maintained due to the crossbreeding with other breeds. In order to characterize the Asil Arabians (less than 2% of all horses accepted internationally as *purebred Arabians*) and to safeguard them in the interests of their preservation, we now use the term 'Asil Arabian'."

These four groups shared a common goal of preservation of the *purebred Arabian* horse, but each group had its own opinion, each varying slightly, about what was to be preserved, and why. None of these groups was able to establish the value its pedigree claims or authority with any degree of certainty or unanimity. The breeding of Arabian horses was becoming compartmentalized as each group promoted its definition of the desert horse. Each had its enthusiastic adherents, but it became clear as time passed that the failure to agree on a definition was the result of failure to agree on the reliability of the historical record and subsequent disagreements over the validity of pedigrees.

Each group started out with the same purpose and same raw material, the mass of written documentation dating back several centuries which claimed to be a true and honest account of each of the horses in question. Each group analyzed this data and arrived at differing interpretations. There was, on the whole, agreement about a core of Egyptian horses which all of the groups considered to be asil, but there were enough horses with questionable and debatable documentation that full agreement remained elusive. Each group drew its own "line of preservation" around those horses which fulfilled its criteria for inclusion.

The international market was in a similar state of disarray. Many countries had a centralized national Arab horse registry, and each registry considered itself fully knowledgeable on the question of what does and does not constitute a *purebred Arabian horse*. Each registry regarded itself as the true and valid protector of the *purebred Arabian* and regarded all other registries who disagreed with them as suspect or questionable. There was, however, no international agreement on the meaning of the term "*purebred Arabian horse*". Many of the fraternal organizations avoided the use of the term entirely, aware of its vexatious nature.

There was little reciprocity in transferring the registration of Arabian horses from one country to another. Each registry practiced a form of economic protectionism, eager to assist in the sale of horses to other countries but less inclined to cooperate with the importation of foreign Arab horses into its own country. There was inevitable conflict and acrimony. For many years, the AHRA did not accept horses registered with the Canadian Registry. The Arab horse market was globalizing, but the registry dilemma was becoming a serious impediment.

By the end of the 1960s, the fragmentation of the international registry system and the growing compartmentalization of the various breed groups set the stage for a new and unprecedented movement to unify the world Arab horse community. Breeders from around the world were looking for a way to improve the reputation and the recognition of the Arabian horse throughout the world. They sought to create a renewed sense of respect and broader acceptance for the horse in the public eye. However, differences among Arab horse breeders in terms of culture, religion and political background were a serious impediment to global cooperation in the marketplace.

The growing compartmentalization, registration and "definition" battles in the international Arabian horse community did not end here.

THE RIDDLE OF THE SPHINX

Figure 54 - *The riddle of the Sphinx; what is a purebred Arabian horse?*

One group of Arabian horse breeders was more commercially motivated, and had a broader vision of the future. They were men who wanted to sell horses globally. They realized early on that the single greatest impediment to the cultivation of new markets for horses was the growing furor and debate over the dividing line between the identification of the *purebred Arab* and the "impure" or part-bred Arab horse. This group of horse breeders was not concerned with preservation of any specific subset of Arabians, but rather was concerned with facilitating the commercial aspects of the buying and selling of the horse that was generically known around the world as an Arabian.

They were to champion the generic Arabian horse.

Arguments within the newly formed group, an organization that would change the face of the Arabian horse forever, arose over the precise meaning of the terms "*purebred*, desert bred, part-bred, pure blood, and alien blood". Pedigree debates raged. Most notably, however, the members of the group were distressed over the lack of Arabian horse registry reciprocity throughout the world. This was the key point. An Arabian horse had no commercial value if it was not registered as a *purebred* Arabian horse with the national registry in the county where it was kept. No two national registries agreed on the purity of a standard set of horses. There was general agreement on a core set of historically *purebred* horses, but there was no agreement on the rest of the world's Arabian horse population. The AHRA was singled out as the most troublesome, having refused to accept many foreign horses for registration. Its refusal to accept Spanish Arabian horses was the most explosive issue.

This group of ambitious forward thinking breeders wanted universal reciprocity.

They were also greatly distressed over the absence of breed standards, unreliable pedigree authentication and the failure of existing registries to monitor and enforce these standards.

The awareness that the Arabian Desert was entirely devoid of high caste Arabian horses was apparent to everyone. European Arabs at the time were a mixture of Polish, German, English, and Egyptian Arabians, most often the result of crosses to European stock, notably the English Thoroughbred. There was a desire among Arab horse breeders to bring to an end the constant and divisive international bickering that had been caused by the many unsuccessful attempts to create a common and universally acceptable definition of the "*purebred Arabian horse*".

Around the world, Arabian horse breeders had enough; it was time for a revolution. And it was not long before a group of revolutionaries coalesced around a powerful leader.

The international market in Arabian horses was in a state of disarray. Many nations had a centralized Arab horse registry, and each registry considered itself fully knowledgeable on the question of what does and does not constitute a *purebred Arabian horse*. Each registry regarded itself as the true and valid protector of the *purebred Arabian* and regarded all other registries who disagreed with them as suspect or questionable. There was no agreement on the meaning of the term "*purebred Arabian horse*". Many of the fraternal organizations avoided the use of the term entirely, aware of its vexatious nature.

There was little reciprocity in transferring the registration of Arabian horses from one country to another. Each registry practiced a form of economic protectionism, eager to assist in the sale of horses to other countries but less inclined to cooperate with the importation of foreign Arab horses into its own country. There was inevitable conflict and acrimony. For many years the AHRA did not accept horses registered with the Canadian Registry. The Arab horse market was globalizing, but the registry dilemma was becoming a serious impediment.

By the end of the 1960s, the fragmentation of the international registry system and the growing compartmentalization of the various breed groups set the stage for a new and unprecedented movement to unify the world Arab horse community. Breeders from around the world were looking for a way to improve the reputation and the recognition of the horse throughout the world. They sought to create a renewed sense of respect and broader acceptance for the horse in the public eye. Differences among Arab horse breeders in terms of culture, religion and political background were a serious impediment to global cooperation in the marketplace.

However, globalization of the Arabian horse marketplace would not be impeded.

One group of Arabian horse breeders realized early on that the single greatest impediment to the cultivation of new markets was the growing furor and debate over the dividing line between the identification of the *purebred Arab* and the "impure" or part-bred Arab horse. Arguments over the precise meaning of the terms "*purebred,* desert bred, part-bred, pure blood, and alien blood" appeared endless, and pedigree debates raged. Most notably they were distressed over the lack of Arabian horse registry reciprocity throughout the world.

They were also greatly distressed over the absence of breed standards, unreliable pedigree authentication and the failure of existing registries to monitor and enforce these standards.

The awareness that the Arabian Desert was entirely devoid of high caste Arabian horses was apparent to everyone. European Arabs at the time were a mixture of Polish, German, English, and Egyptian Arabians, most often the result of crosses to European stock, notably the English Thoroughbred. There was a desire among Arab horse breeders to bring to an end the constant and divisive international bickering that had been caused by the many unsuccessful attempts to create a common and universally acceptable definition of the "*purebred Arabian horse*".

Against this background, a movement composed of pedigree skeptics began to emerge. These individuals were commercially inclined Arab horse breeders with their own herds to promote, and they were stridently opposed to regulations and judgments imposed by any Arabian horse registry which cast aspersions on the purity of their horses. The principle caster of aspersions that especially rankled these men was the AHRA. Purity, said the Arabian Horse Registry of America, had been decided long ago; purity in the Arabian horse breed was determined by the AHRA. It was, after all, very old and revered, especially among its own officers.

**As a calf falls, more knives are drawn.*

www.ingramcontent.com/pod-product-compliance
Lightning Source LLC
Chambersburg PA
CBHW051559190326
41458CB00029B/6484
9780988293946